James C Wilcocks

The Sea-Fisherman

Comprising the Chief Methods of Hook and Line Fishing in the British and Other

Seas

James C Wilcocks

The Sea-Fisherman
Comprising the Chief Methods of Hook and Line Fishing in the British and Other Seas

ISBN/EAN: 9783337412869

Printed in Europe, USA, Canada, Australia, Japan

Cover: Foto ©berggeist007 / pixelio.de

More available books at **www.hansebooks.com**

THE
SEA-FISHERMAN:

COMPRISING

THE CHIEF METHODS OF HOOK AND
LINE FISHING IN THE BRITISH AND OTHER SEAS, AND
REMARKS ON NETS, BOATS, AND BOATING.

BY

J. C. WILCOCKS

PLYMOUTH (LATE OF GUERNSEY):
WINNER OF THE PRIZE OF ONE HUNDRED POUNDS AT THE INTERNATIONAL FISHERIES
EXHIBITION OF 1883 FOR THE BEST ESSAY ON IMPROVED FISHERY HARBOUR
ACCOMMODATION FOR GREAT BRITAIN AND IRELAND, ETC. ETC.
AND AUTHOR OF SIX OTHER PRIZE ESSAYS ON FISHERY
SUBJECTS AT THE FISHERIES EXHIBITIONS AT
NORWICH 1881 AND EDINBURGH 1882.

*PROFUSELY ILLUSTRATED WITH WOODCUTS OF LEADS,
BAITED HOOKS, KNOTS, NETS, BOATS, &c. AND
DETAILED DESCRIPTIONS OF THE SAME.*

'I cast my line in Largo Bay,
And hauled up fishes nine;
They're three to boil, and three to fry,
And three to bait the line.
The boatie rows' &c. (*auld Song*).

FOURTH EDITION,
MUCH ENLARGED AND ALMOST ENTIRELY REWRITTEN.

LONDON:
LONGMANS, GREEN, AND CO.
1884.

PREFACE
TO
THE FOURTH EDITION.

THE SUCCESSIVE EDITIONS of this work have been carefully revised, with many augmentations, both of the matter and in the number of illustrations.

The latter have been done under the eye of the author, by a friend, Mr. G. LE MASURIER of Guernsey, in a manner calculated to convey to the amateur faithful representations.

J. C. WILCOCKS.

CONTENTS.

	PAGE
Introduction	1
Marks, and How to Take Them	4
Methods of Fishing	8
The Possible Bag	10
Shifting Leads for Sea-side Visitors	10
Sea-fishing and Hydrography	11
Gear or Tackle required in Sea-fishing	12
The South-West Coast of England, Channel Islands, &c.	13
The Whiting	31
Ground-fishing Gear	33
Lines	47
Barking or Tanning Lines	48
Baits	48
Drift-lines on the Whiting-ground	51
Disgorgers	53
How to Cure Whiting	53
Drifting on the Whiting-ground	55
A Day's Whiting-fishing	55
The Pollack or Whiting-Cole or Coal	60
Drift or Tideway Fishing	61
Baits	63

	PAGE
How to Bait with living Sand-Eels	63
The Courge or Sand-Eel Basket	66
How to make Courges or Sand-Eel Baskets	67
How to Bait with Rag-Worms	68
How to Bait with living Shrimps	69
Horse-hair Lines, and How to Make Them	69
Pipe-leads and Moulds	72
To Knot the Links together	73
A Day's Drift-line Fishing off Guernsey	74
Streaming for Pollack	78
Whiffing	79
Lines for Whiffing	79
The Cornish Whiffing Line	81
Baits for Whiffing, Natural and Artificial	82, 83
Fly-fishing at Sea, the Rod	86
Use of the Gear or Tackle	90
Boat and Net	91
Sheaf-fishing	92
Rod-fishing for Pollack from Shore	93
Floats	94

CONTENTS.

	PAGE
Rod-fishing with a Light Line, the Pater-Noster Line	95
The Floating Trot	96
The Coal-fish	98
The Whiting-Pout	99
The Power or Poor-Cod	101
A Day's Ground-fishing	101
The Dab	115
The Flounder or Fluke	119
Flounder-spearing, the Fork	121
The Fluking-pick or Pike	122
The Halibut	122
The Mackerel, Sailing-boat for Mackerel-fishing	123
Lines, Leads	124
Baits	126
Booms or Bobbers for Mackerel-fishing, Artificial Baits	127
Ground Mackerel-fishing	128
A Day with the Mackerel	129
Mackerel-fishing at Anchor	134
The Scad or Horse-Mackerel	136
The Bass	137
Ground-fishing for Bass	139
The Bulter, Trot, or Spiller	142
Trot Basket and Hook Holder	143
'The Outhaul' Bulter	143
Harbour Ground-fishing from a Boat	144
Drift-line Fishing for Bass in Bar-harbours	145
Canvas Bucket for Bait	148
Whiffing for Bass	150
Angling for Bass	151
The Dory	152
The Grey Mullet	153

	PAGE
The Chervin, or Shrimp Ground-bait	154
The Smelt and Sand-Smelt, or Atherine	155
The Cod	158
The Newfoundland Cod-fishery	159
The Ling	164
The Sea-Loach or Rock-Ling, the Haddock	165
The Hake	165
The Sole	167
The Plaice	168
The Turbot and Drift-Trot	169
The Brill, the Wrasse or Rock-fish, the Sea-Bream	170
The Braize, or Becker	173
The Gar-fish, or Long-Nose	174
The Red Mullet	174
The Gurnard or Gurnet	175
The Conger	176
The Skate	181
Dog-fish and Sharks	182
The Herring, Pilchard, Sprat, and White-bait	182
The Freshwater-Eel, Bobbing	183
Clotting	184
Hook and Line	185
General Baits for Sea-fish, the Mussel	187
The Lug-Worm, the Rag-Worm, Rock or Mud-Worm	188
The White Sand-Worm	189
The Varm or Sea Tape-Worm, Earth and Lob-Worm	190
The Sand-Eel and Launce	190
Dipping Sand-Eels on the Surface	191

CONTENTS.

	PAGE
Freshwater-Eels	191
The Lampern or Lesser Lamprey, the Limpet	192
The Whelk, and Cuttle-fish	193
The Squid	194
The Sucker or Poulpe	195
Shrimps and Prawns	196
The Common Green Crab	196
The Hermit or Soldier Crab	197
The Solen or Razor-fish	197
Fish baits, Artificial Baits, Knots, Splices, and Bends	199
The Bowline Knot and Timber Hitch	200
The Killick or Sling-stone, and Yoke Anchor	201
Eye-splice and Short Splice with Two-strand Line	202
Crowning or Scowing the Anchor	202
Slipping the Cable, Belaying Thwart, the Anchor Bend	203
Hooks, Whipping and Bending on Hooks	204
Reels, Gaffs, and Bait Tray, and Table of Hooks	212
Gaffs, the Short-handled Gaff	213
Landing Net, the Bait Tray	214
The Fish-Basket	215
The Nossil Cock or Fisherman's Spinner	216
The Lester Cock Trot, and Sunken Lester Cock	218
The Otter	218
Crabs, Lobsters, Cray-fish, and Clothing	219
Cork Seats, Ocean-fishing	220
'The Grains'	221

	PAGE
The Triangle Net	222
Remarks on Nets, the Trammel	222
The Seine and Sand-Eel Seine	228
The Night or Small Seine	231
The Mackerel Seine	232
The Pilchard Seine	235
The Trawl	237
Size of Mesh	239
The Otter Trawl	240
The Dredge	241
Shrimp and Prawn Nets, the Pool Net	242
The Strand Net	243
The Baited Prawn Net	246
Drift-Nets	247
Moored Herring-Nets, Peter-Nets, Drum-Net, Tanning Nets	249
Boats and Boating	250
Spritsail Boat with Mizen	254
Guernsey Spritsail Boat	255
Lugsail Boat	257
The Yawl or Dandy	258
The Itchen River Rig	260
Prices of Boats	261
Centre-Board Boats	262
Remarks on Beach or Surf Boating	262
Launching from a Steep Shingle Beach	262
Launching from a Flat Sandy Shore	263
Beaching or Landing	264
Harbour Boating	265
General Remarks	266
Staying, Wearing, Belaying the Main-sheet	267

CONTENTS.

	PAGE
Beating to Windward and Scudding or Running	268
The Coble, Safety Fishing-boats	269-277
Improvement of Fishing-boats	270, 273-277
On the Management of Open Rowing-boats in a Surf, &c.	277
Practical Hints for the Consideration of Seamen and Others, &c.	283

APPENDIX.

Instructions for Saving Drowning Persons	285
Directions for Restoring the Apparently Drowned	286
INDEX	293

LIST OF ILLUSTRATIONS.

FIG.		PAGE
	Pilchard Bait for Hake, Cod, or Conger . . .	*Frontispiece*
1.	Chart of Part of Guernsey, illustrating Marks and How to Take Them	6
2.	*a.* The Southampton Rig . .	35
2.	*b.* The Guernsey Rig . . .	35
3.	The Kentish Rig . .	38
4.	The Dartmouth Rig	39
5.	The Grapnel or Creeper Sinker . .	40
6.	The Newfoundland or Banker's Lead	41
7.	Boat-shaped Rig. Lead and Section with brass wire, &c.	42
8.	Mould for Boat-shaped Leads	44
9.	,, ,, Longitudinal Section .	44
10.	,, ,, Cross Section	44
11.	Illustration of Lines	48
12.	Half-Pilchard marked diagonally for cutting into Bait	49
13.	Hook baited with piece of Pilchard, Mackerel, or Herring, &c., for Whiting	49
14.	Mussel Bait, first insertion of Hook	50
15.	,, ,, Hook completely baited . . .	50
16.	Living Sand-Eel Bait in tide-way	64
17.	,, ,, baited for slack tide (recommended by the late P. le Noury	64
18.	Living Sand-Eel Bait; ordinary method at slack tide	65
19.	The Courge or Sand-Eel Basket	66
20.	,, in Tow	66

LIST OF ILLUSTRATIONS.

FIG.		PAGE
21.	Hook baited with Rag-Worms	68
22.	,, ,, Living Shrimp	68
23.	Spinning Machine or Jack	70
24.	Knotting Hair-links	73
25.	Freshwater-Eel (dead bait for Whiffing)	82
26.	River Lamprey (dead bait for Whiffing)	82
	India-Rubber baits, consisting of—	
27.	Rubber Band Imitation Rag-Worm	83
28.	Brooks's Double-Twist Spinning Eel or Lug-Worm	83
29.	Hearder's Captain Tom's Spinning Sand-Eel	83
30.	Tail part of an Eel (Whiffing bait)	84
31.	Rag-Worm when fish are shy	85
32.	Earth-worm baited for Whiffing	85
33.	Flies and Feather Baits	89
34.	Paternoster and Pipe-lead and Trace for Rod-fishing	95
35.	Floating Trot	97
36.	Mode of spreading Lines round the Boat in Dab or Flounder-fishing	117
37.	Flounder Fork	121
38.	,, Fluking-Pick or Pike	122
39.	Plummet-lead and Revolving Chopstick	124
40.	Baited Hook for Mackerel-railing	126
41.	Tobacco-pipe Bait for Mackerel	127
42.	Red Sand-Eel, cut in two for bait	135
43.	Leger-lead, Trace, and baited Hooks	140
44.	Bulter, Trot, or Spiller, for Cod, Conger, &c.	142
45.	Pater-Noster, for Smelts or Mullet	156
46.	Boat-shaped Lead and Trace with Copper Swivels, and Swivel the actual size	177
47.	Conger Hook, with Snood traced over with green hemp	178
48.	Hook baited with the tail half of a small Whiting, Mackerel, or other fish; for Hake, Cod, or Conger	179
49.	Lead and Clot of Worms for bobbing from a boat	183

LIST OF ILLUSTRATIONS.

FIG.		PAGE
50.	Baited Clotting-Pole	184
51.	The Lug-Worm	188
52.	The Razor-fish Spear	198
53.	The Overhand or Common Knot	199
54.	The Bend	200
55.	The Bowline Knot	200
56.	The Timber Hitch	200
57.	Stone Killick or Sling-stone	201
58.	a. Eye-Splice commenced	202
58.	b. Short two-strand Splice	202
58.	c. Anchor scowed	202
59.	The Anchor Bend	203
60.	Whipping and Bending on Hooks	206
61.	Whipping and Fastening off a Hook	208
62.	Exeter Round-bend Hooks, Nos. 1 to 7	210
63.	,, ,, ,, ,, 8 to 14, &c.	211
64.	Reel, Gaffs, and Bait-Tray	213
65.	Guernsey Fish-Basket	215
66.	The Nossil-cock, or Fisherman's Spinning Machine	216
67.	Pork-skin bait for Ocean-fishing	221
68.	Side View of Trammel-net	224
69.	End View of Trammel-net	226
70.	Sand-Eel Seine	229
71.	The Trawl	237
72.	Trawl, with Beam and Irons	238
73.	The Otter Trawl	240
74.	The Pool Shrimp or Prawn-net	243
75.	The Strand-net	244
76.	The Baited Prawn-net	246
77.	Midship Section of Beach Boat	251
78.	,, ,, Harbour Boat	251
79.	Spritsail Boat with Mizen	254

LIST OF ILLUSTRATIONS.

FIG.		PAGE
80.	Guernsey Spritsail Boat	255
81.	Lugsail Boat with Jib and Mizen	258
82.	The Yawl or Dandy	259
83.	The Itchen River Rig	261
84.	Boat Carriage for a flat sandy shore	264
85.	Belaying the Main-sheet	268
86 to 93.	The Safety Fishing Boat of the Royal National Life-boat Institution	273, 274
94 to 97.	Four Illustrations of Directions for Restoring the Apparently Drowned	288–291

THE
SEA-FISHERMAN.

INTRODUCTION.

AMONGST all the useful arts, there is probably not one of which the knowledge has been so much confined to those who make it their vocation, as Sea-Fishing. Any detailed accurate accounts of the modes of capture of our best known fish—such modes of capture as the amateur would wish to avail himself of—were formerly unknown, the difficulty of accumulating facts having always rendered the progress of information slow. The requisite practical knowledge cannot be acquired by a short visit to the sea-side and an occasional day's fishing; neither must it be thought that the methods of one particular locality will suffice for all others : on the contrary, it is only by a long residence at different points of the coast, combined with the practice of sea-fishing as a pursuit, that the required information can be collected regarding the various methods called into action by the varying circumstances of different localities. There can be little doubt that the greater facilities of transit afforded us by the extension of the railway system—inasmuch as they have vastly increased, and in many cases created a great passenger traffic to the coast—have been the means of causing a demand for information on the subject to which this work is devoted. Sea-side visitors must find some occupation *pour passer le temps* : the monotony of the marine parade becomes wearisome, notwithstanding the attempts at its alleviation by calling in the aid of the last new novel from the library—but they cannot read all day ; walking in the heat is not agreeable ; everyone is not interested in the study of marine zoology, fashionable as it has become of late years, and deservedly so, although the *furore* has now somewhat abated ; the common objects of the sea-shore fail to afford amusement ; they have had a complete surfeit of German bands ; mind and body are alike satiated with that fashionable and intellectual amusement so perseveringly followed by beach haunters of all ages, to wit, pelting Father Neptune with the pebbly shingle where there is any to the fore ; even the row or the sail require some additional zest. What, then, is there at hand to supply the *desideratum*? Nothing—positively nothing but Sea-fishing, which in its various phases, afloat or from the shore, affords a field of observation and occupation of which the world at large can have but small conception. Considering the variety of sport to be derived from sea-fishing, its votaries have not been many in number until the last thirty years ; latterly they have much increased, and sufficiently to induce a few among them to indite their experiences for the instruction of their fellow-sportsmen. The author has frequently observed how very partial is the knowledge of many amateurs of sea-fishing, and how much sport they lose in consequence of their limited acquaintance with the subject ; he has endeavoured, therefore, in this work to supply the *desideratum*, feeling he might venture so to do from the life-long experience he has had on various parts of the coast of England and the Channel Islands. Regarding the fishing of these islands, the methods of taking, preserving, and using the living Sand-Eel for bait, are so excellent, and have been hitherto so little known on the British side of the Channel, that the author has given them a prominent position in both description and illustration, and he is happy to say that former residents in the islands are following out these methods on the British coasts, over the whole extent of which they ought to be disseminated, as superior to all other methods of coast-fishing for Mackerel, Bass, and Pollack in particular, and eminently useful when applied to the capture of other fish.

INTRODUCTION.

The intention of the author is that this work should be found a truthful description of Sea-fishing, and a *vade mecum* for reference and instruction, both for the British and other seas; for, rely on it, any man who is practically acquainted with the varieties of sea-fishing round our own coasts will be able to take fish on the coasts of all the world. This is not a mere assertion of the author, but the result of the experience of British fishermen who have found their knowledge available from Newfoundland to the Antipodes; and in the latter region particularly on the coasts of Australia and New Zealand, where large quantities of fish have been taken by the same methods.

A careful inspection of the gear in the Great International Exhibition of Fisheries held in London in 1883 has quite confirmed the author's views as here stated.

Throughout this work the author has endeavoured to keep before him the necessity of describing and depicting the different kinds of gear and baited hooks, with which, above all things, the sea-fisherman ought to be familiar. Another matter to which particular attention must be paid is the *habitat* of the fish, as well as that of the various worms, shell-fish, &c. used as bait, which will be of great service in discovering their haunts at any particular locality. By aid of a chart he should endeavour to ascertain the nature of the bottom, positions of rocks &c. which practice has been specially recommended in an article devoted to the subject. In the following article on 'Marks and how to take them,' accompanied by an illustrative chart, plain directions for fixing the positions of fishing-grounds in order to find them on a future occasion have been laid before the reader. Proficiency in any art cannot be attained without considerable attention and practice, to which Sea-fishing is certainly no exception; and when to this is added the necessity of an acquaintance with the management of a boat, if one really wishes to go thoroughly into the subject, the reader will perceive he has need to exercise no small amount of forethought, as well as bring into play a not inconsiderable amount of personal activity, if he would be anything more than a stern-sheet fisherman, an animal guiltless of letting go, or getting up an anchor, spritting a sail, bowsing on a tackle-fall, *et hoc genus omne*, added to which he stows himself abaft the after-thwart to eat, drink, and—yes, oh! climax of effeminacy!—to have his hooks baited!

Marks and How to Take Them.

Fishermen, from time immemorial, have, in the exercise of their vocation—namely, in trawling, dredging, setting Crab and Lobster pots, hand or long-line fishing—discovered certain 'grounds,' as they are termed, frequented by different varieties of fish, according to the nature of the bottom. It was of course desirable that these discoveries should not be lost; fishermen, therefore, have contrived methods of taking 'marks,' that these positions may be revisited when required. A boat or vessel is placed in any one of these positions by aid of two imaginary lines drawn from objects on shore, and crossing each other at the said boat or vessel. Two objects are chosen for each imaginary line, which two objects being seen on with, or over each other in one direction, and two other objects being seen likewise over each other in another direction, the position of any particular fishing-ground may be determined with facility, and revisited as occasion may require. As great precision is necessary in some kinds of fishing, the marks selected should not be too close to each other, neither should the lines of direction be too oblique. If too close together, their relative positions will not appear to alter rapidly enough to enable you to detect a change in the position of the boat instantly, and the result will be similar if the lines of direction are too oblique, because you will be too long passing the intersection of the cross bearings. Some degree of obliquity is generally unavoidable; but where a choice of objects exists, select those which will afford you as near an approach as possible to a right angle, or, as fishermen term it, take your marks 'upon the square.' By adhering to these cautions, your marks will be 'quick marks,' as they are called, and if well selected will alter even by turning your head right or left. The smaller the objects, the nearer they may be to each other; the larger they are, the greater should be the distance between them. If the distant object be of considerable magnitude, and the nearer object of small size, it will be necessary for the observer to bring the smaller object in front of the centre or either end of the larger object, in order to attain the necessary precision. The objects most commonly used are of course the most conspicuous which can be rendered available from the position—always remembering the foregoing cautions in selection—and consist of remarkable houses, chimneys, towers, castles, churches, obelisks, windmills, trees, rocks, small islands, points, headlands, hills, mountains, flagstaffs, beacons, or lighthouses. We have, up to the present time, been dealing with theory, but, with the reader's permission, we will endeavour to reduce our theory to practice, that is, as far as we are able to do so on paper.

For this purpose I have appended the accompanying small chart, which will, I think, render my instructions sufficiently clear to those unacquainted with the subject, whilst the experienced will, I trust, therein recognise the practical nature of the instruction conveyed. This chart (fig. 1) represents a portion of the sea coast adjacent to the harbour of St. Peter Port, Guernsey, with a set of marks and lines of sight in use for finding a certain fishing-ground, on a sufficiently large scale (four inches to a mile) to avoid confusion of lines. It is required at half an hour before low water spring tides, direction of the stream SW., wind NNE., to place a boat at a certain position according to the following marks. The De Lancey Obelisk, M, its own breadth open east of the breakwater lighthouse, and the gap in the back-land touching the sharp edge of the eastern cliff of the shingle bay. A is the line of direction of the obelisk, M, passing outside the lighthouse on the breakwater, and forming the long mark ; B the line of direction given by the gap in the back-land with the edge of the cliff, cutting the line A at the boat's required position at F. After rounding the breakwater, steer to the southward, getting on the mark A ; keep on this line of direction until the mark B is discovered, which is the cross mark, and shows you to have arrived at the spot indicated. The killick, or mooring stone, must not be let go here exactly on the spot, for when any amount of rope were paid out, the boat would of course be much beyond the required

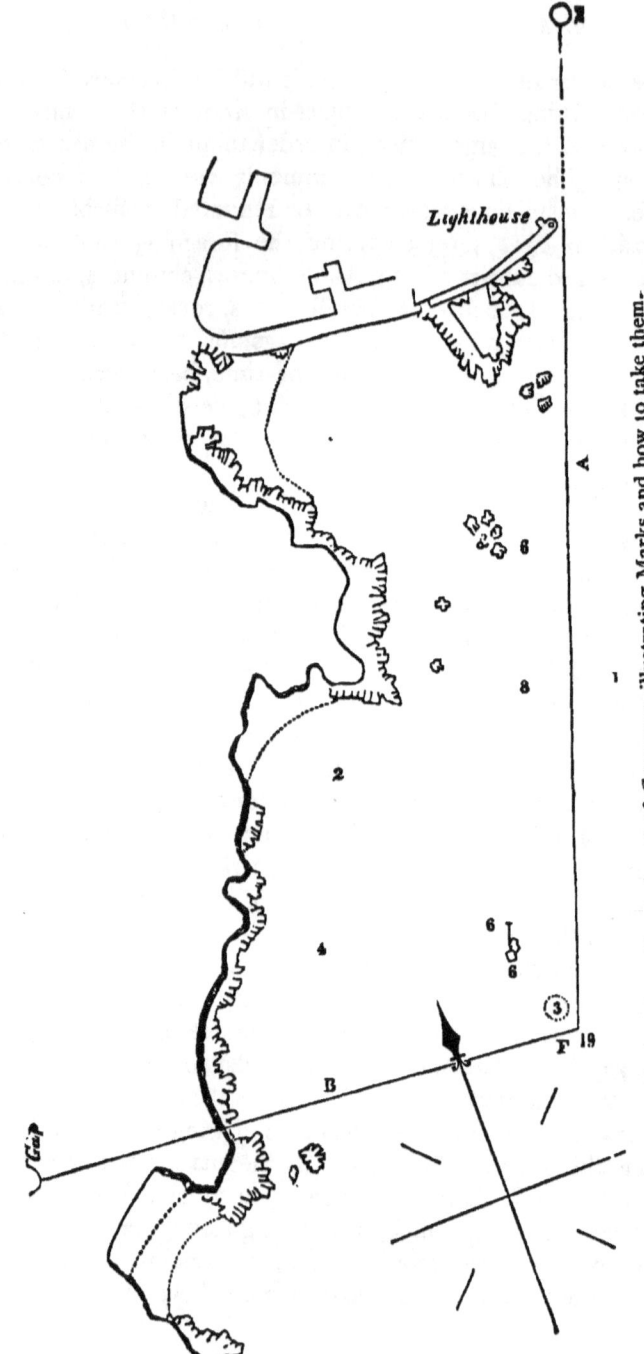

FIG. 1.—Chart of part of Guernsey, illustrating Marks and how to take them.

position, but instead of so doing, the boat should be rowed back again a short distance, until the cross marks are a little open of each other, or one of them is hidden, as the case may be, taking into consideration the direction of the wind and tide, and then drop the killick, or sling-stone, veering out the cable until the marks are correct. A little practice will soon enable anyone accustomed to sea-boating to attain the necessary precision. In order to bring up conveniently, your cable should certainly be half as long again as the depth of the water where you purpose to fish, that you may have sufficient scope to veer and haul upon; but in bringing up at a place which requires great precision, veer as little as will suffice to hold, as you will not then swing so far out of the spot when the boat sheers on one side by the force of the current or flaws of wind. The depth of water being marked at nineteen fathoms, use a rope not less than thirty in length, as much scope is sometimes required if the wind freshens; on sand it is of course best to bring up with an anchor. On a rocky bottom, a sling-stone or killick should always be used in lieu of an anchor, which frequently gets irrecoverably hooked in some projection or crack in the rock (see fig. 57, p. 201). If the ground be of a mixed character, the anchor may be 'scowed' (see the illustration, fig. 58, p. 202). After the boat is brought up, if you find you are somewhat to the left of your position, make fast your cable about two feet from the stem of the boat on the port side, which will cause the boat to tend to the right; but if you are to the right of the required position, make fast the cable on the right or starboard bow, which will cause her to move to the left. This is termed 'putting a boat on the sheer,' and in a tideway the helm may be lashed sometimes with advantage, as an additional aid. Another method of taking marks, but less commonly used, consists in seeing one object over another at a considerable angle, such as the top of a hill over a narrow dip or depression in the edge of a cliff, or the top of a tower, summit of a steeple, vanes of a windmill, or base of any building, seen or just hidden as the case may be; for instance, near Budleigh Salterton, Devon, is a fishing-ground known as 'Two Stones,' the mark for which is the high hill called Shaldon Beacon, beyond Teignmouth, just visible in the dip of the land at the outer part of Strait Point. Other instances of combinations of this kind will be from time to time met with, and are also useful for keeping vessels clear of dangers near the coast. A third method, when great precision is not requisite, is to set two objects on the land by compass : that is to say, to place the compass in a convenient position, and to find one object bearing east and another north, or on any other bearings, so that they may form with each other a right angle, or as near an approach thereto as possible, or contain seven or eight points of the compass between them ; thus taking the bearings on or nearly on the square. This method can be adopted by anyone at all familiar with a mariner's compass, who may be a stranger to the locality, when in case of obtaining sport, marks may be taken for future use. It will answer well where the ground is of the same nature for a considerable distance, as is often the case on Whiting and Dab grounds, or where Cod and Haddock abound, but rarely in Pollack, Bream, or Pout-fishing, where very great accuracy is often necessary.

METHODS OF FISHING.

The different methods of taking sea-fish may be briefly comprehended under the following heads, namely : hand-line fishing, long-lining or trotting, net-fishing and spearing. There is also a method of taking fish by weirs, gradually becoming obsolete. Hand-line fishing has two great divisions ; namely, at anchor and in motion. Fishing at anchor, or moored with a stone killick, subdivides itself into ground and drift-line fishing, of which ground-fishing may be defined to consist in using a single heavy lead as a sinker at the end of the line, on or close to the bottom ; whilst in drift-line fishing, leads at intervals are placed as sinkers, usually at the distance of two fathoms, and from twelve to forty to the pound ; besides which, drift-lines are also used entirely without lead, at the stern of the boat ; and sometimes with a cork float to assist in taking out the line when little tide is running, a method much in vogue at Plymouth, in which case one light lead only is used. Fishing in motion includes reeling or railing, and whiffing, both which signify the act of towing lines after the boat; the former when under sail, the latter when sculling or pulling slowly. Whiffing is in some parts also applied in the same sense as reeling or railing, but it would be better to confine it to fishing in motion, by rowing or sculling, to avoid confusion of terms. Reeling or railing may be defined to consist in towing a lead, of a pound weight and over, after a sailing-boat.

We may also include under hand-line fishing, the method of throwing out a leger-line with a weight at the end from the shore, as from piers, quays, shelving beaches and steep rocks; together with angling with a long strong rod from any of these positions or from a boat. The use of trots, bulters, long-lines, or spillers, which are in fact synonymous terms, is another kind of ground-fishing in which the lines are left to themselves for a greater or less time, according to circumstances, sometimes being shot over night and hauled next morning, the management of which is so different from other lines that we may very well class it by itself. The distinguishing characteristic of this kind of line is that it consists of a long line or back with snoods and hooks at intervals, the whole affair being sunk and moored by stones or anchors. In another part of this work descriptions and illustrations are given. There yet remains net-fishing, and the use of spears and harpoons; of the former I have made mention under the heading of 'Remarks on Nets,' and of the latter, under the 'Flounder' (pp. 121, 122), and the 'Grains' (p. 222). The usual fish taken with ground-lines are Pout, Whiting, Haddock, Cod, Ling, Bream, Dabs, Flounders, Plaice, Gurnards, Eels, and Congers; in drift-line fishing, Pollack, Sea-Bream, Mackerel, Bass, and Gar-fish or Long-Noses; in reeling and whiffing, the same as in drift-line fishing, the Sea-Bream, however, only exceptionally.

In leger-fishing or throwing out a line from the shore, Bass, Congers, Pout, Pollack, and Wrasse, with Eels and Flounders from certain quays and piers, and exceptionally Red Mullet. In angling with rod and line, the same as with the leger-line, also, Atherine and Coal-fish, and in some places off steep rocks, Sea-Bream, Mackerel, and Horse-Mackerel or Scad. On trots or bulters, every kind of ground-fish, including Turbot; and on the floating-trot, Pollack and Gar-fish, and sometimes Bass. Every variety of fish found in the British seas is taken at some time or other in nets; and the Flounder and Eels by spearing. The Grey Mullet, although very abundant round our southern coasts, are seldom taken in any quantity with rod and line. The exceptional places which have come to my knowledge are Dover, Southampton, Plymouth, Penzance, Jersey, and near Bideford, and Barnstaple, at all of which excellent catches are occasionally made. They are rarely taken by spearing, this amusement being more followed in warm climates, particularly in the Mediterranean, where it is carried on by torchlight. The Bonita, Dolphin, Albicore, and King-fish are often struck with the 'Grains' in warm latitudes; and the harpoon is brought into requisition for the Shark, the Porpoise, and the Whale. The Shark, as is well known, is also often taken with a chain-hook, and is sometimes noosed in a bowline-knot, being coaxed to swim through it by a bait held in front. The chain must have a very strong swivel at the top.

THE POSSIBLE BAG

is very useful for travellers, who naturally wish to add as little as possible to their *impedimenta*. Let it contain a brass tobacco-box, with loose hooks and swivels, two or three hanks of snooding wound on cotton reels, a hank of gut, also a skein of thread, and a bit of shoemaker's wax in leather. These and some other matters the sea-side visitor will find mentioned under 'Gear or Tackle required in Sea-fishing,' pp. 12 and 13, for carrying which the bag is very useful.

SHIFTING LEADS FOR SEA-SIDE VISITORS.

Some kinds of ground-fishing leads are not calculated from their shape for railing or towing after a boat, but railing or reeling plummets may be used for ground-fishing, especially in the Kentish Rig (fig. 3, p. 38). The object in shifting is to avoid the weight of two sets of leads for sea-side visitors, and the exchange of chopsticks and snoodings as well as the strops is soon effected by casting off the line from the strops, drawing off the chopsticks, and substituting the other strop with the required chopstick and snood for the first form of gear, when the line is to be bent on as before.

Each form of gear must be in itself complete, having its own strop, for the Kentish Rig requires a double strop, the railing or reeling gear a single one, and a single strop would allow too much play to the Kentish Rig, whilst the double strop would be too large to pass through the hole of the reeling chopstick. See cuts, fig. 3, p. 38, and fig. 39, p. 124.

SEA-FISHING AND HYDROGRAPHY.

As the nature of the soundings generally determines the kind of fish to be taken off any coast, an acquaintance with local Hydrography will be of great service in the absence of information which the old hands alone possess, and are sometimes rather chary of affording to strangers. I need not inform the experienced that Pollack are found off the rocky points and on or close to foul ground, Whiting usually on soft ground, and Dabs on both oozy bottoms and hard sand, Mullet and Sand-Smelts at the mouth of streams or drains, &c.; but to all, whether previously well acquainted with general sea-fishing or not, I strongly recommend the purchase of the chart or plan of that particular harbour, arm of the sea, or section of coast they may visit, on as large a scale as possible, as a useful preliminary to the acquisition of the necessary local knowledge. The Admiralty charts are recommended, and are sold by Mr. J. D. Potter, 31 Poultry, and Mr. E. Stanford, Charing Cross. They run from one to four shillings each, according to extent. For a certain kind of information you must rely on residents of the locality, such as the time of tide to fish certain spots, where to procure bait, and the very precise marks at times requisite for Pout-fishing. This, however, should not prevent your informing yourself as to the general features of the harbour, coast-line, and offing, which the chart or plan will afford, thus gradually rendering you independent of external aid, and bringing you acquainted with various ground which might otherwise escape your notice. Let us suppose a stranger visiting Plymouth, tolerably acquainted with sea-fishing and desirous of taking Pollack ; on referring to his chart, he will find many suitable grounds, namely, for shore-fishing in the east corner of Millbay, at the mouth of the stream for Smelts, off the pier or pontoon for Pollack and Mullet, and at Flat Rock for Pollack on flowing tide.

The entrance gates of the Great Western Docks are much frequented for Grey Mullet angling.

For boat-fishing, round the Cobbler Buoy at the entrance of Catwater, either whiffing or moored, for Pollack ; and other places too numerous to mention. (See Plymouth, p. 22.)

The advantages of an intimate acquaintance with local charts are incalculable, for having first made yourself acquainted with the habits of fish, you will thereby be enabled to determine their haunts.

Gear or Tackle Required in Sea-fishing.

This varies much in different localities ; instead therefore of giving a long list of articles, I shall just mention a few lines which will be useful everywhere, leaving my readers to gather from the following pages what will meet their requirements, according to the kinds of fish of which they may go in quest. The few following lines &c. are far from expensive, and may be obtained from any sea-fishing tackle manufacturers. The materials of sea-fishing gear, viz. lines, snoods, and hooks, can be purchased at almost all sea-side towns and villages, but places where really good gear is sold ready fitted for use are yet very few and far between, practical information on this subject being far from as widely diffused as is desirable.

The amateur had better therefore provide two lines for

ground-fishing, one with a sinker of one pound, and another with a sinker of two pounds' weight, either of the boat-shape (fig. 7, p. 42), or Kentish Rig (fig. 3, p. 38). Also a pair of two-pound Mackerel leads, with two fine Whiting lines, each thirty fathoms, on reels six inches by five, for convenience in packing. The lines can be attached to either the Mackerel plummets or ground leads and shifted at pleasure, a great convenience for tourists; but residents have each line for its own work generally—much the preferable method of the two. A few spare snoods, a hank of strong gut, twenty-five white tinned hooks, No. 11, for either Pollack or Mackerel; twenty-five No. 13, for Pout, Dabs, or Flounders. (See the cut of hooks, fig. 63, p. 211). Two horse-hair lines, lightly leaded, as described under 'Pollack,' on reels, six inches by five, for drift-fishing for either Pollack, Mackerel, or Sea-Bream, which may also be used whiffing under oars, but not under sail from the great risk of breakage. The two first-mentioned lines, intended to answer a double purpose, should be of two strands only, and marked at the required length, as described under 'Mackerel.' The stock of lines &c. can always be increased as the fisherman becomes familiar with his sport, which will be a better course than encumbering himself with a large quantity of gear which he may not require at the locality he may visit.

THE SOUTH-WEST COAST OF ENGLAND, CHANNEL ISLANDS, &C.

In the present article I have mentioned places with which I am personally familiar on portions of our coasts, where all the various methods of taking fish in the British and in many foreign waters, are in practice.

Portsmouth.—Reeling or sailing for Mackerel is much followed by the Gosport fishermen in June, July, August, and September, at the east end of the Isle of Wight, near the Nab Light Ship, opposite Bembridge, which is as near to Portsmouth as any luck can be expected, as it is quite an exception to take any number of Mackerel closer to the shore.

Whiting-Pout near the Buoy of the Boyne, outside Southsea Castle and in various other spots. In the harbour at the moorings of the 'Boys' Brig,' but here usually they are very small.

Whiting at Spithead, likewise small as compared with fish farther west: Bass are sometimes taken off Blockhouse Point Beach by line, and by shooting a trot in the mouth of Haslar Creek. Best bait, Cuttle-fish. Sand-Smelts in Portsmouth Harbour.

Isle of Wight.—Whiting-Pollack (here called Whiting-Cole) off Sea-View, near Bembridge Ledge, and also off other rocky points at the back of the Isle of Wight, with Mackerel anywhere off the last mentioned in the summer months. Good long-lining or trotting off Sea-View, lug-worms abundant. A Sand-Eel seine is kept at Sandown, but is seldom used after July.

The Solent.—The fishing in the Solent is not of much account at the present day, whatever it might have been formerly, and consists of some Pouting and Whiting catching, and trotting for small Conger &c., with Smelt-Fishing, in the various rivers emptying themselves therein, in which also Flounders, Eels, and Mullet will be found. Bass at the mouth of Lymington Creek, and other harbours.

Southampton.—Sand-Smelts and Mullet in the docks and from piers and quays on the Itchen River. Whiting very numerous but small, at the lower portion of Southampton Water and near the Bramble Shoal; also off Hythe. Trawling in Southampton Water. Many fine Eels are taken both by hook and line, bobbing and spearing. No bait is equal to the soft Crab for line-fishing, for Eels and Flounders in a tideway.

Christchurch and Poole.—Christchurch Bay affords Mackerel-fishing, which may be understood to be almost universal on the coast westwards during the summer months, and the rocky ground of Christchurch Ledge will furnish Whiting-Cole; in this and Poole harbours, Sand-Smelts are very numerous, with Eels, Flounders, and Grey Mullet. Wild-fowl shooting is good here in hard winters. Bass in the mouths of both these harbours.

Swanage.—Round Peverel Ledge, and Durlestone Head to

St. Alban's—a corruption of St. Aldheim, is good Pollack-fishing. Eels for bait procurable in the brook. Living Shrimps much used as bait. Bass also here.

Weymouth and Portland.—Between Poole and Weymouth, Mackerel, and near the rocky ground Whiting-Pollack, and Pout; and Whiting in the offing. The same in Portland Roads, and the Pout are taken from the breakwater and its immediate vicinity. The wreck of the old Indiaman 'Abergavenny' is also much frequented for ground fish. In Weymouth Harbour, Eels, Sand-Smelts, and Flounders, with Bass occasionally off the jetty end in rough weather. By watching your opportunity, you may frequently procure a Squid or Cuttle-fish when the Mackerel seines are drawn on shore, on any beaches between Portland and the Start Point. Off Chesil beach, which connects Portland with the mainland, Mackerel are sometimes taken in large quantities, particularly in easterly winds, which make smooth water on this part of the coast. Bass may be taken by a ground line thrown off the beach, when there is too much sea to go afloat. In the Fleet or backwater behind, running from Abbotsbury, Eels, Flounders &c. abound, and Bass may be met with at the bridge. At the opening in Portland Breakwater, Pollack, Pout, Bream, Congers, and Sand-Smelts may be caught with rod or line, with or without a boat, and sometimes Bass with spinning bait. Between Portland Castle and Fleete Bridge, large Bass can be taken on the grass banks with a fly, and fine Red Mullet in trammels at the anchorage.

Just outside the light of Weymouth Harbour is a rocky shoal, the Mixon. Pollack are taken here, but there is better fishing along the north shore on the ledges towards St. Alban's Head. Close at the back of Weymouth Jetty or north pier, good Bass fishing has been obtained, and angling for both Bass and Pollack may here be followed.

Bridport.—Mackerel-fishing in the offing, and Bass from the pier-heads, either with a ground-line or the rod; Eels and Flounders, also Whiting-Pollack on rocky ground off this harbour, from which one spot receives the name of Pollack stone. Whiting-Pout, called provincially Blains, and Dabs on sandy ground. A few miles in the offing Whiting both numerous and large in autumn and early winter, with some Cod &c. ; in fact an increasing improvement going westwards from the Bill of Portland. There is a little Trout-fishing in the River Bride, which river is used to scour the harbour, and small freshwater Eels for Pollack-fishing may be here obtained.

Lyme.—Sea-fishing as off Bridport. Good Pollack-fishing along shore between Lyme and Seaton.

Seaton.—General sea-fishing good. At Axmouth Harbour, at the east end of Seaton Beach, ground-fishing for Bass and rod-fishing from the ruined pier and quay. Trout-fishing in this river, which has still Salmon, Salmon-peel, and Sea-trout. Squid for Bass to be procured from the Mackerel fishermen or Beer trawlers. In the harbour soft Crabs are sometimes obtainable.

Beer.—A fishing village. Pollack off Beer Head and variety of fishing in the offing.

Sidmouth.—Sea-fishing as before mentioned, but a very wild shore, and shallow at low water, which, when the wind blows from seaward, causes a dangerous surf. The little river Side has Trout, Eels, and Lampreys.

Budleigh Salterton.—Large quantities of Mackerel are sometimes taken in the bay between Sidmouth and Budleigh Salterton in the summer and Whiting in winter. Here I once took in half a day 412 Mackerel with hook and line, and near a thousand each were taken by boats fishing the whole day. Whiting have also been taken more than two feet in length and in large quantities, but of late years the fishing has a great deal deteriorated. Dabs are found on sandy ground in this neighbourhood, and at Budleigh Salterton on the rocky ground opposite the mouth of the Otter River, Whiting-Pollack ; and shoals of Bass are seen sporting, but cannot often be taken with the fly here, as in boisterous weather, most favourable for the fish to feed, the sea runs very high on this open coast : at such times, however, I have had good sport with the ground lines, throwing them off the beach in the river's mouth, and sometimes just westward of it ; the fish varying from six to twelve pounds' weight each. Rag-worms for Pollack procurable in the mud of the Otter estuary, and Mussels for ground fishing are brought from Lympstone. Fisherman, John Middleton.

Whiting-Pout—here called Blains or Blinns—are frequently abundant on the rocky ground in front of Budleigh Salterton, a few minutes' row from the shore, also east and west of it; and Dabs are caught on sandy ground along this part of the coast.

The Trout-fishing in the river is strictly preserved until five miles above the tide, after which permission may be obtained.

Eels, Mullet, and Flounders are found in the river, and in the ponds inside the embankment.

Exmouth.—Off Straight Point, between Salterton and Exmouth, large Whiting-Pollack are taken with a dead Sand or Freshwater Eel, or with flies &c. Good sport has been had here with the living Sand-Eel (*à la mode de Guernesey*), which may sometimes be obtained at Exmouth: there being two Sand-Eel seines, but not regularly worked.

Whiting-Pout frequent this locality also, and Dabs the sandy ground to the westward and eastward. Flounders are taken in great numbers in Exmouth harbour with the soft Crab, and Bass are frequently plentiful. During the summer the Mackerel-fishing is much followed with hook and line off Exmouth, and with seines on the whole extent of the coast-line, from Portland to the Start Point, in suitable situations. The best Mussels on this coast are procured in Exmouth and Teignmouth harbours, and every yachtsman should secure a peck or two at least, as they are such excellent bait for all kinds of ground fish. Beds are laid down at Lympstone, but boats are often dredging near the coast-guard station at Exmouth. Good whiffing under sail for Bass. I have taken numbers of Bass from a boat moored to the jetty using drift lines with living Sand-Eel, and occasionally both Pollack and Bass from the jetty with a rod. Fisherman, H. Parker.

Dawlish.—Dawlish is the next watering-place, off which is good fishing for Dabs, Mackerel, and Pollack &c., and with westerly winds and moderate weather, good sport is frequently obtained. There is no harbour here, but the railway breakwaters afford some shelter from the swell, and its position towards the west land favours it.

White sand-worms for Pollack-fishing are obtained from the beach and from Exmouth harbour, and small Mussels for Dab and Pout-fishing on Dawlish rocks. A Sand-Eel seine is kept here. The brook, in common with others on the coast, contains Trout up the valley, and small Eels for Pollack. Leger-fishing for Bass may be practised off the breakwater pier, at the mouth of the brook, with Squid or Cuttle bait from the Mackerel nets.

Teignmouth.—Teignmouth is a bar-harbour, in common with many others, the resort of numerous Bass, which formerly afforded excellent sport by fly-fishing for them. Of late years, however, they have not shown themselves so much on the surface, and this method of taking them has been almost given up, and has been replaced by drift line-fishing with the living Sand-Eel, fully described, and illustrated in this work. Fisherman, John Cox. At the Ness at anchor fish at spring tides, from two hours before until high water, outside east or west of Teignmouth, on falling tide. Straight Point, near Exmouth, six miles off Teignmouth, may be visited in a large boat, towing a basket of living Sand-Eels, with much success, for fine Pollack, with a northerly wind. Dawlish Pollack ground is only two miles from Teignmouth. Pollack begin to be met with 150 yards from shore abreast East Teignmouth Church.

A special article on 'Drift-Line Fishing for Bass in Bar-Harbours' has been added to this work, page 145, which see. Pollack of moderate size are caught at the Ness, and larger fish are met with nearer Hole Head, and towards Dawlish, and in the deeper water at Maidencombe close under the cliffs, two miles to the south-west, towards Babbacombe.

Mackerel, Pollack, and Dab-fishing can be followed at various spots in this neighbourhood; but the Silver Whiting proper are not to be obtained without going a great distance seaward, with the exception of small fish to be taken a mile outside the Ness. Late in autumn, during the Herring season, good sized Whiting may, however, be caught, especially during light evenings when drifting with Herring nets.

Mussels can be dredged in the harbour.

Nets are kept here for taking Sand-Eels, and much sport is now had in using them alive for Pollack or Bass, which method seems comparatively little known in England—although the chief one adopted in the Channel Islands, where it can be seen in perfection, particularly in Guernsey—all the spring and until the end of June, after which the Pollack are less plentiful. Sand-Eels are commonly taken just outside the harbour's mouth, at the back of a sandy shoal, and just before low water several boats may generally be seen proceeding to the spot ; also just before high water at Ferry Point. I have procured them from the fishermen, providing a Sand-Eel basket to tow them in after the boat. They will not remain sufficiently lively if placed in a bucket. See article, 'Courge,' page 66, under 'Whiting-Pollack,' where full directions as to lines and their use are set forth. Fish on the flood, abreast the Ferry Point. Sand-Smelts, Flounders, and Eels can be taken from the quay, and from a boat. Fly-fishing for Trout, Salmon, and Sea-Trout commences about eight miles above Teignmouth.

Babbacombe.—West of Teignmouth, off Babbacombe is the same kind of sea-fishing, and round the Orestone or Big Rock, and between the other insulated rocks at the entrance of Torbay, large Whiting-Pollack are found. The living Sand-Eel method has been tried here, as it is scarce an hour's run from Teignmouth with a fair wind. At Hope's Nose there is angling from the shore. Sport excellent at times, but best from a boat. Fisherman, Thomas, Anstey's Cove.

Torquay.—In Torbay, off Torquay, Paignton, and Brixham, the usual sea-fishing for Mackerel, Pollack, Dabs, &c. is attainable, and as the bay is sheltered by its western horn of Berry Head, the water is on the average smoother here than further east, and the three above-mentioned harbours render landing and embarking easy. Off Corban, Livermead, and Paignton Heads are often many Pollack, until middle of July, by whiffing, fish on falling water. The land here trending fast seawards, the Whiting ground is more accessible, and may be reached in from five or six to ten miles' distance, according to the time of the year, for, towards autumn, fish approach the shore more closely than in the early part of the season. The early part of summer has here been barren of Whiting for some years. Whiting are sometimes taken in abundance in

Torbay, but this is not the rule, it is the exception, as they are generally found in deeper water. A Sand-Eel seine is kept at Paignton, a good position to start from in consequence. Pollack and other fish may be caught by angling from the end of Torquay new pier, depth 16ft. at low water. There is also a good spot half a mile to the left of Torquay, close to the perforated rock known as London Bridge, the quarry rocks there being perpendicular, with deep water at the foot.

I have had good Pollack-fishing with drift lines close inside Thatcher Rock, with living Sand-Eel bait at spring tides, from two hours' flood to one hour's ebb. Fisherman, Cumming.

Brixham is one of the chief fishing towns of England, and employs a large number of trawling vessels and fishing boats. Whiting-Pout are often plentiful in Brixham Roads, commonly called 'The Sedge;' and Pollack and Bass towards Berry Head. Many Bass caught also at the trawlers' moorings in the Sedge. All along the coast between Brixham and Dartmouth, from the shore itself to a mile distance, excellent fishing is obtainable, on account of the deep water running so closely up to the land. Close to Berry Head itself there is great facility for rod fishing from the shore, for the rocks rise from the water like a wall, and admit vessels to lie alongside and load with limestone for days together during fine weather. Both Pollack and Bass range along these rocks, and can be taken by angling with or without a float, baiting with rag-worms. For Bass, Soft Crab, Cuttle Fish, Pilchard, and Squid are recommended. From the commanding positions which can be taken up on these rocks, spinning natural or artificial baits may be tried without difficulty. Abundance of Squid and Pilchards are brought into Brixham by the trawlers and seine fishermen. Immediately after rounding Berry Head, a large rock, called the Eastern Cod, is seen a short distance from the land. Abundance of fish, such as Bass, Pollack, Conger, &c., are found here. Bass have been seen in very large shoals on the surface on the flowing tide, and excellent sport has often been obtained, by casting with the fly, and with artificial bait, under favourable circumstances of wind and weather. Also from high water as the tide ebbs.

There is much rocky ground between this place and Dartmouth, affording Conger, Pollack, and Bream-fishing, and there are also two or three bays, with a smooth bottom, abounding in Dabs and other fish. Pollack, and sometimes Bass, may be met with off every rocky point between Berry Head and Dartmouth.

Dartmouth.—Dartmouth is a fine deep harbour, and sea-fish enter it in large numbers. Mackerel are found in it, and Whiting-Pout, and also small Whiting late in autumn ; outside, in the neighbourhood of the Mewstone, Blackstone, and other rocks, Pollack may be taken, also at and just inside the East Ledge Buoy and Homestone. With living Sand-Eels from Slapton good sport has been obtained at Dartmouth. At the inner part of the Mewstone several rocks above water jut out to the westward ; here the shoals of Bass are found on the flood-tide ; on the ebb they frequent the north-east side of this rock. Two-thirds of a mile more east is the Eastern Blackstone, much frequented by shoals of Bass ; you will see the gulls hovering when fish are abundant. During the summer and autumn months, good fishing for Pollack, Bass, or Mackerel has often been had by bringing the boat up, on the ebb, about fifteen yards inside and abreast of this rock, and fishing with drift lines and living Sand-Eel bait. When the Mackerel are plentiful they will take Pilchard gut well on a light line inside the harbour. Mussels for ground-fishing at a shilling a basket are obtained from Dittisham men, who bring them in their boats to the upper New Ground steps, generally in the afternoon in summer. A mile and a half outside the harbour, abreast of Blackpool Beach, there is Whiting and other ground-fishing ; but the chief ground is in the offing, the harbour entrance north, north by east, or north by west, and Prawle Point, just open of the Start. Fish at neap tides. Scarcely any Whiting fishing until middle of July. This fishing-ground extends many miles. Pollack at ebb tide at Kingswear jetty.

Slapton Lake.—Jack and Perch-fishing.—This lea or lake is five miles by sea from Dartmouth, and between it and the sea is a comfortable inn, the Sand's Hotel, the landlord of which keeps boats on the lake, and supplies tackle and bait to visitors.

In the neighbouring Start Bay all the usual sea-fishing is attainable, and large quantities of Sand-Eels are taken in nets by the fishermen, which would afford opportunity to try them alive, as bait for sea-fishing. Steam yachts from Dartmouth, calling at Slapton for Sand-Eel bait, have had good Pollack-fishing at the Start. Teignmouth, Paignton, and this are the last places on the coast, I believe, where nets are kept for the purpose, in Devon.

Although so plentiful, they are not used alive, but dead, and for Whiting catching. Slapton is one of the most eligible stations on the coast for a visitor, embracing both salt and freshwater-fishing within a stone's throw of each other, and a fine beach several miles in extent, with good sea-bathing. In the small brooks near, some Trout are found. The hotel at Torcross has been well spoken of by visitors. A new lake for Pike, Perch, and Rudd has been created at Torcross, where good sport is also obtained.

Start Point.—There is good angling for Pollack in Start Cove, and Whiting-Pollack &c. are to be caught near the Start Point and rocks, and near Peartree Head and its insulated rocks, and over the foul ground extending to Prawle Point.

Salcombe Harbour affords Mackerel and Pollack, which are also found in the Range or roadstead outside; Smelts, Mullet, and Bass are plentiful. Half-way up to Kingsbridge on the right is the Salstone Rock, close to which fishing from half flood until high water spring tides, myself and friends have taken from a few up to thirty Bass at a time, running up to six pounds each. Bait, living Sand-Eels from South Sands, used on drift lines at anchor.

To the westward is the usual fishing in the large bay of Bigbury, into which the two rivers Erme and Avon discharge their waters. In these rivers there is fly-fishing for Trout. They produce also Salmon and Truff, or Sea-trout.

Plymouth.—Plymouth Sound affords various fishing; Pouting near the Hoe by the Mallard Black Buoy: Pouting are found also near the White and Red Buoys, and in Cawsand Bay. Whiting-Pollack on the shoal water between Drake's Island and the main, at the Cobbler Buoy off Mount Batten, the Leek Beds near Bovisand, at the east end and back of the Breakwater, and the Red Buoy between the west end and the land of Mount Edgecumbe.

In the middle of Cawsand Bay I have taken a good many Dabs and an occasional Whiting, and late in autumn you may meet with a chance Cod. Off Penlee Point, between the shore and the Buoy, is the best Pollack-fishing in this neighbourhood, either moored or moving, and it is also a good spot for fly-fishing for Bass : there is sufficient shelter here from north-westerly winds. At Penlee Point you always try on the ebb tide, for on the flood the fish do not appear to frequent the locality. Whiting-Pollack are also found at the Knap Black Buoy, and further towards the Mewstone you will take them between the White Buoys of the Tinker shoal, also close under and inside the Mewstone and near the Shagstone, which last-mentioned spots are best in easterly winds. Between the Knap Buoy and Penlee Point is good cruising ground for Mackerel, which may there be taken before they have become plentiful inside the breakwater ; and late in the season Mackerel are found near the vessel in which the workmen live, and may be there taken with the inside of Pilchard for bait, whilst at anchor, and also in Batten Bay. The living Sand-Eels would answer well for Plymouth fishing, but no nets are, I believe, kept for their capture : the usual bait here is the smaller rag-worm, which can be procured from boys in Plymouth who collect them for sale. In this bay, and also near the Breakwater, much ground fishing for Mackerel is done from the middle of July until October. See 'Ground Mackerel Fishing,' page 128.

Mussels can be obtained from Saltash, and Pilchards from the fishermen in Plymouth Pool at the Barbican.

Sand-Eels are found at Cawsand and Bovisand Bays, at Yealm River Bar, and in Catwater ; they might be taken by a seine provided for the purpose.

Yealm harbour, six miles from Plymouth, and east of the Mewstone, is much frequented by Bass. Shoals often show up from half ebb to low water, in the Narrows inside the Bar, and at the Bar on young flood. Season, April to November ; Sand-Eels here. Net-fishing is private ; fly-fishing answers well here from a boat. It is best always at spring tides, in common with Bass fishing by other methods.

The ground for Whiting is from seven to ten miles outside Plymouth, more or less according to season, and boats are generally to be found there. Cod, Ling, Haddock, Gurnards, Bream, &c. are taken amongst them, and Pilchards and Hake at night. Hake-fishing has much fallen off of late years. July, August, September, and October are usually the best fishing months. Bass are also taken with the fly between Drake's Island and Mount Edgecumbe in the summer; and further up the harbour opposite Devonport Dockyard, near the West Mud, both Mackerel and Bass, in July, August, and September.

Shore Fishing.—Fishing from shore with rod and line from the following spots: Millbay Pier for Pollack and Mullet Sand-Smelts at the mouth of the Mill-stream falling into the east corner of Millbay, and Whiting-Pout from a boat twenty yards off the pier. Whiting-Pollack may be also taken off any steep rock found available.

Millbay is a convenient place to keep a boat, as from the dock wall it may be hauled in and out at a mooring. As the docks are owned by a company a small charge is made of seven or eight shillings per annum, which gives you the *entrée* at all times.

Some spots have been frequented for angling from the shore from time immemorial: and the favourite place of this neighbourhood is at Devil's Point, Stonehouse, where the hard limestone has been rendered quite smooth by the feet of successive generations. The water is here deep close to the rocks; in fact, there are twenty fathoms within a short distance. It is the custom to fish here for Pollack from low water until half an hour after high water, but it is next to useless on the increase of the ebb stream, as the line cannot be kept clear of the rocks, owing to the set of the current on the Point. A Government boat-house having been built here, the public are excluded from the greater part of this stand.

At Flat Rock, under Long Room Barracks, is another spot for Pollack-angling on the flowing tide. A float is commonly used here.

Looe.—On the rocky grounds off the coast of Cornwall, from two to six miles, the Pollack run very large, and are often abundant. Between Plymouth and Looe, Whiting, Mackerel, Dabs, &c. are caught the whole distance, according to the nature of the ground; numerous Whiting-Pollack are taken just outside the harbour's mouth, near Looe Island, and at other spots; Mussels are found in the river. Sand-Eels abundant here. Above the reach of the tide, and in the back streams, Trout may be taken.

Polperro.—Polperro is the next harbour, off which the fishing is much the same as at Looe, which, with every other small village or town on this coast, participates in the Pilchard trade.

The Udder Rock, a sunken danger half a mile off shore between Polperro and Fowey, is a noted place for Pollack, and has sometimes shoals of Bass, which latter here rarely take a bait.

Fowey, an excellent Station.—Fowey is the best and most extensive harbour on this coast, except Plymouth and Falmouth, of the same character as Dartmouth, but with less depth of water.

The railway station at Par is only four miles from Fowey, and an extension is now open to Fowey.

The inner Whiting ground is not more than three miles off the harbour, which is a great advantage. Mackerel and Whiting-Pollack are often numerous both inside and outside the harbour. Near the Cannis Rock, by Gribben Head—on which the sea-mark is erected—the latter are often very fine.

Trout are numerous in the streams tributary to the Fowey River, and in the Leryn about six miles up: taken altogether, I consider Fowey to be a very desirable station, ranking with Plymouth and Slapton, without doubt the best of all in the district, from the east end of the Isle of Wight to Falmouth.

The result of one day's fishing off Fowey, in July 1862, was as follows: one hundred red Gurnards, sixty Sea-Bream, forty Whiting, and a large Pollack about ten pounds' weight. On another occasion, about one hundred and twenty Whiting and thirty Gurnards. There is special streaming fishing for large Pollack in the offing. Fisherman, C. Pill, Polruan.

From the beginning of July to the end of the season, Pilchards for bait may frequently be procured in the morning from the drift-boats, in default of which Mussels can be obtained by dredging higher up the river.

The outer Whiting ground is easily found by steering out S. by W. or SSW., until the Gull Rock midway between the Deadman Point and Falmouth is visibly open of the Deadman or Dodman. This rock is also called the Gray.

Mevagissey.—The general sea-fishing on this coast will be found in the bays of St. Austell and Mevagissey. At Mevagissey there is an extensive Pilchard fishery. Dabs may be taken over the side of a vessel at anchor almost everywhere in the bay. I have gathered Mussels for bait off the dock entrance walls at Pentewan, the next place NE. of Mevagissey. They should not be used as food. There is Pollack ground by Chapel Point, and round the Gwinges Rock off it, also under the Deadman. NE. of Gwinges, 720 feet, is a small rock, the *Yaw*, appearing at very low tides. Gwineas or Gwinges Rock is also noted in summer for large shoals of Bass.

Yachts crossing the bay between the Deadman and Falmouth, will, just before coming up with its western horn, called the Zone or Zoze, meet with some Crab-pot buoys, on a shoal called the 'Bizzies.' This is considered good Pollack ground.

Falmouth.—Before entering this harbour is an excellent spot for Pollack and other fish, known as the 'Old Wall,' a pinnacle rock with $26\frac{1}{2}$ feet over it, and 5 or 6 fathoms around. It is S. by compass, $1\frac{1}{4}$ mile from St. Anthony's lighthouse, and I mention it because an old Cornish fisherman informed me it was an unusual place for fish. The marks are— Restronguet smelting-chimney up the creek with east end of the broken rocks at St. Anthony's Point (the lighthouse) and Greeb Point next north of Killygerran Head showing east of it. Another mark is Flushing Mound seen over the rising ground at the north end of Pendennis land. Very large Pollack in the offing, on other grounds well known to fishermen of St. Mawes. When running along the coast, weather permitting, I generally tow a line with a Mackerel bait, spinner, or flies, round the headlands, *en passant*, and not unfrequently pick up a Pollack or two, for these localities are the strongholds of these fish, as well as insulated rocky grounds at a distance from the shore. Round the Black Rock in the harbour entrance, and under Pendennis Castle, some Pollack may be taken. Mackerel enter Falmouth, in common with other extensive arms of the sea, and some few Whiting are at times to be caught in Carrick Roads. In the outer roads off Swanpool Beach there is ground-fishing for the usual fish on the coast, and on all the more western portions of Cornwall, off Penzance &c. is excellent fishing : Pilchards can generally be obtained for bait. There is some pier-fishing from the new works.

Helford.—Four miles beyond Falmouth the River Hel falls into the sea, forming the harbour of Helford. It is a very snug anchorage, and abounds with fish. On a visit to this harbour in a yacht we took one day, in a trammel, twenty-two Red Mullet and five Soles, besides some Plaice, Flounders, Grey Mullet, and Bass, as well as three Lobsters. Pollack, Mackerel, and Congers can also be caught here with hook and line, without going outside the harbour entrance. About 2½ miles south-east of Helford the Falmouth boats often catch abundance of Whiting and Bream, and the latter fish may be taken at a much less distance from the harbour. At the Manacles Rocks, beyond Helford, very large Pollack are often met with by the fishermen ; shoals of Bass also here.

Isles of Scilly.—Both in the various harbours, and in the open sea outside, the Pollack-fishing is remarkably good. Without leaving the harbours, a hundred or more Pollack may be taken in a day, either by whiffing with the natural or artificial baits, or by fishing at anchor with rag-worms dug from the disintegrated granite of the beaches, as in the Channel Islands, or with the living Sand-Eel, which can be caught with seines, by those provided with them. At New Grimsby anchorage I have had excellent fishing, and by Shipman Head, and the Kettle Rock, just off the entrance, the fish are larger. The sea being often heavy at the last named spot, caution must be used. At the Seven Stones Rocks, Pollack are very large.

The Island of Guernsey is worthy a visit by anyone taking an interest in sea-fishing, particularly on account of the living Sand-Eel method, which can be here seen in perfection in summer, and would, I am confident, if adopted generally, yield astonishing results.   There is good fishing also round Alderney, but at Jersey it is not considered so good ; in fact, that island receives much of its fish from Guernsey; however, the fishing of Jersey is not fully developed.   I would recommend visitors to these islands to take with them on their return at least a couple of hair-lines with pipe-leads, and a Sand-Eel basket, there called a 'courge.' The fishing of Guernsey has much deteriorated during the last fifteen years.

Alderney.—Visitors to Alderney will find numerous Pollack in the rocky bays under the fort, and close to the shore whiffing with flies or worms, particularly just before and after sunset. Also Bream, by mooring with a stone, and fishing with lightly-leaded lines, or lines without lead. The soft part of Limpets is a very good bait for Bream (see.'Sea-Bream,' p. 171). Red Mullet may be caught by a trammel. Lug-worms are dug out of the sand near the old pier. Good sized Pouting are taken near the end of the breakwater ; and near the Island of Burhou, on the other side of the Singe Passage, there is excellent fishing, provided you go with an island fisherman, without whom it would not be prudent for any stranger to venture outside the breakwater, on account of the various settings and rapidity of the tides amongst the rocks. The rock or mud-worm is here found by digging under stones &c. Sea-fish of all kinds may be taken from the breakwater, or any steep rock found available, with the rod or throw-out lines. Off the breakwater a floating trot is found killing. Large Bass are taken in the Swinge, or Singe Passage, by trolling under sail with a Sand or Freshwater Eel bait ; under favourable circumstances the fly might be used.

The Isle of Man also affords good fishing, the Bahama Bank being a celebrated ground for Mackerel and Plaice &c. Sand-Eels are very abundant in Ramsay Bay, but I have not heard that nets are kept for taking them ; on the contrary, I am informed they are dug out of the sand with forks. The

Bahama Bank is six miles from Ramsay Pier, and between it and the shore is some stony ground where Pollack or Lythe are very large. There is good fishing also in Douglas and other bays, and Cod and Haddock are taken in the offing. Small Freshwater Eels and worms are procurable in the mouths of the rivers which form the various harbours of the island, but for ground fishing Herring is chiefly used.

Filey, Yorkshire.—Although my remarks on fishing stations have chiefly been confined to those I have myself visited, Filey has been so frequently recommended by correspondents of the 'Field' newspaper, that I think it well to mention it here. It offers special advantages for both fishing from the shore with rod and line, and also for boat-fishing. At times many Pollack and Coal-fish are caught from the Brig, which is a reef of rocks running out into the sea a considerable distance, and forming as it were a kind of natural jetty or pier, with deep water alongside. The fish often run very large, and extra strong tackle is necessary, with fifty or sixty yards of running line. All the rocky ground about the Brig affords good Pollack-fishing from a boat by the various methods described in this work, with either natural or artificial baits. The method of fishing with the living Sand-Eel, if procurable, would afford sport unattainable by other baits.

Tynemouth.—Good pier-fishing, also in the offing.

Bridlington.—Sea-fishing good.

Flamborough Head.—Large Whiting-Pollack, with spinning baits or flies, from the rocks, or better from a boat.

Good fishing from the piers and in the offing, at both Scarborough and Whitby.

Yarmouth and Lowestoft.—Fishing from the beaches and piers, with throw-out lines, and variety of fishing in boats, Whiting plentiful, Cod and Codlings late in autumn and early winter.

Dover.—Mullet, Pout, Smelts, Whiting-Coal, Codlings, and occasional Bass from the jetties, by angling with lug-worm bait, found between the rocks under the cliff. Whiting also from the end of the Admiralty Pier; Mullet, Flounders, and Eels in the harbour. Whiffing with flies along the breakwater or over the rocky ground, for Whiting-Coal, Bass, and sometimes for Mackerel.

Margate.—Bass sometimes taken with artificial spinning bait, or by angling from the piers or jetty with lug-worm or cuttle bait. Local name Sea-Dace, or Salmon-Dace. Flat-fish and small Whiting by line or rod, with boiled Shrimp, from the jetty end. Grey Mullet also met with. Codlings, Cod, Whiting, and sometimes flat-fish in the offing. For Mullet use a portion of the inside of Skate, in appearance like sweet-bread.

Ramsgate.—Same kind of fishing as at Margate, but fish in greater quantity.

Bass are met with in the mouth of Sandwich Haven in summer; fish where any flock of gulls may be seen in a state of great excitement, with fly or spinning bait. Dabs, Whiting, Codlings, and Cod in the offing.

Aberffraw, Anglesey.—Excellent sea-fishing, especially for Pollack. Fish run large. The landlord of the 'Prince Llewellyn' keeps a boat for visitors. Mackerel, large Gurnards, Conger, Cod, and Bream are also here taken.

Carnarvon.—Bass and other sea-fishing.

Beaumaris.—Bass and Pollack.

Barmouth.—Bass at the harbour entrance.

Aberdovey.—Bass-fishing.

Milford.—Good sea-fishing. Pollack round Sheep, Thorn, Stack, and the other islands. Very large fish round the outside islands towards the Smalls.

Tenby.—Bass and Pollack round Caldy Island, and the Woolhouse Rocks. Trotting or long lining for Cod in the bay, in which is good trawling.

Ilfracombe.—Bass, Pollack, and ground-fishing. Licensed pilot and boatman, Buckingham, familiarly known as the 'Duke.'

Appledore, North Devon.—Bass-fishing in summer and autumn. Bideford Bay is best fished from Clovelly, the tide being so very strong, and the passing of the bar being attended often with much risk.

Clovelly.—Bass and Pollack fishing more or less along the shore eastward to Rock's Nose, and north-westward to Hartland. Also round Lundy Island.

Hartland.—Most excellent Bass-fishing with Pollack, off the point; between forty and seventy fish have been taken at a time, from three to ten pounds weight, with spinning Red Eel. Good Trout-fishing in neighbouring streams.

Padstow.—Good Bass-fishing at times in summer and autumn. Pollack under the cliffs towards Stepper Point, round the Gull and other rocks and Pentire Point. Sand-Eels at the Dumbar Sand.

St. Ives.—Pollack on all the rocky coast and grounds, flat-fish in the bay, and Bass at Hayle Bar.

Penzance.—Angling from the piers and quay, for Pollack and Grey Mullet. Pollack immediately outside the entrance at the Geer Rocks, on which is a beacon, and on all the rocky grounds of the bay. Pilchard gut on a small hook, very excellent bait for the Grey Mullet. Much fish collects at the drain outfall, at the back of the east pier.

The Lizard.—Plenty of large Pollack round the Stags, and Bass are sometimes taken by spinning off the rocks, with india-rubber eel.

The Whiting.

(Merlangus vulgaris.)

There are several fish known under the generic term of Whiting, and as they are frequently confounded with each other, it has been deemed advisable to mention the different varieties, as well as their provincial names, in order to prevent mistakes.

First, in order and quality, is the common Whiting (*Merlangus vulgaris*), called also the Silver Whiting.

Second, the Whiting-Pollack (*Merlangus pollachius*) or Whiting-Coal, by which name it is known in Hampshire and the Isle of Wight &c., and is, as its name implies, of a darker hue.

Third, the Coal-fish or Sillock (*Merlangus carbonarius*). This is the Race or Rauning (an old Cornish word for ravening or ravenous) Pollack, and is of a much darker green on the back, and of a rounder form of body generally than the preceding classes. This fish is more abundant on the northern than on our southern shores, although found everywhere. It is numerous and large in Norway and North America.

The common or Silver Whiting (*Merlangus vulgaris*) abounds upon our coasts, where the bottom is sufficiently soft or oozy for the abode of the worms on which it is supposed to feed, and sometimes grows to four or five pounds in weight, but is usually much smaller. The finest run of Whiting I have seen have been caught on the southern coasts of Devon and Cornwall, in the fall of the year, the average size of which certainly is not more than two pounds in weight. It is true that numbers of a much larger size are taken, but they must be regarded as above the average.

In early summer the fish are much smaller than at the latter part of the year, when, after having enjoyed their summer's food, they are in the best condition for the table, being light and nutritious.

The depth of water in which the greater part of the Whiting are caught near Plymouth is from 28 to 32 fathoms, and between the distances of from $2\frac{1}{2}$ to 6 miles from the shore; that is to say, between the Eddystone and the Rame Head, the westernmost headland of Plymouth Sound. They are by no means confined to these limits, as both east and west of this great quantities are taken, as well as over nearly the whole circuit of our coasts; and, as the season advances, they approach the shore to within 1 and $1\frac{1}{2}$ mile, and small sections enter the different bays along the coast, and the deep water harbours of Plymouth and Falmouth &c., and proceed up some miles, there being, on the average, 10 or 12 fathoms' water at a considerable distance up these arms of the sea; this, however, is not the case with the main body of the fish, which remains in the deep water, for in 30 fathoms Whiting may be taken nearly the whole year round in the English Channel.

In moderate depths of water at the distance of one or two miles from the shore, Whiting are rarely to be taken in any quantity before June. Although they are to be met with in congenial localities all round the coasts of the kingdom, there are considerable intervals either entirely destitute of this fish or only occasionally visited by them.

Where a large extent of the sea-bottom exists frequented by this fish during the chief part of the year, it has received the name of a 'Whiting Ground,' from time immemorial, amongst fishermen.

Ground-fishing Gear.

I have placed the article on ground-fishing gear, to follow that on the Whiting, that being the most important fish taken at the bottom. In treating of other fish I refer to this article, mentioning such modifications regarding sizes of lines, or weight of leads &c., as I have found necessary.

The methods of fitting gear for ground-fishing differ much in different localities, much more, in fact, than necessity requires, the truth being simply this—that any tackle which will work clear, and place the bait fairly within reach of the fish, will be sure to catch them, the form of the gear being bound, as far as success is concerned, by no arbitrary rules ; but to fulfil its requirements, whatever be its form, the tackle must be the best of its kind, or it will prove a failure. In such deep water as that to which I have just referred at the end of the previous article, heavy sinkers are of course required, and even then it is impossible to keep the baited hooks near the bottom during the strength of the stream of tide at the springs, which is necessary, for although the fish sometimes are found at a less depth, yet this is quite an exception to the general rule, and the baits should therefore be kept just clear of the ground. Off jutting points and headlands heavy gear is also necessary, even in a moderate depth of water ; but in bays and off parts of the coast, where depths and streams of tide are not beyond the average, much lighter sinkers may be used, provided they will keep the ground. The most common error is to use a stout line where a fine one would be just as effectual ; this by necessity involves a heavy sinker, because the stout line holds so much tide that bottom cannot be kept with a

D

light one, that is to say, with a sinker of moderate weight. Although the varieties of lines figured and described in this article are fitted out with intention of taking Whiting, the captures are by no means confined to this fish, but grey and red Gurnards, Sea-Bream, Cod, Haddock, Ling, and Dabs &c., to say nothing of those plagues, the Dog-fish, are constantly caught therewith.

In describing the various kinds of ground-lines, their differences and relative advantages will be pointed out, and the variety of form of sinker and method of fitting will be rendered evident from an inspection of the accompanying cuts. Leads of one and of two pounds' weight are very useful for most kinds of ground-fishing alongshore, and in common with others of greater weight are always fitted with booms or bearers-off, to keep the hooks from twisting round the lines, which booms, whether they consist of wire, wood, or whalebone, are known by the general name of chopsticks. Some are fixed, others revolving.

Snooding.—The material to which hooks are attached is generally known as snood or snooding, sneads or sids, and consists of either twisted white hair, single or twisted gut, silk, hemp, or flax. The snood is either long or short, according to the kind of gear, and of which it forms a very important part. (See also p. 205.) Two-stranded snooding is far preferable to three-stranded, being stiffer and less liable to foul.

Dip Leads are of varied form, sometimes of a sugar-loaf or true conical shape, and frequently more elongated, with eight sides like a ship's hand lead; others like a rifle bullet of the Enfield make. In fig. 2, *a*, in the accompanying woodcut, it is represented.

The Southampton Rig, consisting of a sinker one pound in weight and in form an octagonal cone, with a piece of brass wire (having eyes turned at the ends) cast into its base. Size of wire ———. A loop of fine line or snooding being worked into the eye at either end, the hook-link is attached in the usual manner known as the loop-slip. A correspondent suggests small brass swivels between the loops and hook-links, to prevent the kinking of the snoods; this is an excellent addition to the gear. The hook-links must always be of such a length that the two hooks when brought towards, will not touch each other, and the line having been securely fastened to the loop in the top, the tackle, with the addition of the bait, is complete. As soon as the boat is anchored, the lead being dropped overboard, is allowed to find the bottom, and then raised a few inches until a bite is felt, when the fish is to be hooked with a slight jerk, and to be drawn on board as speedily as your tackle will permit, without endangering its breaking.

This kind of tackle will answer best where the depth is not more than 10 or 15 fathoms; the cross piece of brass wire or

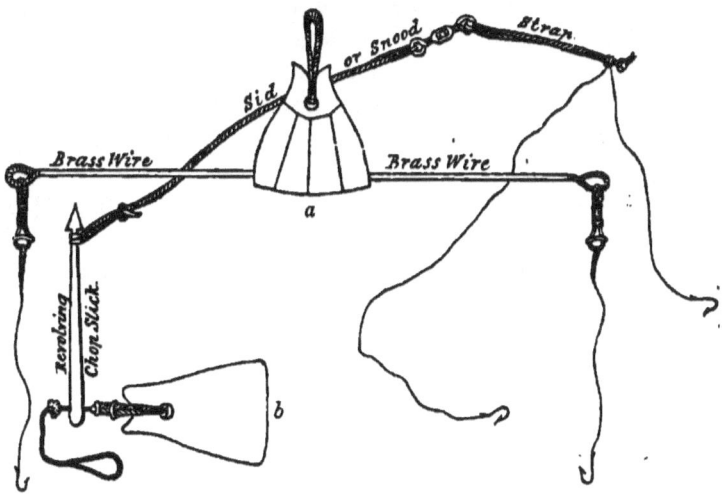

FIG. 2.

whalebone, however, offers a very great resistance to the passage of your tackle through the water, and the shortness of your snooding, which cannot be more than 10 or 12 inches long, the length of the wire being hardly two feet, is a great objection, particularly if the fish are shy; if the united length of your two snoods exceeds the length of the wire they will be continually fouling, which is both unpleasant and unprofitable. I mention this form of ground-tackle because it has been much used at Southampton, and at other localities, where fishing for ground-fish is followed in comparatively shallow water. I consider it however, a very poor form of tackle, both cumbrous to use and troublesome to stow, the long wire or whalebone spreader being much in the way, as it is longer than the reel can accommodate either along the sides or diagonally.

The Guernsey Rig (fig. 2, *b*), for ground-fishing, consists of a dip-lead with long snoods, a portion of line known as the sid-strap, brass swivel and two hooks, the whole attached to a 5-inch elder-wood revolving chopstick, having one end dart-shaped, and the other wider and flatter to admit of a hole being burned out with a hot wire. Ten inches of the end of the line should be served over with waxed thread or tarred sail-twine, to prevent chafing, and the piece of line being put through the hole, a knot is made both above and below to keep the chopstick in its place. A loop is also shown in the cut, useful for bending or unbending the line. Sometimes a sling of stout upper leather is used in lieu of a piece of line. For offing-fishing in 20 fathoms and upwards, 9 feet of sid-strap and 3 feet of snood will be a good length; but for Whiting, Dab, Pout, or other ground-fishing near the shore, both snood and sid-strap together should not exceed 5 feet in length. This form of gear, fitted with snoods, swivels, and hooks of proportionate strength, is used in Guernsey for Congering. For Whiting-fishing, the bottom being sounded, haul up equal to the length of the snood if the tide is slack, but only just clear of the bottom if much current.

The Portsmouth Rig consists of from two to three chopsticks revolving on the line at intervals of one, two, or three feet, and sometimes less, above the dip-lead, the chopsticks of whalebone, about 8 inches long. The chopsticks are often fixed and not revolving, being attached to the line by a clove hitch only, over a notch at one end. Two chopsticks are better than more, and the snoods should not be longer than the chopsticks, or much fouling will ensue. Let a swivel be fastened to each, to which loop on the hook-link. On clean ground, with this gear, the lead is allowed to rest on the bottom. A neat way of fitting is to make the chopsticks of brass wire, taking four turns at one end to make a spiral coil through which to reeve the line, answering to the hole in the wooden chopstick. The chopsticks are kept in place by knots above and below as before-mentioned, and about three inches of the line at each chopstick must be served over with waxed thread, or the line will soon chafe and break. These lines are not favourites of mine for boat-fishing, as the shortness of the snoods gives so little liberty of action whilst unhooking the fish, but for throwing out from a pier or rock they are the best which can be used in my opinion. N.B.—One evil of this form and of all other gear with short snoods, is the liability to hook your fingers, by the lead falling off the seat of the boat, when placed thereon for the convenience of fresh baiting, or unhooking.

The Sprool Rig consists of a dip-lead of from two to five pounds' weight, of the shape of a ship's hand-lead, having a hole an inch or so below the hole in the top. Through the lower hole a bow of three-eighths galvanised iron is thrust, which measures from 20 inches to 2 feet across ; and loops being lashed on to the ends, the snood and hooks are fastened thereto. The bow of iron is considerably arched—almost, in fact, half a circle—and as it is not a fixture the ends turn upwards as the gear descends, and drop downwards when the lead reaches the bottom, maintaining the same position in ascending. The lead is prevented sliding beyond the centre of the bow by two leather washers of the size of a sixpence, which are themselves kept in their places on either side of the lead by collars of sail-twine, lashed tightly round the bow outside the lead sufficiently to allow room for it to work. It is much used in Cod, Haddock, and Whiting-fishing in the North Sea. A snood of from 4 to 6 feet is attached to either end of the bow, and the manner of using is to sound and then haul up just clear of the bottom. It should be fitted with swivels of brass. The Cod smacks are hove to, and, as they slowly drift along, six or eight lines are worked from the weather side. For light fishing one and two pound sinkers are sufficiently heavy, and the bow may be used one foot across.

The Kentish Rig (fig. 3) is a dip-lead slung with a piece of double upper-leather, over which a widely-forked brass wire is passed, the middle of which has been twice turned round a half-inch bar of iron, to receive the leather. The arms are from 6 to 10 inches long, and have eyes at their ends for the snood and hooks, with the swivels. The angular bending of the wire makes it hang true with the tide, and sometimes the wire is cast into the lead and bent to the required angle afterwards. Fishing for Whiting or for Pout with this gear, raise the lead one to two feet from the bottom after sounding; but for Dabs, Plaice, or Flounders it should be just off the ground. The length of the snoods may be 2 or 2½ feet, and the most useful weights for leads one and two pounds. For offing-fishing quite double are requisite, and frequently heavier. I consider this to be the best of all the chopstick methods, and prefer the wire to work on the leather to being cast into the lead. For beginners I recommend the snoods to be only a trifle longer than one arm of the wire, but after a little use they may be lengthened to 2 or 2½ feet.

FIG. 3.—The Kentish Rig.

The Dartmouth Rig (fig. 4) consists of a conical dip-lead of from two to five pounds' weight, slung by a piece of Cod-line having an eye spliced at either end, to which is lashed a spreader of either whalebone, brass, or galvanised iron wire, the two latter to be preferred. For the heavier weight three-eighths wire 18 inches long, for the smaller quarter-inch 1 ft. in length. The illustration represents the gear in the act of descending, and the use of the spreader or chopstick will be rendered evident at a glance in keeping the snood and hooks apart from the line. When the bottom is sounded the lead is raised sufficiently to keep the hooks just off the ground, the requisite height depending on the strength of the tide, which accordingly streams out the snood and hooks more or less as the case may be. If a good stream of tide is running,

THE DARTMOUTH RIG.

the lead should be lifted just off the bottom ; if moderate, a little higher ; if slack, the height of the whole length of the trace.

For Whiting or Haddock-fishing in the offing 9 ft. will be found a very useful length for the trace ; but near the shore 5 ft. is sufficient from the lead to the lowest hook. Place a brass swivel in the trace, as shown in the woodcut, and loop on the snood and hooks over the knot. The sid-strap or upper portion of the trace may be of thirty-six hairs for fishing alongshore, but for the offing forty-eight. Many Devon and Cornish fishermen use a plaited sid-strap of twine, which is less liable to kink than three-stranded fishing line. I have myself been trying some two-stranded sid-straps of double snooding, and find them answer well, although nothing is superior to white horsehair, often very difficult to obtain. This arrangement of snoods and sid-straps is alike common to the Dartmouth dip as well as the creeper or grapnel sinker, the Newfoundland or Bankers, and boat-shaped leads, about to be described.

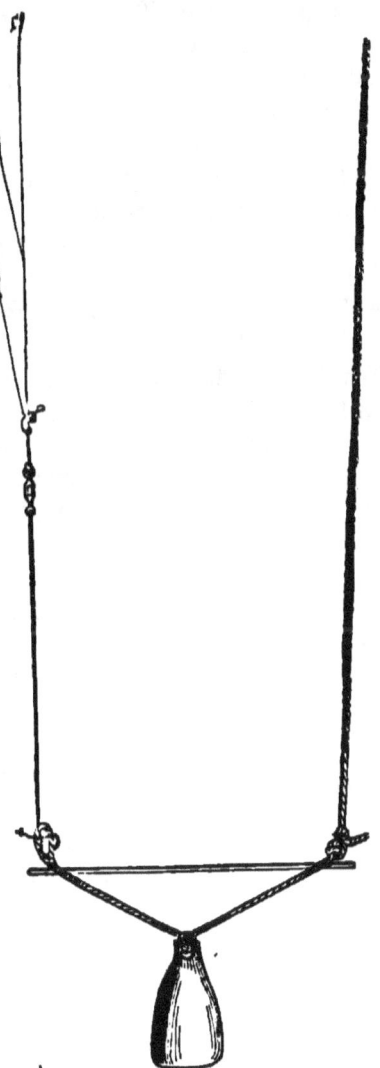

FIG. 4.—The Dartmouth Rig.

In the illustration three hooks only are represented, which will generally be found sufficient for any kind of hand-line ground-fishing, and if two pairs of lines be diligently worked, the result will generally be fully equal and often greater than if more hooks were used, which are often a positive hindrance, from the increased liability to entanglement, and loss of time incurred in clearing away before the gear is again in working order. If from the strength of the tide you are compelled to use two lines only with the 'Grapnel or Creeper Sinker,' the number of hooks may be doubled, but six on each I consider quite sufficient, even under these circumstances. At Plymouth three form the complement of hooks on one line; at Salterton, Sidmouth, and Beer, two only; yet these fishermen will take as many fish on the average as those at Dartmouth, and have often made catches of from thirty to seventy dozen Whiting in a day's fishing, facts the reader will coincide, I think, with myself in esteeming pretty conclusive.

The Grapnel or Creeper Sinker (fig. 5) is much used off Dartmouth and Start Bay, on account of the strength of the tidal currents in the offing, at and about spring tides. These creepers have five claws, are about 18 inches long, and are provided with a conical piece of lead of two or three pounds' weight, cast on to the shank as close to the eye as possible, in order that the ring end may not rise from the bottom, and the claws thereby break out of the ground. A thimble of the kind used by sailmakers to insert in the bolt-rope of sails is closed on the eye of the grapnel, and a piece of three-eighths galvanised wire is slung to the thimble by passing a piece of Cod-line through the ring and round the circumference of the thimble, when both ends of the piece of Cod-line, on each of which an eye has been spliced, are lashed close to the ends of the wire. The snood and hooks, sometimes amounting to more than a dozen, and the line, are bent on as in the ordinary Dartmouth Rig. The weight of the

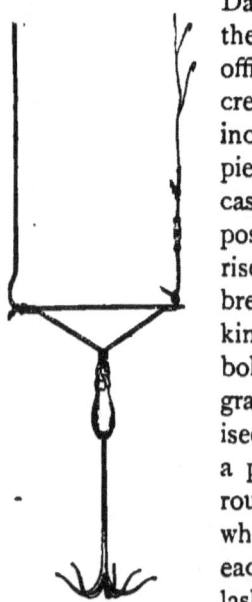

FIG. 5.—The Grapnel or Creeper Sinker.

Creepers complete is about eight pounds each, and by their aid bottom can be kept in a considerable tide when at anchor. They are of course only used on ground known to be free of rocks. I have only met with them at Dartmouth, but they would answer anywhere under similar circumstances. The whole affair should be galvanised.

The **Newfoundland** or **Banker's Lead** (fig. 6) is a favourite sinker in the Cod fishery, of a sugar-loaf form, with the base cut off at an angle. It works with a long snood like the Dartmouth Dip, and in Cod-fishing with two hooks only. At each end holes are made to receive a ganging or stiffening of served yarn, and the line being bent on to the ganging of the sharp end the snood is attached to the other. It is the favourite rig with the Teignmouth men for Whiting and general ground-fishing. By thrusting a sugar-loaf piece of wood on a slant into moulders' sand the angular base is readily formed.

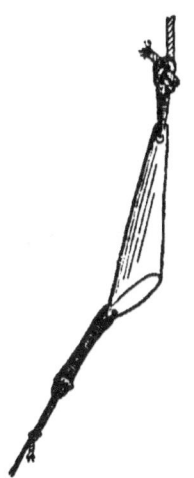

FIG. 6.
The Newfoundland or Banker's Lead.

The **Mould for Dip-Leads** is easily made for conical sinkers by sharpening a piece of wood and pressing it into a pot of sand the required depth. Use moulders' sand from a foundry if procurable. A cone $2\frac{1}{2}$ inches wide at the base and 3 high will weigh about 3 pounds; one of 2 inches at the base and $2\frac{1}{2}$ high about 2 pounds. A mould for a ship's hand-lead shape is made by planing a piece of wood with $1\frac{1}{2}$ inch to each side of a square, and then taking off the angles until you have eight equal sides. One inch at the top should be reduced to $\frac{3}{8}$ thickness, which when pressed into the sand will form the top for boring the hole to sling it, for which a bit of stout upper leather is most durable. The wooden model should taper slightly. Moulds may also be made in freestone or chalk.

The **Boat-shaped Rig** (fig. 7).—This lead derives its name from its form, that of a whaleboat or canoe, and is much used on the south-west coast of England and elsewhere. The boat-shaped sinker has probably descended to us from our Scandinavian ancestors, for varieties of it were numerous in the Norwegian, Swedish, and Danish departments of the International Fisheries Exhibition of 1883, held at South Kensington. This form has not been chosen without design. A boat at anchor is liable to be moved to the right or left of a right line by force of wind or stream of tide, and when a lead is cast overboard it is always likely from this cause to get foul of the lines already down, which hang under the boat at an angle when the boat is thus deflected from the line of her anchor. The boat-shaped lead, when cast overboard, as it strikes the water, sheers out

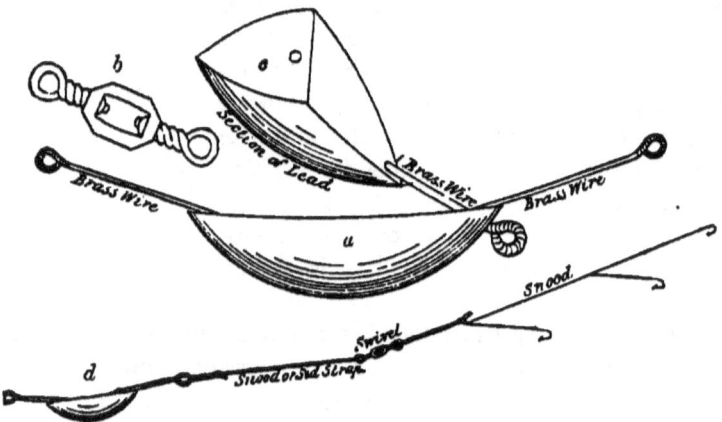

FIG. 7.—Boat-shaped Rig. Lead and Section with brass wire, and Cross Section showing form of cavity.

from the side of the boat from which it is thrown, and consequently has a better chance of avoiding the lines on the other side of the boat than if it descended perpendicularly. In the woodcut, fig. 7, *a* represents the lead with a brass wire cast into it projecting from both ends and answering to the bow-sprit and mizen-boom of a boat, which serve to keep snood and line apart in descending, as in the Dartmouth Rig, and thereby to prevent fouling. A swivel is shown at *b* (two-thirds length) of sufficient strength for offing Whiting-fishing, and a section of the lead *c* to show its form, whilst in *d* we have lead, brass wire, snood or sid-strap, brass swivel, and the snood itself with three hooks looped on over the knot below the swivel, wanting only the line to be bent on to the upper eye of the brass wire. This is on a smaller scale in order to show it entire on the same page. The fishermen dispense with wire, and use served rope yarn instead, worked in through holes in the ends of the lead. Wire is the simplest and most speedy method. This wire or ganging is absolutely necessary, or snood and line would twist together. The knot at the end of the sid-strap is most useful, as you can immediately detach the hooks, either if you have the ill-luck to get foul by a fish sheering across the other lines, or from any other cause; this removal of the hooks of course facilitates greatly the clearing of an entanglement, and I recommend it for all ground-fishing gear for hand-lining. In fig. 7, d the lead is represented in the position it would occupy when in use (holding the book sideways), and as it is drawn to the surface in the same position, the wire of course offers little or no resistance in passing through the water, and the tackle is drawn up with comparative ease; but in the Southampton Rig, the wire or whalebone spreader running through the lead at right angles to its upward course, nearly doubles the labour, a matter of no small moment in offing-fishing, although alongshore, using light leads, it does not matter so much. To use this gear, the boat having been anchored, first sound the bottom, and if a good stream of tide is running lift the lead just off the ground; if moderate, a little higher; if dead slack, the height of the whole length of the trace, by which I mean all the tackle below the lead. The sid-strap should never be quite as strong as the line, in order that it may break first, by which you will save your lead if you get fast in the bottom. For offing work use twelve feet from lead to lowest hook, alongshore four or five feet.

Moulds for Boat-Leads can be made in wood, freestone, and ordinary, or moulders' sand from a foundry. The dimensions of a mould for a lead of one pound weight will be $4\frac{1}{2}$ inches long, $1\frac{1}{4}$ wide, by $1\frac{1}{8}$ deep; for one and a half pound, $5\frac{1}{4}$ inches long, $1\frac{1}{2}$ wide, and $1\frac{1}{4}$ deep, and others in proportion of greater weight.

Here are three illustrations of the mould for boat-shaped leads in wood; first, the mould screwed together (fig. 8); second, half the mould showing the position of the brass wire (fig. 9); third, a cross section, showing the form of the cavity (fig. 10).

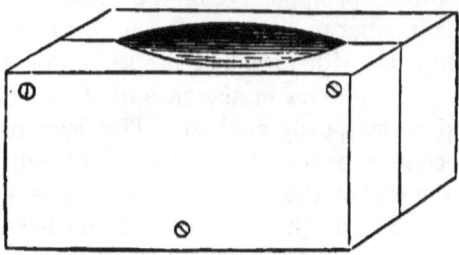

FIG. 8.—Mould for Boat-shaped Leads.

The moulds may be made of wood, as follows : take a piece of fir or deal, three inches square and a foot long, saw it through the middle, and with a spokeshave take out of the inner edges half the dimen-

FIG. 9.—Longitudinal Section with brass wire.

sions for the lead, and screw the two pieces tightly together.

If of freestone, rule a straight line on it, and, marking the shape of the lead, scoop out the mould with a knife and gouge, or get a stonemason to prepare it for you.

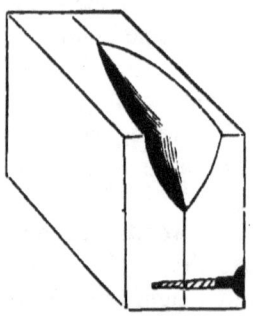

FIG. 10.—Cross Section, showing form of cavity.

You may cast twenty to thirty leads in the wooden mould, but that of freestone will never wear out, although it may break through the heat of the molten metal, to prevent which fill it with hot ashes previously. The charring of wood moulds may be hindered by oiling them, but when they begin to burn they should be lined with thin sheet iron, which when first put on should be painted with

MOULDS FOR BOAT-LEADS.

lamp-black and water, or the lead may adhere. If in sand, make a model in wood first, and press it tightly down into the sand, having first driven a nail, screw, or gimlet, into the top or flat, to enable you to raise it without breaking the edges of the mould. Get a piece of brass wire three-sixteenths of an inch thick, and long enough to project beyond the ends of the mould in the same proportion as shown in the illustrations, whatever be the weight of the lead, and having made an eye in each end and bent it into the form of a bow, place it in the mould, securing it in the wooden or stone mould by a lump of putty or clay, but in the sand mould by a little of the same material. Now pour in the molten lead, and allow it at least three minutes to become solid, or it may divide in moving the mould. To cast smaller leads is easily done by pressing one already made into some sand a less depth, not of course forgetting the wire. Tackle-makers and plumbers keep moulds for leads in wood or metal, but amateurs will find the wood or sand the easier plan. A sharp-edged oval pebble from the beach is often used where procurable, but a model may be made out of a piece of deal in a few minutes. For a three-pound lead the wire should be a quarter of an inch in diameter. I have made moulds out of a Bath brick, but find them very liable to break. Good moulds are also made out of thin sheet copper.

Boat-leads are sometimes made of a half-breadth form that they may work out from the side of the boat more than those of ordinary shape, which seem, however, to answer every purpose. The half-breadths must, of course, each be kept to its own side of the boat, or their very property of sheering will destroy its object, by fouling each other. This applies to those leads made with a long arm only at one end, and an eye at the other, but when the wire projects equally from each end, the leads can be used on either side, always remembering to so fix them to the line, as to keep the *convexity of the lead towards* the side of the boat. They are much used now by the Plymouth fishermen, and have nearly superseded all other forms for ground-fishing. They are known as 'sheer' leads, from their action in the water, and are being also used for Mackerel-fishing under sail. The neighbouring fishing villages, however, adhere to the whole breadth boat-lead.

Weight of Sinkers.—No arbitrary rule can be made for this, local requirements varying so much. Those of one and two pounds' weight are of general utility and quite heavy enough for sea-side visitors; but if fishing in twenty to thirty fathoms, and where anything of a strong tide runs, your lightest leads, of which two are used, one on each side aft, cannot be less than two pounds and a half, and these can only be used at neap tides; your forward leads, of which two are also used, must be at least four pounds in weight, the additional weight being requisite to keep these forward lines more nearly perpendicular, and consequently clear of the after lines, which, from their leads being lighter, will tail away much farther astern. During spring tides you will hardly keep a line down with leads of less than four and seven pounds' weight, and there is but a short slackening of either the flood or ebb stream at the springs; it is better, therefore, at this time to pay your respects to the Whiting-Pollack &c. (of which hereafter), as they are to be taken in shoal water, and close to the shore, in congenial situations.

Snatch-blocks for Deep Sea-Fishing.—As the hauling of a line of thirty or forty fathoms in length with a heavy sinker, by which I mean anything from two to ten pounds' weight, involves considerable labour, consequent upon the friction of the line on the gunwale of the vessel or boat, snatch-blocks through which to lead the lines will be found very convenient in lightening the haulage. These blocks have an opening at the upper end into which you can snatch the line as you begin to haul, and being provided with a short piece of sinnet or plaited line, tapered in the plaiting and known amongst seamen by the odd name of a *monkey's tail*, may be thereby attached to any part of the rigging of the boat or vessel at the required height. A four-inch block is a convenient size, and thus fitted is better than a davit, as it at once adapts itself to the inclination of the line in a tide-way. For the two after lines, crane-davits of galvanised iron, in shape of the ordinary boat-davits, the crane part about six inches across, turned into an eye at the end to receive the tail of the snatch-block, will be found very convenient; by being made with a bend they might ship into staples in the sides of a yacht's stanchions to avoid making holes in the top-rail. The block should hang about shoulder high for easy hauling; length of davit according to height of vessel's bulwarks —in boats, about two feet would be sufficiently long. The North Sea Cod-smacks have an arrangement for this purpose. With light gear and a moderate depth of water, it is of course superfluous, but either for Congering, Cod-fishing, Haddock, or Whiting catching in deep water I recommend this arrangement with confidence, as I find it a great improvement on working the lines over the gunwale or top-rail, for it not only lightens the labour, but also saves rail and gunwale from being cut into deep notches by the friction of the lines.

Lines.—For heavy ground-fishing and Whiting catching, hemp is the material in general use; but for a moderate depth of water and run of tide I prefer a much finer line made from a strong twine in two strands, which I find much less liable to foul than the usual threefold make.

Plaited lines are very good, but much too expensive for general ground-fishing, yet make excellent rod-lines if well dressed with a stiffening solution, of which I do not find, on the whole, anything more effectual or lasting than coal-tar and turpentine. Bridport is the chief seat of net, twine, and line manufacturers, and has been so for centuries. Cotton lines are also used, and large quantities are exported to Newfoundland and neighbouring colonies for use in the Cod fisheries.

In the woodcut fig. 11, page 48, lines of different sizes are shown for various fish, and, by the numbers attached, they can be selected or ordered from any tackle-maker's or manufacturer's stock, personally or by letter.

No. 1 is well fitted for a bulter or long line for Congers; it is also a good size for hand-line fishing with the grapnel or creeper-sinker as used off Dartmouth. No. 2 is a good size, with leads from three to six pounds, for the offing Whiting-fishing, for hand-line Congering, or for a trot or bulter for Bass in a harbour. No. 3 is a suitable line for inshore Whiting, Pout, and Dab-fishing. No. 4 is the double flax or cotton line for Mackerel; and No. 5 is snooding for Whiting in the offing, and Eels and Flounders in harbour. When anything finer is requisite, we mostly use gut, double or single.

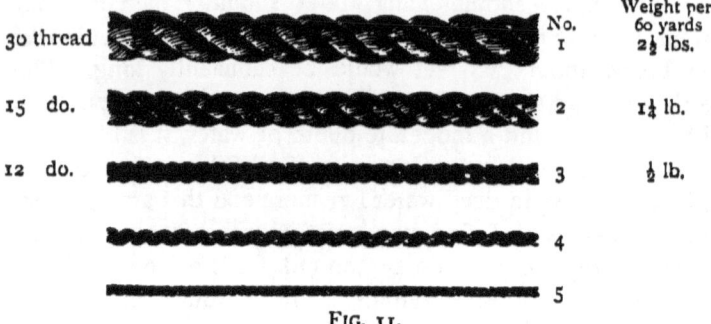

FIG. 11.

Barking or **Tanning Lines.**—Catechu and oak-bark are used for this purpose : half a pound of catechu to three pints of water and a handful of oak-bark, if you can get it ; pulverise and boil until the catechu is well dissolved, then pour over the lines in a basin or tub and allow to soak twenty-four hours. There is no objection to using oak-bark exclusively if you are near a tan-yard or bark-mill, but a gallon of oak-bark has not more power of tanning than half a pound of catechu. In a town catechu is often procurable when oak-bark is not to be obtained.

Baits.—In the choice of bait for Whiting-fishing much must depend on the particular locality. Some of the following are to be had almost everywhere: viz., fresh Pilchards, fresh Mackerel, Mussels and Lug-worms, Gar-fish, also called Long-Nose, Snipe-Eel or Scoot, and fresh Herrings.

Fresh Pilchards are undoubtedly the most attractive bait (salt ones are also sometimes used), being very full of oil, so full indeed that pellicles of it may be seen to float away on the surface when you drop your bait in the water; this no doubt circulates through the water, as smoke through the air, and serves as a clue to guide the fish to the bait. The intestine of the Pilchard should always be saved, as it forms perhaps the best bait of the whole : Mackerel is also very good, for although it is not as rich as Pilchard, it is very much tougher, and will therefore remain on the hook a longer time ; if you are provided with them, use them both, to which end make it a rule to tow two Mackerel lines or more, both in going to and returning from the Whiting-ground. Full directions will be found under the article 'Mackerel,' as to the gear necessary for their capture. Long-Nose and Herrings are but passable, and may be considered as supplementary aids, if your supply of Pilchards or Mackerel runs short. To prepare either of the above-mentioned for bait, having first scraped off the scales (if a Pilchard or Herring), lay the fish flat on your bait-tray (see fig. 64 b, p. 213), enter your knife at the tail, and cut up along the backbone towards the head, when of course the fish will be split in half; now remove the backbone by inserting the knife under it, and, dividing the bone from the flesh, turn the silvery side downwards, and cut it up diagonally in pieces three-quarters of an inch in width, and each side will make about eight or nine baits, as shown in fig. 12.

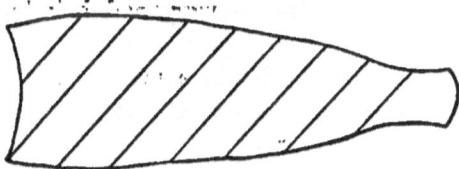

FIG. 12.—Half Pilchard marked diagonally for cutting into bait.

To bait the hooks, take a piece of the fish and enter the point of the hook in the fleshy side of the blue end of the bait, pass it through, and, turning it over, hook it through the silvery side. (See fig. 13.)

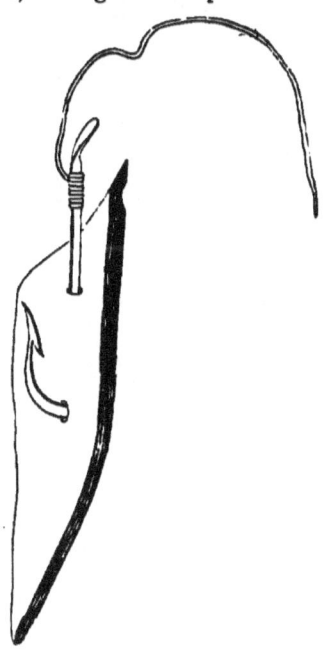

FIG. 13.—Hook baited with piece of Pilchard, Mackerel, or Herring &c. for Whiting.

Memorandum. — As the fleshy side of a Mackerel is rather too thick, and would clog the hook, before dividing it

E

into baits slice off about a third of the thickness from the inside.

To bait with Snipe-Eel or Long-Nose use the same method as for Pilchard.

Mussels also are very good bait, but do not hold on the hook as well as the preceding; they are very plentiful on the gravel and mud-banks in most of the large tidal rivers of the kingdom, and are very generally used as food; but are often

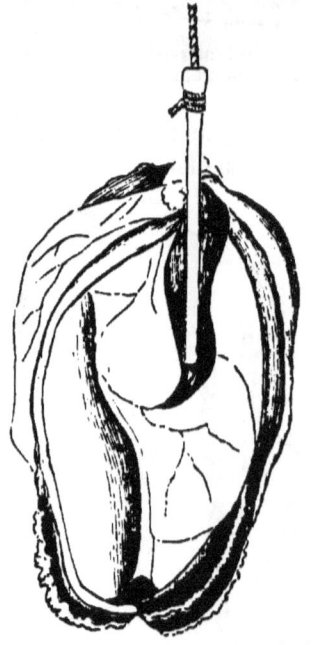

FIG. 14.—First insertion of hook.   FIG. 15.—Hook completely baited.

the cause of serious illness. I should therefore advise my readers never to venture on this perilous shell-fish. They are occasionally met with on the rocks of the open shore, but of a smaller size than those in rivers; it being apparently necessary that they should have a mixture of fresh with salt water, for in such situations they are of greater magnitude and much more abundant. To this rule there are, however, some exceptions.

For Whiting catching they certainly should not be less than

about two inches and a quarter in length, unless the fish run small. (To open Mussels, see p. 187).

To bait with a Mussel (for which see figs. 14 and 15), having first taken it out of the shell, spread it open in your hand, when you will at once discover the tongue, through which you are to pass the hook (see fig. 14); then closing it as you would a book, turn it over and hook it through the round gristly part, by which it attaches itself to the shell (see fig. 15).

Note.—Mussels may be kept alive three or four weeks, by hanging them over the side of a boat in the water in a basket or net, or in a closely made Crab-pot, or in a sheltered pool amongst the rocks : this renders them very valuable to the fisherman, who, in some parts where bait is naturally scarce, sends as far as twenty miles for them.

Lug-worms are also in very general use. The largest of these are about the thickness of the little finger, and have a great number of small legs, somewhat like those of a large caterpillar, whilst the tail part is very small ; they are to be obtained by digging in the sand in the sheltered bays of the sea-shore, and in most harbours where the ground is congenial : their presence may always be detected by the little hillocks of sand and marks on the surface, having the appearance of worms ; hook them through two or three times. These will be found a very good bait when used quite fresh, but are almost useless after the second day. (See also p. 188.)

Note.—Whilst fishing for Whiting you may also take fish of the following varieties : namely, Cod-fish, Haddock, Bream (sometimes called Chad-Bream), red and grey Gurnards, Dabs, Ray or Skate, Hake, Pollack, &c., and occasionally a large Mackerel.

Drift-lines on the Whiting-Ground.—It is a good plan to put out a drift-line without lead, having a large hook (No. 4 in the cut of hooks, fig. 62, p. 210) baited with the side of a red or grey Gurnard of seven or eight inches in length ; a piece of a Long-Nose or Gar-fish will also answer very well ; as by this method you will be likely to take very large Pollack from ten to twenty pounds' weight.

These baits are to be hooked on by passing the hook once

only through one end, in order that the baits may have as much freedom of action in the water as possible, which renders them all the more attractive. Some fishermen use two hooks on one snood, one a foot above the other.

To this line you should fasten an old tin pot, a piece of wood or other object; this will make a noise and attract your attention (if engaged with your Whiting lines) by the violent struggles of the fish to escape.

N.B.—Half the side of a Pilchard is a most excellent bait, or a quarter of one added to a bait of a piece of Bream or Mackerel.

In Cornwall it is the custom to have several spare leads on board, and to attach one to these lines, and place it on the gunwale. This retains the line from running out, but on a fish seizing the bait, the lead falls overboard, and calls the attention of the fisherman by the splash.

It is also well to keep out a fine line with a small hook, baited with a strip of any bright shining fish, which will frequently catch you a Long-Nose or Snipe-Eel, very useful when the other bait runs short, and the Whiting are well on the feed.

Note.—In taking the Whiting off the hook, be careful of your fingers, as the teeth of this fish are very numerous and sharp; and should the blue Dog-fish pay you a visit, you must handle them with circumspection, as they are furnished with two sharp spurs close to each dorsal fin, very similar to those of a cock, and with which in their struggles they are capable of inflicting a severe wound. As a precautionary measure I usually set both feet firmly on the head and back of the fish, and divide the vertebræ at the neck before attempting to unhook them.

Whiting bite best in the early morning or just before and after sunset, but do not ordinarily feed in the night, unless occasionally when the moon shines brightly.

In calm weather, during the full blaze of a noon-tide sun, they usually cease feeding, but in cloudy weather and a fresh breeze may be often taken whilst the sun is high. On some parts of the coast it is the custom to set trots or long-lines for Whiting.

Disgorgers will be found very useful, especially to those accustomed to angling. The handle of an old tooth-brush makes a good disgorger by cutting off the bristles and putting a notch in the end. This will do for ordinary fish, but for Congers make one out of a bit of ash or oak or beech, 18 inches long and of the thickness of a policeman's staff, plane one end chisel-fashion, and cut a notch in it $\frac{1}{2}$ an inch deep. To extract a hook, push the forked end down on the bend of the hook, take a turn of the snood round the disgorger, and force it down until the hook relinquishes its hold, when you will withdraw it easily. For Conger it is often necessary to screw round the disgorger until you can clear the hook *vi et armis* ; this the strength of hook and snood allows in Conger-gear. The teeth of Silver or common Whiting are very numerous and sharp, and the fingers may be easily badly torn in taking them off the hook if the fish has pouched the bait. The Pollack also has numerous teeth, smaller than the Silver Whiting, yet sufficiently large to scratch the fingers considerably, and I have constantly got my hands into a very disagreeable state after a day's fishing from extracting the hook, which they often gorge as deeply as a Pike, particularly if you are baiting with living Sand-Eels, after which they are perfectly voracious.

How to Cure Whiting.

The following extract from the 'Field,' contributed by W. T., Isle of Wight, is so thoroughly to the purpose that I make no apology for its introduction in the present work ; it is both short and simple :—' As soon as possible, clean, and then with a sharp short knife, split the fish from throat to tail, taking care in so doing that the knife feels its way by pressing gently along the backbone, thus making a neat cut and avoiding ragging the meat of the fish. By the aid of the knife dissect out the backbone to two-thirds of its length towards the tail, and break it off.

'Sprinkle salt on the inner side of the fish, and lay one over the other in piles of about three dozen.

'In an hour, if the fish are small, in two hours if they prove large, they will be sufficiently salted, when they may be placed on a grating, or hung in the air to dry.

'When perfectly free from all moisture—say in four or five days' time—they may be lashed up in bundles of a couple of dozen each. While drying, place them under cover at night or the dews will considerably retard the progress of the work.

'The process above described is not as tedious as it reads; two or three hands can, after a little practice, clean, split, and salt many dozens in an hour.

'If, when very lightly broiled and well peppered, a piece of butter is rubbed over them, and they are dished hot for breakfast, they will prove as delicious, delicate, and appetising as when, with their tails through their eyes, they are served in their pretty pale brown crumb-and-egg jackets.'

The method of curing Buckhorn (which is West of England vernacular for dried Whiting) is almost identical with the foregoing directions, the chief difference consisting in merely opening the fish through the back and top of the head, instead of through the throat and belly. By this method the fish fold up like a book and have possibly a more sightly appearance to connoisseurs of Buckhorn; but the fish are equally as palatable if the above method be followed, which I have, in my own case, found to afford greater despatch.

Whiting-Pollack I also find very good salted and dried, and the larger Whiting-Pout from ten inches and upwards in length, although they are not equal to the real Whiting, which is scarcely surpassed in delicacy by any sea-fish.

A clean beach of pebbly shingle is an excellent place to dry fish, for the heat rising from beneath and given off from the stones, will, on a fine warm day, dry the fish (by turning them occasionally) in a very few hours.

A fish-stick is often seen at a cottage door in the West of England, and consists generally of a young holly bush deprived of its bark, and the branches left about a foot in length at bottom, diminishing to six inches at the top, the fish being thrust on through a hole in the tail part; they drain well and soon become firm when thus suspended.

If fish are found to have become unpleasantly salt, they

should be toasted first, and being placed in a basin and kept at the bottom, let boiling water be poured on them for two minutes, after which a little butter should be rubbed over them immediately, which method will extract any superabundance of salt.

In drying fish the great difficulty is to avoid the depredations of cats, to do which I sometimes suspend them on a line stretched across the garden with hooks at intervals of nine inches, trot fashion. Another good plan is to have a light wood frame, about 4 ft. by 5 ft., with cross bars at intervals of a foot, into which tenter-hooks are to be driven to receive the tails of the fish. It can be suspended at any height, as most convenient. The fish should not overlap each other.

Drifting on the Whiting-Ground.—In a calm or laid to, Whiting are often caught as a vessel drifts with the tide. It is constantly practised by the North Sea Cod-smacks; and the Orkney and Shetlanders use a drift-sail in the water as a floating anchor. It is not much followed by boats and small craft, but sometimes by bending on a piece of chain to the cable to check the drift. If laid to, fish on the weather side. As a rule, it is best to anchor.

A Day's Whiting-fishing.

One fine morning in July I threw up my window and looked out across Cawsand Bay and the entrance of Plymouth Sound. In front of me rode tranquilly at her moorings my little 'Fairy,'—a dandy-rigged yacht of seven tons—and several fishing-boats, some of which had already slipped their buoys, were getting up their canvas as they slowly drifted, rather than sailed, out of the bay before the light morning breeze. A shower of gravel at the staircase-window told me that my old man-of-war's man was at the door; and hastening below, I gave him a fish-box to carry, told him to launch the punt, then rushed upstairs to make a hasty toilette, and, taking with me a basket containing some provision for the day, I was soon on the beach.

It was but four o'clock, consequently much too early for breakfast; and, besides, the time is much better spent in sailing towards the 'Whiting-ground,' which the fish frequent at a greater or less distance from the shore according to the time of the year, keeping well in the offing abreast or outside the Eddystone early in the season, and nearing the shore as summer merges into autumn. As we pulled out to the little bark, several boats returning from their night's fishing were passing the Breakwater lighthouse, and from one near us we procured a supply of Pilchards, which is the chief bait used on this coast from July to December, but in the earlier part of the season Mackerel and Mussels are in general alone procurable.

Having arrived on board, and our canvas being set and sheets eased off, the bachelor's kettle is put in commission. 'Coffee's ready, sir,' says Hannibal (for such is the 'Christian' name in which my man rejoices), and, giving him the helm, I adjourned to the little cabin to take my early meal. We had long since passed Penlee Point, abreast of which, as we hauled our wind, we set the foresail, and the breeze came down with increased freshness from the high land of the peninsula terminated by Rame Head, the outermost point of Plymouth Sound.

Thinking it was as well to try for a Mackerel or two, I put out a couple of my heaviest lines fitted with 3-lb. leads; and as I had not then met with Hearder's plano-convex spinner, I used as a bait a bit of tobacco pipe about an inch and a half in length, cutting off the snood from the hook, and, after threading on the pipe stem, making fast the snood again to the hook. One line, however, I kept on board to receive a bait from the tail of a fish if I succeeded in taking one, which happened in ten minutes after putting out the first line, when I immediately baited and veered out a second, first giving the helm to Hannibal whilst I cut the bait. To do this with facility both hands are needed; and it is best to kill the fish, if only to avoid cruelty, by dislocating the vertebræ at the neck, bending back the head for that purpose. From the Rame Head we kept our course into the offing about forty minutes; by which time we were well on the ground, and nearly up with several boats which had probably arrived at daylight. We had caught only our Mackerel on our way; these, however, were very acceptable, for if our Pilchards ran short one or two of them would make excellent bait for the Whiting. We had lowered our topsail some time previously, and now took in the foresail, intending to pick up a berth at a reasonable distance from a boat we were approaching. As we ran past the stern of this boat, which belonged to our bay, Hannibal hailed with, 'Well, skipper, how d'ye rise 'em?'

'With their heads upwards, Hannibal, with their heads upwards,' exclaimed he, making use of a standing joke amongst fishermen. 'Us an't a din (done) so bad; about five dizzen, I reckon,' looking down into the boat's bottom, and still hauling away at his line in that steady business-like manner which is so characteristic of an old fisherman when he feels he has his hooks loaded.

Just as we were passing out of hail I observed the lead come on board; a couple of fathoms more, and he swung into the boat three fine Whiting (one on each hook), fresh baited, hove his lead clear of the side of the boat, and turned to tend his other line, on which he found a pair of fish on getting it on board.

'That looks well, Hannibal,' said I; 'I think we shall have something more than a water-haul to-day.'

'No doubt o' that, sir,' rejoined Hannibal; 'there's a fine school of fish on the coast; but I think we're a brave berth from the other boat now.'

'Well, then,' I observed, 'down jib, and let go the anchor; I'll see the cable clear, and you can clap on a bit of service before coming aft to lower the mainsail.' The foresail had been taken in two or three minutes previously.

As soon as we were fairly brought up, and the sails stowed and coated, we got out our ground-lines—a pair of which are worked by each man when two are in the boat—the after leads each of 2 lbs. weight, the forward pair nearly 4 lbs., in order that the lines may tail away at a different angle, and not become entangled. Off Plymouth they commonly use the boat-shaped sinker, with a stiffened tail at each end equal to half the length of the lead—which we may, to carry out the similitude, term a bowsprit and outrigger, for these, projecting from either end of the lead, keep the line and snood apart in descending, without which they would twist round each other and foul constantly. My old man-of-war's man used a sid-strap—or, as in freshwater parlance it would be termed, a trace—12 feet in length, to which he bent on a snood with three hooks, after the manner of a bottom or collar used in fly-fishing; this added another 3 ft., making in the whole a length of 15 ft. below the lead. Such an extraordinary length of trace and snood is quite unnecessary; but, as it is the custom in these parts, it is simply absurd to argue the contrary, as many old hands can never be persuaded that any improvement may be made in the form or arrangement of tackle or gear to which they have been accustomed from their early years. Having been used to fish more to the eastward, I had fixed my own gear with shorter snoodings, namely a 9 ft. sid-strap, with 3 ft. of snood and three hooks, which is certainly long enough in all reason, making 12 ft., but in fishing alongshore I never use more than 5 ft. Our bait was Pilchards, which Hannibal prepared by scaling them carefully, to avoid breaking the skin, splitting them in two from head to tail, dissecting out the backbone, and then cutting them up diagonally into pieces nearly three-quarters of an inch wide, so that each side made about seven or eight baits. The hook is baited by passing the point through the fleshy side at the blue end of the bait, and then forcing it through the silvery side.

The baits are now cast clear of the boat, the lead being thrown into the water away from them; for if dropped on the snood, as I have seen done by many unthinking persons who have only been accustomed to chopstick lines, a foul is the natural result. All four lines are now out, and are trimmed by sounding the bottom, and then raising the lead about a yard clear of the ground, if there be any tide running; but if not, just sufficiently to keep the hooks from touching, as they are more visible to the fish than if allowed to fall on the ground, where the bait is likely to be taken off by Crabs or Star-fish.

Directly we had put out our second lines we tried our first, and feeling fish, we both commenced hauling, and took a pair of Whiting. Having readjusted the bait, we turned to tend our lines on the other side of the boat, first carefully putting out those which we had just hauled in. The lines on the other side we also found loaded; and as the Whiting were now evidently well on the feed, we had no occasion to stop to feel them, but dragged away as a matter of course, first on one side, then on the other, until the boxes began to make a very respectable show of fish. Having taken four Mackerel on our way out, we cut up a couple of them for bait, as the skin is much tougher than the Pilchard, and the hooks are consequently not so often robbed.

After we had continued fishing some time, rather a lull occurred in our sport, and I determined to put out a drift-line for Pollack, which are occasionally taken of great size on the Whiting-ground. Desiring Hannibal, therefore, to split a Gurnard in two (of which fish we had taken six or eight), I took one side, and, hooking it through one end with a medium-sized Cod-hook, cast it overboard, paying out about twenty-five fathoms. I attached to the bight of the line a small tin bailer I happened to have on board, so that my attention would be attracted if a fish should seize the bait whilst I was occupied with my Whiting lines.

I had again been hauling Whiting for some minutes when suddenly the tin bailer struck the gunwale and bounded overboard; and on taking hold of the line I found I had hooked a fine fish, which bore down hard, so that I could not venture to haul on him but with care and caution, for the strength of a large Pollack is very great when he chooses to exert it. I brought him steadily upwards some distance, when he started off again, and required some little amount of coaxing before I could get him alongside; but Hannibal, standing ready with the gaff, at length hooked him under the jowl, and lifted him on board. He was a famous Pollack, weighing fully twelve pounds, but they are sometimes taken of as much as eighteen.

I put out also a drift-line, having a small Mackerel hook baited with the ordinary last or slip of skin cut from the tail of that fish, and with this took a Scoot—the West-country appellation of the Long-Nose or Gar-fish. This as well as the four Mackerel are cut up as supplementary bait for the

Whiting. Included in our catch was a Hake, which in playing about amongst the lines, had become hooked close by the ventral fin, and sheering round, got three of the lines in a considerable foul. This gave us some trouble to clear; but such things will happen, and as the fish weighed seven or eight pounds, we might consider he atoned for the entanglement.

As our boxes were now full, I determined to cease fishing; we therefore got up our anchor, and, setting our canvas, an hour's sailing saw my little 'Fairy' again riding at her moorings in Cawsand Bay. Our catch consisted of twelve dozen Whiting, one Hake, one Pollack, a dozen Bream, and a dozen red and grey Gurnards—a fair, although not a large catch. For this fishing a small yacht, or large waterman's boat, may be hired at 12s. 6d. or a pound a day, under the Hoe, at the Barbican, or if a visitor elects to lodge at Cawsand, he may arrange excursions from thence. N.B.—Cawsand is three miles on the way to the Whiting-ground, and about a mile from Penlee Point.

THE POLLACK OR WHITING-COLE OR COAL.

(Merlangus pollachius.)

This fish, although somewhat similar in shape, is of quite a different colour from the genuine Whiting, being of an olive brown on the back, the sides shading off to a yellowish white; it attains a very much larger size, sometimes even 15 or 20 lbs.

It differs quite as much in its habits as in its colour from the true Whiting, which prefers a soft bottom, whilst the Pollack frequents the rocky ground, particularly off headlands where the rocks run a long distance seaward, over which the tide sets strongly. Wherever an extent of sunken rocks exists at a distance from shore, there is a stronghold of the large Pollack in all the British and Northern Seas, as, for instance, round the Eddystone, off Plymouth &c. On rocky ground, or close thereto, they are more or less to be found at all seasons of the year; they occasionally, however, disperse in pursuit of the Sand-Eels or Launce, of which they are immoderately fond, and after remaining any length of time on the sandy

THE POLLACK OR WHITING-COLE OR COAL.

ground become of a much lighter colour, and improve in quality for the table, so that any of these fish over the weight of 6 or 7 lbs. nearly equal Cod in firmness and flavour.

My best sport in Pollack-fishing has always been on the SW. coast of England from March to the end of June, after which I have found them less numerous, although you can rarely try without some success. In Guernsey I find it the same. In the deep water of the offing however, there is good fishing for the larger Pollack, all through the summer and autumn, and sometimes even up to Christmas. The Pollack feeds at all depths, but much more above than on the bottom; in fact, is often seen in large numbers on the surface in pursuit of Britt and small fry. Differing so much in its habits from the real Whiting, very different tackle is required for its capture. No gear is equal to the pipe-lead horse-hair lines for drift line fishing, fully described (p. 69), which are much superior to hemp lines for this fishing, being light, elastic, and, from their stiffness and mode of manufacture combined, rarely become entangled.

The Pollack is both active and voracious, and as it takes the baited hook well, may almost be considered the staple of amateur sea-fishermen on such parts of the coast as are congenial to its habits. It approaches the shore very closely, and enters the harbours in considerable numbers, when it may be taken either by fishing from a boat with hook and line, or from the piers or any projecting rock with the angling rod. This fish, as well as Mackerel and Bass, will rise very freely at a fly, which method of taking them has become very general within the last few years, and will therefore be treated of in due course. It takes the fly also under water.

Drift or Tide-way Fishing.—There are two methods of hand-line fishing for Pollack: the one when the boat is fixed in the berth or position chosen by an anchor or stone killick, thence termed 'Drift or Tide-way Fishing' (because the lines are lightly leaded or without lead, and consequently drift or stray out with the current); the other termed 'Whiffing' or towing, because the boat is kept in motion by either sailing, sculling, pulling, or rowing slowly over and around rocky ground, ledges or insulated heads of rock, or other haunts of the fish. We have first to describe the former, which with certain baits is the more killing of the two and more convenient, as the boat requires no attention, which can be given solely to the lines, more of which can be put out at once. If unacquainted with the depth, sound with a ground-line as soon as the boat is fixed in position, and pay out line accordingly, recollecting the leads are placed at two fathom intervals along the line. Sounding and finding, for instance, six fathoms with little tide, put over two and keep the third lead in-board, the snood allowing the bait to descend sufficiently beyond the bottom lead; when the tide strengthens put out the third and keep the fourth in-board; when stronger one or two more may be veered, the length required depending on the angle the line takes from the speed of the current: the faculty of estimating this will soon be acquired by observing these directions in practice.

Two hair-lines with leads, and one without, are quite sufficient to occupy one person when the fish feed well, but if two hands are in the boat, another pair may be used amidships, the leads of which must be twice the weight of those at the stern.

Previous, however, to obtaining familiarity with the use of the lines, even if two are in the boat, one line apiece will find them full occupation, supposing the fish are well on the feed.

The lines at the stern without leads may be made of hair, cotton, or flax or hemp, and I often use a pair, where the water is not deep, or the tide too strong, as I find them very killing for both Bass and Pollack. I begin fishing with these lines at the commencement or at the slackening of the tidal current, and frequently take with them Mackerel, Bream, and Gar-fish, in addition to Pollack and Bass. What is required in these light lines is an entire absence of any tendency to kink, and become entangled, only attainable by care in the manufacture, and also by being dressed with a stiffening solution. I have for several seasons used lines manufactured by the Manchester Cotton Twine Spinning Company, and have found them answer remarkably well. Each of these lines is 30 yards or 15 fathoms long, and at one end I splice a brass swivel $\frac{3}{4}$ of an inch in length. To this I add six yards or three fathoms of snooding, three yards of it consisting of dressed plaited line, two yards of fine cotton, and one yard of double gut twisted. I sometimes use two lengths of the strongest single gut at the end, especially for Bass-fishing. The line itself is of the size No. 3, plaited line No. 4, and the cotton No. 5. (See fig. 11, p. 48.) I use these lines also in whiffing for Pollack over shallow rocky ground, and they are equally useful for Bass when whiffing under sail with dead bait.

At some localities, or at certain times, there is so little tide that the line will not stray away from the boat. A float of large Jack size is useful on these occasions, and a lead of two ounces 12 feet above the hook. The float should have neither ring nor cap, merely a stick through it, over which the line may be attached by a clove-hitch top and bottom. The lead should be fitted with wire or a stiffened hemp ganging, projecting from each end, the distance of half its length, on which the snood can be clove-hitched as on the float. Both float and lead can thus be removed at once, when unnecessary through increase of current.

Baits.—No baits for drift-line fishing can equal living Sand-Eels; next rank Rag or Mud-Worms, then living Shrimps; as for dead baits it is exceptionally only that a catch is made with them, when fish may be extraordinarily abundant. The best dead baits I know of are Freshwater Eels, 5 or 6 inches long, small Lampreys, or a strip of Long-Nose the same length and ¾ inch wide, but it is better to use them whiffing. Sand-Eels to be used as living bait should be taken in a seine made for the purpose (fig. 70, p. 229), and afterwards placed in a floating cage or basket, termed a 'courge,' towed astern of the boat (figs. 19 and 20, p. 66).

How to Bait with living Sand-Eels.—When about to use these baits, first dip up 4 or 5 inches depth of water in your boat's bailer, remove the cork from the aperture of the courge, and holding the canvas bucket, pail, or bailer (see p. 148) at the level of the water, pour a dozen or so into it, replace the cork and drop the basket again into the water instantly, for you must be very careful not to keep the courge out of the

BAITING WITH LIVING SAND-EELS.

water longer than necessary, or expose the Sand-Eels to the sun, or you will soon lose all your bait; with ordinary care they will live nearly a week. To bait with a Sand-Eel, hold it between the fingers and thumb of your left hand, throat outwards, put the point of the hook into the mouth, and out of the gills, then turn the hook over and pass it through the throat below the gills, just sufficiently to hold the hook, and throw the bait overboard. A straight hook is alone fit to use, the Kirby

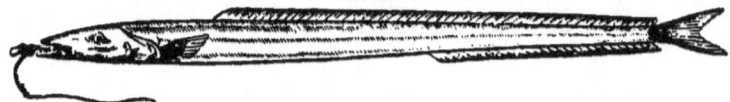

FIG. 16.—Living Sand-Eel bait in tide-way.

bend twisting the head on one side. Fig. 16 shows how to bait 'in a tide-way,' and is also used for whiffing. The boat being moored and the line paid out the required length as previously directed (p. 62), drop the lead you have in your hand in front of one of the thowl-pins of your boat, and when a fish takes the bait your attention will be called to it by the lead rattling over the gunwale, if you are engaged at the other side of the boat, which arrangement will be found preferable to short rods or outriggers. Make fast the reel by passing it once or twice round the thwart of the boat, leaving a couple of

FIG. 17.—Living Sand-Eel baited for slack tide. (Recommended by the late P. le Noury, of Guernsey.)

fathoms or so of slack between it and the lead, that a heavy fish may not bring himself up short and part the line, ere you can attend his summons. Bore a hole in the smaller end of all your thowl-pins, and if the tide or swell of the sea drags the lead overboard, attach one or two to the line close to the lead; this is called a tell-tale, and one should also be fastened to the light stern line.

The second method here shown (fig. 17) is followed when

the stream has nearly ceased to run, or at dead slack, for by being hooked through the nape of the neck, it preserves a more natural position in the water, which would not be so if hooked as in the tide-way. On the other hand, it would be an error to use the slack tide method in the tide-way, because the force of the stream would cause the bait to fall on its side and thus take an unnatural position.

The third illustration (fig. 18) shows the ordinary method of baiting at slack tide, but I prefer that recommended by the late P. le Noury as more calculated to maintain the Sand-Eel in a natural position. Whilst baiting your hook according to either of the methods, be careful not to squeeze the Sand-Eel, for its attractiveness depends on its liveliness remaining unimpaired ; therefore when it becomes feeble put on a fresh one. The living Sand-Eel is such a killing bait that few who have used

FIG. 18.—Living Sand-Eel bait ; ordinary method at slack tide.

it will care to procure any other when provided with it, as it is greedily devoured by sea-fish in general, as well as by Pollack.

The Sand-Eel or Launce is very numerous on many of our own and foreign shores, and particularly at the mouth of those harbours where the sand is of a loose and gravelly nature. They should not be scraped out of the sand, as is the common custom, but be taken in a small seine or net, the bag part or bunt of which must be of very fine netting or of unbleached calico, in order that it may not mesh them (see 'Sand-Eel Seine,' fig. 70, p. 229). After being taken in this they are poured into the baskets about to be described.

There are many localities where Sand-Eels abound, and yet no net is kept for their capture. In such places they can manifestly be alone taken by scraping or digging them from the sand ; in which case provide a small light bucket or bailer to contain salt-water, and throw the Sand-Eels into it instantly on

F

THE COURGE.

capture. When you have a score or so, pour them into your Sand-Eel basket or courge, which should be placed ready in a pool you have dug for it in the sand close at hand.

The Courge or **Sand-Eel Basket** (fig. 19).—These curiously-shaped baskets are made of fine willow or osier twigs, not

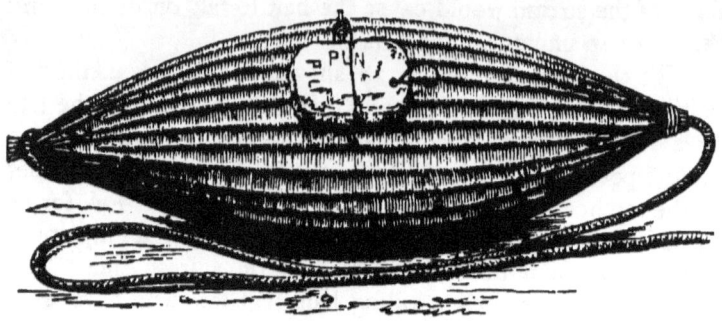

FIG. 19.

more than about one-eighth of an inch in thickness, and are woven sufficiently close to prevent the escape of the Sand-Eels, whilst in the centre an opening is left for their introduction, which is closed by a piece of flat cork accurately fitted to the aperture.

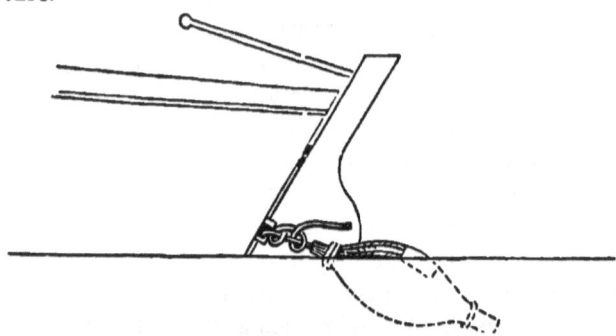

FIG. 20.—Courge in tow.

A piece of small rope having an eye in it is passed over one end and firmly lashed at the other, by which it is towed astern of the boat (fig. 20), the rope having first been made fast through a hole in a small cleet, nailed to the stern of the boat, just above the surface of the water, as shown above.

Of these baskets the most useful size is 2 feet in length, and 7 inches in diameter, and certainly nothing can be better adapted for the end intended, as it is very light, and the water flows easily through it, whilst from its shape it offers less resistance in passing through the water than any other which could be devised. It will be found equally useful as a live-bait cage for Shrimps, Prawns, other crustacea, and small fish in general; it might also be adopted for the same purpose for live-bait in Jack or Perch-fishing on a river or lake. In France these contrivances are made of wood; I have tried them, but the Sand-Eels do not live nearly as long as in the wicker 'Courge.'

How to make Courges or Sand-Eel Baskets.—Procure some very fine small osiers or withies, and soak them a day or two until they are sufficiently supple for weaving, also some others for the framework $\frac{1}{4}$ inch in thickness, and 2 feet in length. Get two hoops of any light and flexible wood of 7 inches diameter, and lash with wax-ends or sail-twine seven of the rods to the outside of one of the hoops, at equal intervals of its circumference, and at 3 inches on one side of the middle of the rods; now insert the second hoop at the distance of 6 inches from the first, and secure it as before. The ends must then be brought together, and, being lashed, will form the figure of two cones joined at the base. Commence weaving at one of the ends, opening the rods by aid of a small marline-spike or pricker, and having done about 4 inches, proceed in like manner with the other end, when the increasing width between the rods will necessitate the insertion of other rods, which are to be secured as at first directed. Continue weaving from either side until within 3 inches of the middle of the rods, leaving here an unwoven space about 6 inches in length and 4 in breadth to pour the Sand-Eels into the basket. This aperture must be closed by an accurately fitting piece of cork when in use. A rod or two will cross the hole, which must be cut off to admit the cork, and the two rods forming the edges of the hole longitudinally must have the osiers doubled round them and the ends of these osiers tucked in.

The cork should be a good thick piece $\frac{3}{4}$ of an inch larger than the hole, and by taking off half its thickness to this width, it will fit well over the edges of the aperture. If a thick piece is not to be had, peg two pieces together. The cork should be attached to the basket by a piece of line, that you may be able to drop it when taking out the bait, and a bit of stick ¾ of an inch thick being thrust through under one of the rods on one side of the cork, a piece of line 18 inches long should be made fast to the rod of the basket nearest the cork on the other side; this being belayed on the short bit of stick serving as a cleet will ensure the cork keeping in its place. Reservoir courges are sometimes 3 feet long; these are not towed but kept on a mooring.

How to Bait with Rag-Worms.—The Mud, Rag, or Rockworm is the next bait, and very valuable, as it can be kept a considerable time with proper care, and is found in the soft mud of harbours, also in sandy gravel and clay formed by the

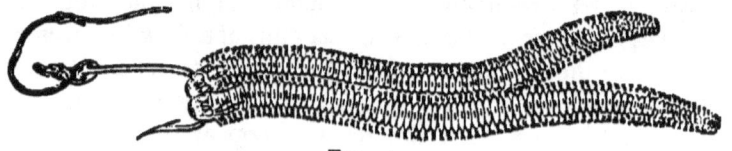

FIG. 21.

decomposition of granite. (See the article, p. 188.) Hook them merely through the head, if small three, if fair sized two,

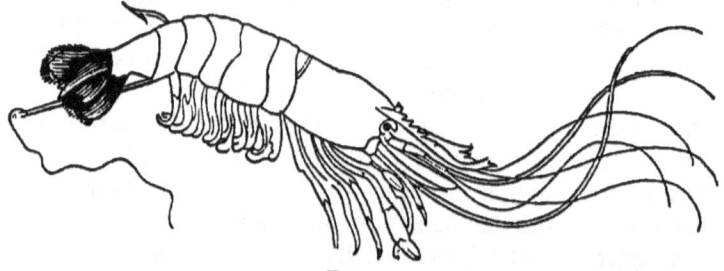

FIG. 22.

at a time, allowing them to hang down their whole length, as fig. 21. A second illustration (fig. 31, p. 85) shows a plan for whiffing hereafter described. These worms should be carried to sea in a wooden box 10 inches by 5 and 2½ deep, the joints pitched, and water ½ an inch deep should be kept in it. Put it out of the sun if possible and on a slant, as the worms like to crawl out of the water for a change.

How to Bait with living Shrimps.—The Shrimps or Prawns are to be kept afloat in a box with holes, or, better still, in a Sand-Eel basket. The hook should be baited, as shown fig. 22, by passing the hook up through the tail. Dead baits are not of much account generally in drift-line fishing : their description under 'Whiffing' will suffice.

HORSE-HAIR LINES, AND HOW TO MAKE THEM.

These lines with pipe-leads at intervals of 12 feet are the best that can be used for Pollack-fishing, Mackerel, Bass, or Bream, when moored, and may also be used when whiffing under oars, but not under sail, as, being valuable lines, the risk of hooking the bottom and consequent breakage is too great. They are more used in the Channel Islands than any other locality I have visited, but I have met with them at Portsmouth, they are well known at Weymouth, and a variety with the lead at the end is used at Plymouth, and another with hemp at the upper end in the Isle of Man ; there are, however, very large districts where they are quite unknown or unused, and here they might be introduced with great advantage. The following is the method of manufacture.

You must provide yourself with a small jack or twisting-engine, also two circular pieces of lead, one of 1 lb. weight for hair, and another of ½ lb. weight for gut, each with a wire hook in the centre. (See fig. 23, p. 70.)

Procure a good long tail (of a horse, not of a mare for obvious reasons), wash and dry it in the open air, and cut a few inches off the end, as it is usually rotten from dirt, &c.; then tie the hair round with twine at the root, in the middle, and at 8 inches from the tail end, place it on a table before you with a heavy book on it or a piece of board, the tail end towards you, and drawing out the longest hairs as they present themselves to the number of twelve or fourteen, according to the thickness of the hair, whether it be coarse or fine, attach three twelves or three fourteens, as the case may be, to the hooks of the spinning machine, as shown in the woodcut, and hanging on the 2 lb. lead, turn the handle briskly round, checking the too rapid twisting of the strands by the scored top held in the left hand, or your fingers, until the twist reaches as far as the top will allow, which is then to be removed, and a knot made on the end to prevent the link untwisting. A three-pronged kitchen dinner fork will answer the purpose of the top, but practice enables everyone to dispense with either, dividing the strands with the fingers of the left hand only. The spinning machine is fitted with a screw spill which you can fasten in a heavy block of wood and place on the table before you, or into a shelf or mantelpiece at a convenient height.

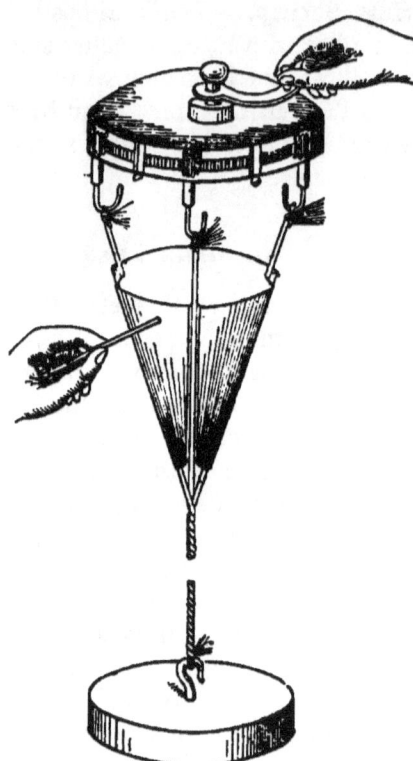

FIG. 23.—Spinning Machine or Jack.

Make the snood part of the line of white hair, three hairs less in each strand, transparent preferable, length 9 feet.

It is a very good plan, when you have selected your hairs sufficient for a strand, to tie one end of them round with fine thread or cotton, and put them to soak in water for half an hour before using. By preparing a number of such strands you will make greater progress in the manufacture of the line. A single hair when doubled should raise a pound and a half weight. If on testing a lot of hair, it will not do this, the hank of hair must be rejected as too weak.

Twist half your links from left to right, the other from right to left, and when knotting them together, be careful to join those of a different twist, for the opposite action of the links will prevent fouling; to make sure of this, tie them up in separate parcels, and label them right and left. In making up these lines, take a link from each parcel alternately. I have dwelt strongly upon this point, as on the careful application of the principle of alternate twist, the useful working of these lines entirely depends, for in this respect they differ from all others.

As the hooks of these twisting machines are sometimes rather widely placed apart, it often entails a loss of two inches on each link, to avoid which take three pieces of thin brass wire, and making an eye at one end turn the other into a hook one inch deep in the bend, passing the eyes over the hooks of the machine, fasten your hair to them, for they will draw together as the hair twists up and prevent waste of material. By having the tail end of the hair towards you, you will use the longest first, shortening in regular succession, and thus put the hair to the best advantage. When the tail is reduced to a foot in length, it is no longer considered worth twisting. Some fishermen prefer the lines two stranded only, which practically are equally as good as three-stranded lines; supposing both well made I cannot see that either has the advantage of the other.

Black hair is in use for the greater part of the line, twenty-two fathoms in length, at the end of which 9 feet of white is attached, the black of two or three strands, of ten or twelve hairs in a strand in the case of three, but of fifteen or eighteen in the case of two strands, the white as forming the greater part of the snood of three hairs less, in order to make it a little finer; whilst for sinkers leads of the shape of a piece of tobacco-pipe, a little barrelled in the middle, half an ounce or more in weight, are placed at every two fathoms, and are of course kept from slipping up or down by the knots which occur in the line.

To the end of the white hair is fastened a yard of white snooding, and to this two lengths of double twisted Salmon gut with the hook. The line thus comprises in all twenty-two fathoms of black hair, with the pipe-leads, to be threaded on whilst knotting the line together, nine feet white hair and three feet white snooding with the two lengths of double gut.

A reel is necessary upon which you will wind your line, and may be either a square of cork of 6 inches to each side, or a frame of like dimensions; if a frame, it may be made 8 inches by 6, or according to the owner's fancy. Many will say that snooding is of itself fine enough, but the majority of fishermen are of a different opinion, unless the fish are very numerous and large; almost all fishermen at the present time use gut both for Pollack and Bass or Mackerel, fineness of tackle in a proportionate degree to size of the fish being quite as necessary as in freshwater fishing. At least four of these lines will be required, and one of the same length without leads; these latter lines are more frequently made of only two strands of hair, but of the same strength, length of snooding, &c., as the preceding. One of these only should be used with the others, and always over the stern of the boat. In the Channel Islands all these lines are known as 'Flottants,' answering to our term 'Drift Lines,' and their use has received due attention under 'Drift or Tideway Fishing,' pages 62 and 76. The various leads used as sinkers can be obtained ready made from the makers (see advertisements at end of the volume), but the method of making them is described in case my readers should like to cast them, or be inconveniently situated for procuring them, at home or abroad.

Pipe-leads and Moulds.—To make the Pipe-leads, take two pieces of freestone exactly alike, and of a square form, and fitting two of the surfaces carefully together, separate them and describe a cross upon the surface of one, and mark on the side the position of the four points of the arms; now place the other piece on it, and with a carpenter's square make also four perpendiculars to the plane of the cross, on the sides of the pieces, which will of course enable you to describe two more crosses on the under and upper surfaces of the portion of freestone. Separate the pieces, and hollow out half the proposed dimensions from each surface, exactly in the centre on the line of the cross, and on the transverse line of the cross on one side make a hole about the thickness of the small end of a tobacco-pipe stem to pour in the lead.

Make also a small furrow lengthways through the mould, and place in it a piece of wire well greased, an eighth of an inch in thickness. You can now pass a few turns of twine round the pieces of freestone, to keep them together, and cast your first lead, and if your work be true, your best plan will be to drill four holes for the insertion of pegs, to hold the pieces in their places. Half-ounce and ounce leads are useful sizes. The pear or fig-shaped plummets for Mackerel-fishing may be made in this manner also ; but still a better plan, if you intend to follow sea-fishing much, is to get these moulds made in iron by an iron-founder, and keep them by you.

In Guernsey, as there is a constant demand for these kinds of leads, they are kept on hand by the plumbers.

To Knot the Links together.—Take two links of a different twist, unlay an inch and a quarter of the ends, drawing them

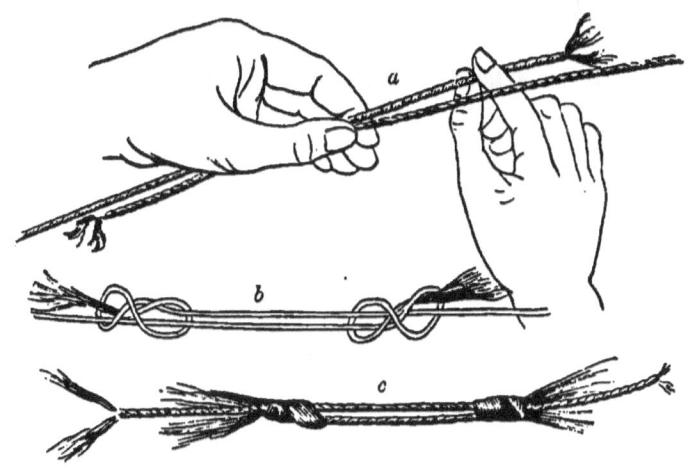

FIG. 24.—Knotting hair links.

through your teeth to flatten out the whole of the twist in order that they may make a snug knot. Place them in the left hand overlapping each other six inches, as shown in fig. 24, *a*, and turn each end twice round the fore-finger and the other link, as shown loosely at *b* ; then put the end through and tighten the knot, not by pulling the end, but by pushing the knot as far as it will slide, as shown at *c*, and having done the same with the end of the other link, draw the two knots together, tightening the knots by taking first one and then the other end between your teeth and pulling with a steady strain on the link, also gently biting the knots to make them draw home in the turns as neatly as possible. Do not cut the ends off too short, but leave them of about two straws' breadth. Measure your twelve feet distances and put on the pipe-leads when required; they should be allowed to run loosely between the knots; if not, the line may rot from damp under them. In addition to these directions, it is well to obtain instruction from old sailors, fishermen, whip or tackle-makers, when opportunity offers.

To Knot Gut Links together.—Previously to either spinning or knotting gut soak it ten minutes in lukewarm water. Spin with the ½-pound lead, and set it revolving by hand, the gut usually having insufficient power to start the lead. In knotting lay the ends to overlap two inches, turn them round your left fore-finger and put the ends twice through, when draw tight and cut off ends. This knot is not so short in the nip as the ordinary one used for gut, and gives more strength in the join.

The following article written for the 'Field' I reproduce as *àpropos* to the subject.

A Day's Drift-line Fishing off Guernsey.

On a fine morning in spring I joined a friend for 'a day's drift-line fishing' off the rocky coast of Guernsey. As we pull out to our fishing-boat we observe the fishermen just starting with their Sand-Eel seine to procure bait for themselves and their *confrères*, who in common with themselves are making for the Mesuriaux rocks, which, on the falling tide, are much frequented by shoals of Sand-Eels.

A run of five or six minutes brings us to the rocks just outside the harbour, where we find eight or nine boats awaiting their supply of bait. The net has been already shot, and is being hauled by the men in two boats, each having a rope attached to one arm of the seine; but the net itself is now

A DAY'S DRIFT-LINE FISHING OFF GUERNSEY. 75

reached, and the crews of both boats get on board one, the cable of which being shifted from the bow to one end of the midship thwart, she swings broadside on, allowing room for four men to haul the head and foot lines at once, whilst a fifth thrusts down an oar continually to drive back the Sand-Eels to the bag of the net. The greater part of the seine having now been boated, the Sand-Eels are seen rushing to and fro in a dark, plum-coloured cloud; and it is very remarkable to observe how the whole shoal turns at one and the same time, as if the act were one of instantaneous volition common to all. The bag of the net is now gathered up closely, and the fishermen hand in their Sand-Eel baskets to be filled in turn, our own being one of the first to receive its quantum of this truly splendid bait. The Sand-Eel basket is secured, and, with our canvas again set, we are off for our fishing ground.

At length we reach the Rousse de Mer (a large insulated rock), and having seen the cable clear I make fast the end on the crown of the anchor, and stop the cable down to the ring with a single turn of rope-yarn, as the bottom is of a somewhat mixed character; by this precaution we may clear the anchor if we hook a rock. We find one or two boats on the ground, and being personally known to all in the neighbourhood, generally give them a greeting.

'Ky b'yottong!' (*quel beau temps!*) exclaims my friend, addressing an old fisherman in the nearest boat who has reversed his sou'-wester, bringing the fantail flap forward over his *os frontis* as an additional protection from the sun, which has lately burst out with great fervour. 'As-tu d'pison?' (*as-tu du poisson?*)—'Pas grammong, m'sieur, pas grammong' (*pas grandement, monsieur*); 'nous avons hallai dow ou tray,' we have hauled two or three—holding up a fine Pollack of about 10 lbs. weight.

Having anchored, we take a dozen or so of the Sand-Eels out of the basket and place them in the boat's bailer, half filled with water, baiting our lines by passing the point of the hook down the throat and out of the gills, then lightly hooking the bait through the skin of the throat just sufficiently to fix the hook; at the same time taking care not to injure the fish, by holding it as carefully as possible. We have here some seven fathoms of water with a nice stream of tide running, and as the lines stray out astern at about an angle of 45 degrees we pay out rather more than the depth of the water, namely, five leads on the lighter lines at the stern, and four on those amidships, there being intervals of two fathoms between the leads. As I am in the midship part of the boat I take the line of the port side in my hand, and having a bite, hook my fish and haul him in, rebait my hook and put out again, when rattle goes my lead and line on the other side of the boat, and beginning to haul I find I have rather a larger fish, for he makes several violent tugs, and I find it necessary to give a little line. I contrive, however, to turn him as he comes rapidly up, and when he is alongside I dip him up in a short-handled landing net, the gaff being chiefly reserved for extra large fish. They are now coming along faster, and it is as much as we can do to tend and fresh bait the hooks. 'More bait, please,' I exclaim, throwing overboard on the line the last in my bailer, and handing it aft to my friend for a fresh supply, who proceeds to replenish from the basket under the stern; but, whilst thus occupied, whizz, rattle, goes his line, and dropping the basket he turns quickly to seize it, but, unmindful of the other little bucket containing his own bait, knocks it over with his knee, and the little silvery fish are all scattered over the stern-sheets.

Meanwhile, as there is evidently a weighty fish on the line, I glance round to see the gaff handy. The fish struggles violently, and it becomes necessary to veer out some of the line; he is, however, soon turned, and comes into view through the clear water, his wide side showing yellow as he struggles head to tide, but to no purpose, for as my friend breaks his sheer once more, he is brought to the surface alongside, and inserting the gaff under his gills, he is safely taken in and done for. The fish proves to be nearly a yard long, and weighs full 12 lbs. This will be a fellow for boiling; we therefore decide on hanging him up with a lump of salt in his head, for with some oyster sauce he'll eat like a Cod-fish.

The tide now began to run considerably stronger, and more

length on the lines was requisite; I therefore prepared to 'rig a soldier.' A soldier-line is one of two-stranded hemp-twine, having for a sinker a two-pound Mackerel plummet, and is made fast to a strong flexible stick about two feet long, which is stuck into one of the thowl-pin holes in the midship part of the boat, that is to say, in front of the drift lines. With this line I chiefly fished during the strength of the tide, substituting it for one of the lighter ones, with fair success; but when it slackened I reeled it up and put out the drift line as before.

But what is this on my line which hauls as dead as if I had hooked a weed? I hope it's a Dory!

This it turned out to be, and I desired my friend to get ready the hand-net, for it often happens that Dories are taken without being hooked at all, but by swallowing a small fish which has previously taken the bait. It is therefore always well to be prepared, and nothing is better than a little handnet, in which our Dory was dipped up, testifying his disapprobation of his entrance into a foreign element by two grunts as he was placed in the basket.

'I think we had better shorten up,' I observed, 'as the stream of tide is fast decreasing, or we shall be hooking up in the long ore weed.' On trying my line, however, I found that the hook was already fast in the uneven bottom. I tried jerking and hauling by turns, but it was of no avail, and putting on a steady strain, I got free, but with the loss of the two lengths of double gut, which form the hook links. Two or three minutes, however, served to repair damages, and I was soon at work again.

'Suppose we put out a light line,' my friend observed; 'I think we may get a Gar-fish or two on the turning round of the tide.'

I accordingly baited a line, without any lead on it, and paid out about ten fathoms. Here I made fast a loose thowl-pin, and then secured the end to the stern ring-bolt, leaving about four or five fathoms of slack line. This line had not been out long before the thowl-pin went overboard with a violent surge, and my friend found the bait had been seized by a good-sized fish, which, after a little humouring, was brought to boat, and weighed about 5 lbs. Four others, all of fair size, followed on this line, and a couple of Long-Noses also, after robbing us of numerous Sand-Eels.

The tide is now dead slack, and the lines are perpendicular. '*Il faut tracher*' (I must whiff) observes a fisherman in a neighbouring boat; and, having wound up all his lines save one, gets up his anchor, and shipping one of his oars in the sculling hole, prepares to whiff, putting out his single line, and sculling along with steady determination to make the most of the slack tide. We, on our part, wind in all our lines save two, being two of us in the boat, and, getting up the anchor, I take the sculls and pull leisurely along, when we pick up half a dozen more; but a nice breeze springing up it becomes inconvenient to whiff any longer, and we make up our minds to return. We therefore wind up our drift-lines, the mast is stepped, sail hoisted, and we head for the harbour. Putting out a couple of Mackerel-lines, we pick up one of these fish (three we took previously on our outward trip), and are only about a quarter of an hour returning, for the inset of the flood is made strongly. On landing we carry our panniers up the beach, and emptying them on the turf, survey our catch, which consists of Mackerel, Pollack, Gar-fish, and the Dory, weighing in all about 50 lbs., the Pollack, as is commonly the case, predominating in number, and varying in size more than either of the other kinds.

Streaming for Pollack is a kind of drift-line fishing which has been followed in Cornwall time out of mind with much success, and is next in efficiency to drift-line fishing with the living Sand-Eel. Its chief feature is a long snood of five fathoms on a stout Whiting or Conger line, and a stout hook, No. 7 size, p. 210. Three lines can be used from one boat, the first line with no lead, the second with a lead of $\frac{1}{2}$ lb., the third with a lead of $1\frac{1}{2}$ lb., of boat shape, all with a five-fathom snood. Being thus of different weights, the lines will stray out at different angles, and thus keep clear of each other.

Bait, half the side of a Pilchard, a piece of Mackerel four inches long, or half the side of a Chad or small Bream scaled.

The snood as hereafter described for whiffing Cornish fashion (p. 81). Anchor on rocky ground, sound the bottom, and haul up sufficiently to keep the hook clear of the ground. Hook the bait on by the small end, and fish in an easy tide. In the offing the largest Pollack are taken in this manner, from 100 to 350 lbs. in a day's fishing. It is usual to make fast the lines to another boat lead, sufficiently heavy to prevent the line dragging it overboard, which happens only when a fish seizes the bait and draws your attention by the splash of the lead in the water if you are engaged with the other lines.

Whiffing

is the process of towing a line lightly leaded, or without lead, after a boat sailing slowly, or gently pulling, or else sculling with an oar, in a semicircular notch in the stern or transom-board. In lake and river-fishing it is termed trailing.

Lines for Whiffing.—The pipe-lead horse-hair lines are well adapted for whiffing, but as they are so much more expensive than hemp or flax lines, and from constantly passing over rocky ground are liable to hook the bottom, when breakage often results, I advise my readers to reserve them for drift-line fishing, unless in very quiet weather when whiffing under oars. A very useful line for whiffing is a 12-fathom Mackerel line of the double Bridport make, as manufactured for Guernsey, with a 4-fathom snood of the same make, and a 5 oz. lead. These lines being two-stranded rarely become entangled, and are yearly becoming more known and appreciated (fig. 11, p. 48, No. 4). They are usually of flax, but there is no objection to twisting up the hemp snooding when extra strong lines are required. Always from their twist retaining a considerable amount of stiffness, they may be used either dressed or not. The end of the snood consists of two lengths of double twisted gut with a hook, No. 11 in the cut of hooks (fig. 63, p. 211), the hook being first provided with an eye of fine snooding tied on Nottingham fashion, which gives the gut much better chance in hooking a large fish, than if whipped directly on to the hook. For whiffing under oars near the shore a lead not less than three or more than five ounces should be used, and the snood may be of gimp four yards size 20, or copper wire of gauge 21. The lead, if of boat form, should be mounted with a piece of double horse-hair thirty-six hairs' thickness, served round with stout waxed thread, projecting from each end half the length of the lead. Join the wire snood to this with an inch swivel, and connect the gut and wire, or gimp, with one of half inch size. A pair of these lines are necessary. The plummet lead with revolving chopstick (fig. 39, p. 124), as for Mackerel, is suitable for Pollacking, but a very considerable latitude is allowable in the shape of leads for whiffing for Pollack, and either the conical, fig-shaped, or boat-shaped, may be used, the form not being of great importance, because the boat should always move more slowly than in Mackerel-fishing, for which the boat leads are not adapted from their tendency to sheer and jump when towed along at any considerable speed. For whiffing under sail one pound will be found a good weight, but in certain positions, as in tide-races, you are compelled to use a two-pound lead, and if of plummet form all must have the revolving chopstick. You may consider you have good 'dray-way' through the water, as the fishermen term it, when your lines form with the surface an angle of about 45 degrees, and you should pull, or set just sufficient sail for the required rate. I recommend the lines to be marked at 3, 6, and 9 fathoms. When wind and tide are in opposition (if both are not so strong as to cause too much sea) is the most favourable opportunity for whiffing under sail, as you can then adjust your canvas to a nicety, but under oars or sculling, a calm or very nearly a calm is most advantageous. In whiffing under oars or sculling almost any punt or sea-going boat will answer, for instance a yacht's dingy 10 to 14 feet long, and one should take the paddles whilst the other tends the two lines. For one person a single line is ample, and the boat may be either sculled or pulled as found most convenient. In Pollacking under sail two lines may be managed by one hand, and any sized boat under 10 tons will answer; but were I to choose I should prefer an Itchen rigged boat of 5 tons, with the addition of a mizen, as it is so very useful in whiffing with the jib or staysail, when the mainsail is not required. The helmsman must be a good pilot, and not become so absorbed in his sport as to forget which way his craft is heading, or he may speedily come to grief on the rocks. The Crab-pot corks must be diligently looked out for or the gear will become unrigged by loss of the hooks and snoods. Amateurs at Plymouth use short rods about a foot long of cane or whalebone stuck into the gunwale, or into little zinc or copper clamps nailed against it; these first yield to a fish, and then allow the line to slip off the top, thereby avoiding breakage. These little rods are called twiggers, and show a bite immediately. The Crab-pot corks are often run under by the tide, which causes great risk to your tackle, and unfortunately they are most frequently placed off the headlands, and in the very spots required to sail over, therefore it is best to do so only at slack tide, when they are all bearing at the surface, and during the continuance of the stream to follow drift-line fishing.

The Cornish Whiffing Line consists of a stout Whiting or Conger line without any weight or sinker, the absence of this being supplied by the length and weight of the line itself. A good method of making the snood is with two fathoms of stout gimp or fine copper wire, secured to a brass swivel $1\frac{1}{2}$ inch long. A loop of double hemp half the thickness of the main line, and 6 inches long, is spliced into the swivel, and looped over a knot on the end of the main line. A second small swivel is attached to the further end of the gimp or wire, and to this 3 to 6 feet of very stout single or double Salmon gut. Ten, fifteen, twenty, or more fathoms of line are used according to the depth of water. This kind of line is much used single-handed when sculling the boat along with one oar over the stern, as being without lead, it does not sink so rapidly as to get foul of the rocky bottom, like a leaded line. Pollack, Mackerel, and Bass are taken with it. It answers well also as a stern-line in a tideway at anchor for these fish, trailing back clear of the other lines. A slip of parchment $\frac{1}{4}$ inch wide $1\frac{1}{2}$ inch long is an additional attraction with Rag-Worm bait, and an imitation indiarubber band in parchment will also catch Pollack.

G

Baits for Whiffing may be used both dead and living, which affords a much wider range for choice than in drift-line fishing. Living and dead Sand-Eels, small dead Freshwater Eels (fig. 25) and Lampreys (fig. 26) from 4 to 6½ inches, Lob or large Earth-Worms, Lugs, Rag, Rock, or Mud-Worms, slips of Long-Nose or Mackerel 3 to 5 inches long and ⅜ to ¾ wide, unsmoked bacon or pork skin, Gurnard or Bass skin, and white leather or rag of the same dimensions, in addition to artificial baits, as spinners and imitative Sand-Eels, with

FIG. 25.—Freshwater Eel (dead bait for Whiffing).

white and red and fancy flies, including a rough kind of palmer or caterpillar made of goat's-hair, which is especially good for Bass. In addition to these, we have three valuable artificial baits for Pollack, which owe their discovery to accident. The first is the red indiarubber band, which was, it seems, tried by an amateur short of bait, who happened to have one of them as a fastening to his pocket-book. After cutting it he hooked on one end, and took several Pollack whiffing. It is the

FIG. 26.—River Lamprey (dead bait for Whiffing).

custom now to whip one end to the head of the hook (fig. 27), instead of baiting it as a living Rag-Worm, to which it bears a remarkable resemblance. The most useful size is about 1½ inch in diameter. There are two kinds of these indiarubber rings, one lies flat when placed on a table, the other rests on its edge. The right kind is that which lies flat like a quoit or Saturn's ring; and if my reader will divide one and hold it up by one end, he will see at once the reason it should be preferred, as it hangs in a curl, which causes a rapid life-like action when drawn through the water, and thus becomes very attractive to fish. The action is so rapid that the eye cannot follow it, and it has the appearance of two, instead of one worm on the hook. The other two baits are Brook's double twist spinning Eel or Lug and Hearder's Captain Tom's spinning Sand-Eel or Lug-Worm (figs. 28 and 29), with which numbers of

INDIARUBBER BAITS.

FIG. 27.
Rubber Band Imitation Rag-Worm.

FIG. 28.
Brook's Double Twist Spinning Eel or Lug-Worm.

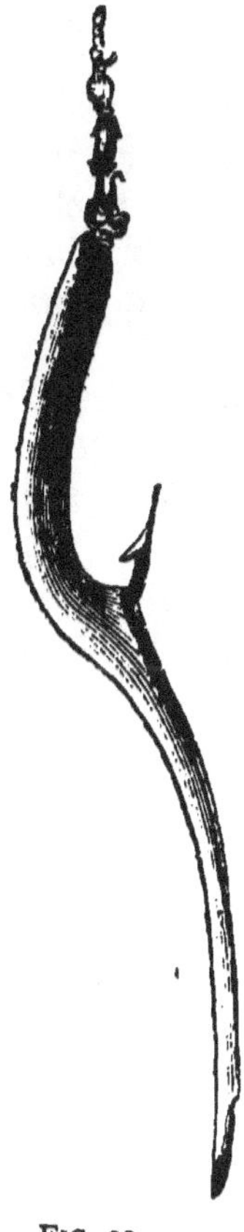

FIG. 29.
Hearder's Captain Tom's Spinning Sand-Eel.

Pollack and Bass have been taken, the so-called Sand-Eel being made of the grey indiarubber pipe placed on the hook so as to form an elbow at the bend, which causes it to spin when drawn through the water, whilst the Lug-Worm is a red bait of the same kind. Brass swivels are attached to the heads of each, to provide for the spinning action. The spinning Sand-Eel and Lugs are taken by both Pollack and Bass, the Grey Eel especially in the dusk, as well as by daylight. With the Red Eel or Lug a friend took off Hartland Point over sixty Bass 3 to 9 lbs. weight on one visit. Having lived on such a variety of coasts, including rocky, sandy, and shingly shores, some abounding in, and others having little or no natural bait, but on many of which much fish could be taken, I have had to try all manner of things, and have given a considerable variety to choose from in the present article. The best natural baits for whiffing are without doubt living Sand-Eels and Rag or Mud-Worms; the living Sand-Eel to be placed on the hook as in the tideway, and the dead as I have here shown in the cut of the Freshwater Eel (fig. 25); the Rag-Worm, when two are used, as at fig. 21, p. 68, and if the fish are shy or bait runs short as at fig. 31, p. 85, to do which enter the

FIG. 30.—Tail part of an Eel (Whiffing bait).

hook about ¼ of an inch below the head, and bring out the point 1½ inch down the worm, and stick the smaller hook through the head.

Lampreys and Freshwater Eels are to be placed on the hook in the same manner, but the point of the large hook should be brought out lower down, as shown in figs. 25 and 26, p. 82. The blind Lamprey or Pride may be baited in the same way, and the three last-named baits should be always killed prior to placing them on the hook. In Cornwall the blind Lamprey or Pride is very much used as a whiffing bait for Pollack, and in baiting it, the point of the hook is brought out through the back, instead of through the belly, and only sufficiently far down to allow of the mouth covering the flattened top of the hook. The mouth is then tied round with thread above the flattened top of the hook to keep the bait straight on the hook in a naturally swimming position. If small Eels are not procurable, and you should have taken one of a foot or 15 inches in length, cut off 6 or 7 inches of the tail part, and having drawn back the skin ¾ of an inch, cut off the flesh and backbone thus far, then drawing the skin back again, tie it round with a bit of thread after having placed it on the hook (fig. 30).

In Ireland it is customary to turn the skin back over the lashing, and then sew it with needle and thread to the skin

FIG. 31.—Rag-Worm when the fish are shy.

behind the lashing. Eels are so tough that a couple or three will last some hours—light coloured are best; if dark, skin and soak them a few hours in milk. Lob or large Earth-Worms, as well as Lug-Worms, are kept in position by the small hook, after having brought out the larger one part way down, as here shown (fig. 32), the point of the larger to be entered ¼ of an inch below the head of the worm. If the worm should turn white before you take a fish, put on a fresh one.

FIG. 32.—Earth-Worm baited for Whiffing.

These worms are by no means equal to the other baits enumerated, and are only to be used in the absence of better; that they will take both Whiting-Pollack as well as Mackerel I have frequently proved, and on one occasion, with a dozen Earth-Worms and one white fly, I took from four to five dozen Whiting-Pollack, and frequently three or four with the same worm, being careful not to spoil the bait in unhooking the fish. It is the most economical way of baiting I have ever met with.

The small hook keeps the bait from sliding down on the

bend of the larger one, which is essential, for were the body to follow the bend of the hook, it would not only have an unnatural appearance, but also would revolve in the water, and cause the tackle to twist. Such little matters as the above, although they may not seem of importance at first sight, I can assure the reader are not to be neglected with impunity. This observation applies to all the baits. Lugs are not usually considered whiffing baits, yet although not equal to Rag-Worms, they will do good service when baited as the Earth-Worm, but being a much shorter bait, the small hook must be tied on to the hook link close to the head of the larger, and a shorter shank hook used than for other baits. The slips of Long-Nose, Mackerel, pork or fish skin, are to be hooked merely once through one end. A strip of Squid or Cuttle-fish is also used, and a horn or arm of the same, or of a sucker or octopus, may be baited like the Eels. The artificial baits may also be often used with success, attached to the whiffing lines, and over rocky ground and shallow water, flies are very killing, particularly after sunset. In moderate weather a rod may be used. In consequence of the success obtained by the previously mentioned indiarubber baits in whiffing, the natural baits are not so much used where they are troublesome or difficult to procure.

Fly-fishing at Sea.

Fly-fishing at sea was formerly confined to the few, but is now very generally practised. In favourable positions it may be followed from the shore, where steep rocks with deep water, or a pier-head, or shingle beach at the mouth of a bar-harbour offer points of vantage, whence the fly may be cast without the embarrassment of a steep wall of cliff; but such are few and far between, and as a rule it is much more successfully carried on from a boat.

The Rod.—For general sea-fishing with a rod, 16 to 18 feet of ash and hickory is a good length, top lancewood, which will answer well in bottom-fishing or angling from the shore, but for boat fly-fishing or throwing off the rocks &c. the long butt should give place to one of 18 inches, which will reduce your rod of course to about 15 feet 6 inches, a much more manageable length. A bamboo rod is the lightest and most agreeable to fish with, but more liable to receive damage from the rough and tumble work which is one of the concomitants of sea-fishing, both in clambering over rocks and in the boat.

Your rod, although strong, should not be unmanageably heavy or stiff, in order to enable you to throw your flies as in Salmon and Trout-fishing, when a shoal of fish is seen leaping on the surface of the water. A stout top 6 inches long to fit into the third joint is very useful for heavy work in the boat. Supposing the rod to be four-jointed, three lengths without any butt may be used for Mullet and Smelt angling from piers and quays. Thus in one rod you will have an instrument useful for all angling in salt water. The rings should stand out as in a trolling rod, but need not be so large, $\frac{3}{8}$ inch in the clear will suffice. It is not in fresh water considered orthodox to use the same rod for both fly and bait-fishing, but in salt-water work we must bend to circumstances for the sake of having only one rod, which can be made to answer all our requirements.

The Winch should be a plain one without any complication whatever: a Nottingham one is preferable, because, from the size of the barrel, the line can be wound up more quickly when a lead is used. As this winch works very freely, to prevent overrunning and thereby entangling the line, a leather washer of proper thickness should be placed between the screw nut and the reel, or between the back cheek and the reel. Any required pressure may thus be obtained, and the principal objection to this class of reel removed. The new Nottingham winches are fitted with a brake or check which prevents their overrunning.

The Line.—Your line should be of white or brown snooding, or cotton, not less than 30 yards in length, with a reel or winch which will run easily when a large fish takes the bait, or some part of the tackle will be carried away; if the fish run very large 60 or 70 yards will not be too much. See fig. 11, p. 48, No. 5, for size of line. The check of the winch therefore should not be heavy.

The eight-plait trolling lines are also very suitable for salt-water angling: in fact, there can be no kind of line better adapted for the purpose; my only reason therefore for recommending the line of snooding or cotton is that it is a very efficient substitute for the plaited line, and may be purchased at about one-sixth of the cost; but whether your line be of snooding or plaited, it will be much the better for being dressed with some stiffening mixture, and I have not found anything more effective than coal-tar and turpentine, the only objection to which is, that it takes a long time to harden, for which purpose it should remain a month in the open air in all weathers, and be occasionally rubbed over with a piece of upper shoe-leather, which will smooth all inequalities. One line will answer both for fly and bait-fishing.

The Collar, Bottom, or Trace, should be 6 feet in length, made two-thirds of double or triple twisted, and one-third of selected single Salmon gut. This if joined together by loops will stand a heavier strain than knotted.

Flies.—For materials the whitest goose feathers to be procured, with such blue, green, red, and yellow feathers and wools as are commonly used to tie flies for Salmon and Salmon Trout; also any of this kind half worn out may be used for sea-fishing, and if of a dark colour a few fibres of white goose feather may be added to the wings with advantage. The following are those most adapted for general use :—

1st. A Salmon fly, red body, gold tinsel, a bushy tail, gaudy and white wings (fig. 33, *a*).

2nd. Irish or Scotch rolled fly, if fly it may be called, either with or without a red worsted body. The feather to be taken from the bottom of the neck of a white goose in the back, just between the wings, then to be rolled up like a leaf between the fingers, and lashed on by the root of the feather to the head of the hook, so that it may point straight back over the hook. This is a very effective bait, and has long been a favourite on the Irish and Scotch coasts, although I have not heard of its use elsewhere (fig. 33, *b*).

3rd. Hearder's fly or feather bait. This is made with two small red or white goose feathers tied flat one on each side of the hook without a body (fig. 33, *c*).

These three flies may be made on hooks Nos. 8, 9, 10, or 11

according to the sized fish expected to be met with. (See cut of hooks, fig. 63, p. 211.)

4th. A plain white fly, consisting of white wings cut from the quill-feathers of a goose or swan on a bare white hook, No. 10.

5th. White fly of a large size, hook No. 8, body red, green or blue wool, ribbed with gold and silver twist or flattened wire,

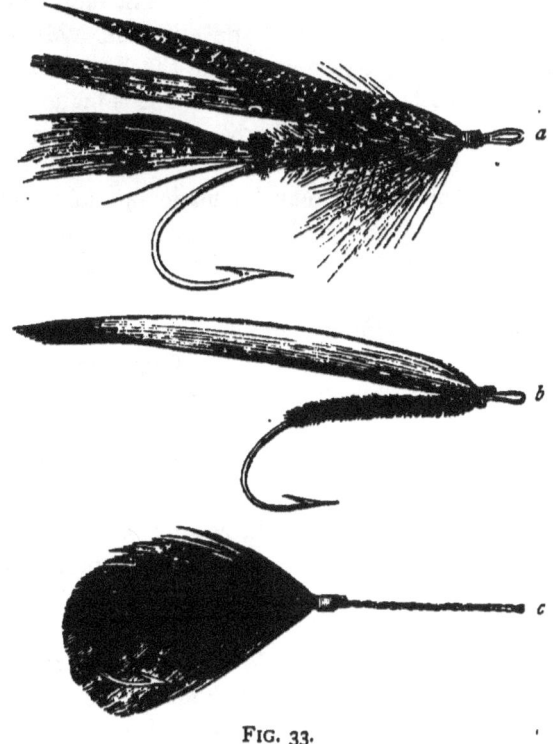

FIG. 33.

or the body red, green, and blue wool in succession, without gold or silver wire.

6th. 'The Shaldon Shiner,' so named from the village of Shaldon on the west side of Teignmouth harbour, close to which this fly was used with great success by the late J. C. Hele, Esq., one of the most expert fly-fishers of his day both in salt and fresh water, and through whom the author was induced to adopt fly-fishing at sea. It is a kind of imitation of the dragon-fly. The body is as thin as possible, being nothing but flattened silver wire, a small brush of scarlet feather for the tail, a little green, blue, and red dubbing out of an old turkey carpet for the shoulders, and bright blue wings, to which add half a dozen fibres of goose feather in front. With this, fishing at the mouth of a river-harbour, or in the pools just inside, you will probably take a Sea Trout or two, or even a Salmon, particularly if you fish at the beginning of the ebb tide. Make it on a 9, 10, or 11 hook (fig. 63, p. 211). In the Taw and Torridge estuary at Instow, N. Devon, the fly in use is made with white and grey feathers and a silver body, and with this great sport is frequently obtained.

A rough but useful fly may be made out of white goat's-hair, with a body of red wool. The beard on account of its length is preferred, but that from the body if long enough will also answer.

Two flies will be quite as many as you can conveniently cast, and more will only embarrass you ; but if not accustomed to fly-fishing in fresh water, as well as to boating at sea, employ a man to pull, for you will find it rather awkward work to manage the rod and keep your footing in the boat at the same time, and had better be satisfied with towing or trailing your line after the boat, until you have practised throwing a fly on shore. The loops at the heads of the flies should be either of fine silk line, fine snooding, or double twisted gut.

Use of the Gear or Tackle.—If alone let out about twenty yards of line and pull leisurely along, resting the rod against a thowl-pin in the after rowlock. If with an assistant, hold the rod in your hand. If you feel a fish or see a rise, strike gently and draw the fish toward you by bending the rod backward until you can take hold of the line, when you may lift the fish on board if of moderate size. If, however, the fish should be large, on no account touch the line, but play him as you would a Salmon or any other large fish, and when exhausted, reel up the line until you have no more out than the length of the rod, then inclining it backwards, strike your gaff into him under the throat if possible, and lift him on board. A landing net with the handle only about a foot long is more convenient than a gaff, where the fish are of a moderate size.

As good a gaff as can be used is a Hake or Bonita hook, 2 inches in width, from point to shank, lashed on to a stick 2 feet in length, as a Salmon gaff is too good for the purpose, and soon gets rusty and spoiled. File off nearly all the barb of the hook.

When employing an assistant from amongst sailors or fishermen, it is very necessary to caution him against catching the line without orders to that effect, as in the excitement caused by seeing a large fish circling round the boat, he is almost certain to do, not reflecting that the elasticity of the rod favours your tackle, and will enable you to kill a much larger fish than if the line were held in the hand.

Boat.—The best boat for this fishing is a yacht's punt, or dinghy, or boat of similar build, not less than twelve, nor more than fifteen feet in length, as these boats are light, steady on the water, and row sufficiently fast for the purpose; but if unacquainted with boating at sea, leave the selection to your assistant.

You may fish under sail in larger boats, but the rod is better dispensed with, and whiffing lines should then alone be used with the fly-trace.

If you prefer a short-handled net for getting your fish on board, procure a forked branch of a tree, of any tough or flexible wood, and bending the ends round, lash them together with waxed twine, leaving the handle about a foot in length; this is much preferable to an iron ring, on account of rust, but if for the sake of portability in travelling you prefer a metal hoop, folding or not, it should be galvanised, unless made of brass, which breaks frequently in the joints and is more expensive. When you see a shoal of fish break the surface of the water, row to windward and cast as for Salmon or Trout. If you wish to trail entirely you may venture on four flies on the trace, but for casting two are quite sufficient, more being difficult to manage. Any artificial baits, such as Spinners, Sand-Eels, or Herring-fry (locally called Brit), may be used in the same manner.

Sheaf-fishing.—This plan of fishing is followed on the Scotch and Irish coasts with severals rods of from 12 to 15 feet in length, and where the fish are plentiful, Pollack and Coal-fish (often known as Lythe and Saythe) are taken in large numbers, and sometimes Mackerel. It is termed sheaf-fishing, either because as many as nine rods are taken on board the boat at one time, and when lashed up into a faggot, form something like a sheaf of straw, or because a sheaf of straw is actually used to retain the rods in their required positions during their use. This sheaf of straw is lashed to a board placed across the boat about three feet in advance of the stern, into which the rods are stuck, spreading out like a fan over the stern and quarters of the boat. The line of strong snooding or horse-hair is used about 18 inches longer than the rods, and the hook, on double or strong single gut, is baited with a particular goose feather, placed on the hook in a peculiar way, not having the appearance of a fly, but rather that of a small fish. This feather is first rolled up as one might roll a small leaf, and firmly lashed on to the head of the hook, at the root part (see fig. 33, b, p. 89). If the line were more than 18 inches longer than the rod, there would be a difficulty in lifting the fish into the boat. The extraordinary number of nine rods is taken into the boat in case of loss or breakage, but half a dozen only are commonly used at once, the other, with lines and hooks wound round them, being kept ready for action at a moment's notice, in order that no time may be lost in the dusk of the evening, when this method of fishing is commonly practised, for it would not do to be refitting the rods when they ought to be at work. These rods are made of good red pine $1\frac{1}{2}$ inch thick at the butt, tapering up to $\frac{5}{8}$ thickness at the top. For safety in case of breaking the rod, the line should come down the whole length of the rod to the butt. The favourite feather used is taken from a white goose at the bottom of the neck on the back, just between the wings. One goose does not afford many of these feathers. The rods when stuck in the sheaf of straw stretch out horizontally, and the tops ought to be 6 to 12 inches above the water. In addition to the board upon which the sheaf of straw is placed, a second is laid across from gunwale to gunwale, a little in advance of the other, for the fisherman to sit on, so as to enable him to look down conveniently upon his rods. The boat is kept in motion by a second fisherman, who pulls along slowly in successive circles of from 50 to 70 yards' diameter. The feather-bait is represented with the flies, any of which can likewise be used with the rods, just as this feather-bait may be used for fly-fishing for Bass or Mackerel on either rods or whiffing lines. (See p. 89.)

Rod-fishing for Pollack from Shore.

Very good sport with the rod, from the steep rocks as well as from the piers of many harbours on the coast, may be had with this and other fish. ·For rod, reel or winch, and line, see pp. 86, 87.

If there is much stream, or a strong eddy, the following tackle will be found to answer well. Take a pipe-lead one ounce in weight, and pass through it a piece of horse-hair line double, of twenty-four hairs thickness and seven inches in length, having previously brought the two ends together and firmly bound them round with waxed thread to within half an inch of each loop, which will stiffen it and prevent the line from twisting round the lead; procure a piece of medium sized gimp four feet in length, such as is used in Jack-fishing, and having made two neat bows with waxed silk at either end, loop the lead on at one, and the hook tied to two links of the strongest Salmon gut at the other; bait with two or three Rag-Worms.

If you cannot obtain gimp, you may use four feet of white horse-hair, twenty-seven hairs thickness, or of the strongest triple Salmon gut twisted, which should be first soaked in warm water, when it will work better; this should also always be done before tying knots with gut, to prevent its breaking. Where fish are very large, six thicknesses of gut, two in each strand, will not be found more than is requisite. In fig. 34, p. 95, a short piece of gimp is only shown, from the limited space available. The hooks should be tied to two lengths of double twisted or single gut, according to the run of the fish on the coast. This tackle with a snood of increased length, nine feet from lead to hook, may also be used in whiffing from a boat pulling, weight of lead four ounces; in tide-way or drift-fishing it will also answer well, and if the current be strong a lead of as much as half a pound will be requisite, beyond which weight the use of a sinker is unpleasant on a rod. Let the line be fine, it will then keep down sufficiently in a considerable tide, if not beyond ten fathoms depth of water. In connection with this gear I have given these instructions for the benefit of anglers who, not being accustomed to sea-fishing, object to the handling of wet lines. The line had better be marked at distances of three, five, and ten fathoms. In harbour fishing No. 13 hook is generally large enough, but off headlands and the open sea you will require No. 11; all should be extra strong, as a large Rockfish, Bass, Bream, or Pollack, will snap off or straighten a weak or brittle hook. (See 'Hooks,' fig. 63, p. 211.)

Floats.—If you are fishing on an open shore where there is much wash, you will find a float useful, but in a strong run of the tide it is of little service. On an open shore the backwash will take out the float and enable you to keep your line clear of the rocks, which without it would be very difficult of accomplishment. From a pier or breakwater head, off which a stream sets, turning round into an extensive eddy, a float may be used with advantage. The old-fashioned pear shape answers well, length 3 inches, diameter $1\frac{3}{4}$ inch, or of the elongated form, length 5 inches, diameter $1\frac{1}{8}$ inch. I prefer them without either ring or caps, merely with a plug through the hole projecting from either end $\frac{3}{4}$ of an inch. Over this, both top and bottom, take two half hitches, which will both make the float secure, and can be cast off instantly, when you desire to dispense with it, which a change of tide or any other circumstance may render desirable. A float may be used either with the lead and trace already described, or with the Pater-Noster line, when the water is inconveniently deep to plumb the bottom. For Pollack-fishing from shore I do not find any other bait worth using except the Rag-Worm, to be stuck on through the head, as shown in fig. 21, p. 68; a piece of Long-Nose 1½ inch long or a bit of parchment is considered an additional attraction.

Rod-fishing with a Light Line.—Off a pier-head, when there is a very little or no tide, or in a moderate tide from a boat, very good fishing may be had with a light line, with no sinker whatever, a collar or bottom of the best Salmon gut, 5 or 6 feet long, being attached to the line. If the water be sufficiently deep, let out about 15 yards of line, use no float, and bait with Rag-Worms; you will chiefly take Pollack, Bream, and Horse-Mackerel.

The Pater-Noster Line (fig. 34).—This kind of line is preferable in fishing off piers &c. in quiet water and gentle eddies, inside a harbour's mouth, or other favourable situation, and enables you to take a greater variety of fish than the other. To fit up this tackle, take 6 feet of double twisted gut and a pipe-lead half an ounce in weight, and having passed through it a piece of double hemp snooding, make a knot below, and you will then have a bow in the top, to which loop on the gut; then take a hook tied to 9 inches of single gut, with a bow at the other end, and fasten it by a sliding loop to the hemp, and at three or four of each of the knots above, place a hook tied to not more than 6 inches of gut; they will now keep clear of each other and stand out from the line, as shown in fig. 34, and you can bait with Rag-Worms as before.

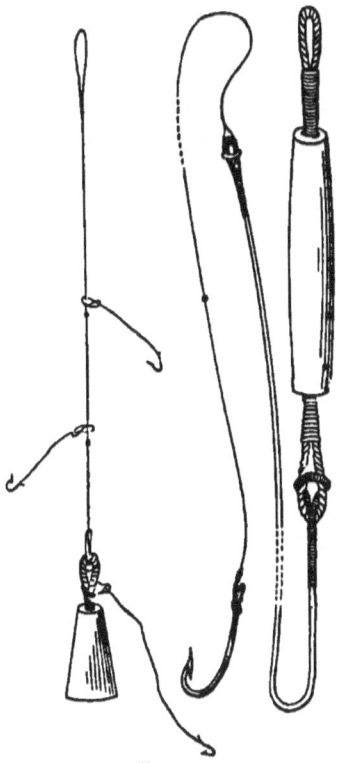

FIG. 34.

Pater-Noster.   Pipe-lead and trace for Rod-fishing.

For open sea-fishing and at the mouths of extensive and deep arms of the sea, stout gimp should be used in place of the twisted gut, and the hooks themselves be tied on twisted triple gut, and of tried strength, for you are always liable in these positions to meet with very large fish, and unless well prepared will have your tackle carried off in the most unceremonious manner by a 10 or 15 lb. Pollack, or even a larger Coal-fish. Three hooks will be quite sufficient for a Pater-Noster for Pollack, and the hooks, supposing you use gimp for the main part of the Pater-Noster, may be kept from slipping down by a lashing of waxed thread on the gimp, at about intervals of 18 inches; if of gut, the knots will be sufficient. Bait with the Rag-Worm.

Plumb the ground with the lead occasionally, and lift it a few inches off the bottom, until you feel a bite.

You may bait the bottom hook with a boiled Shrimp, taken out of the shell, and will take with it Flounders, occasionally Dabs, sometimes called Sole-Dabs, and Freshwater Eels, which are found in most tidal harbours, but use Mud-Worms when procurable.

A landing or Shrimp-net must be provided, with which to dip up the large fish, as your tackle would frequently fail you in attempting to weigh them out. See p. 91, or fig. 74, p. 243.

N.B.—The larger kind of Mud-Worm is frequently known as the Rock-Worm, as it is found in the sand, clay, or gravel, close to rocks, or under large stones; many are also obtained by forcing asunder stones naturally cracked, for the fissures in which they have a great predilection.

In the Channel Islands it is customary to clear a spot of stones, and then to dig in the subsoil with a harpoon-shaped digger of iron, called a 'Petron,' $6\frac{1}{2}$ inches long and $3\frac{1}{2}$ wide, on a $4\frac{1}{2}$-foot handle.

The Floating Trot (fig. 35).—A Trot is a long line with hooks at intervals, and the variety here described is used in Guernsey for Pollack and Gar-fish.

The Floating Trot should consist of a stout Cod-line, well stretched to take out the kinks, having pieces of cork 4 inches long by 2 wide secured flat on the line, at intervals of 2 fathoms, and midway between the spaces bung corks 2 inches across, bored through the centre, and grooved round the edge to receive a 3-foot piece of stout horse-hair line, having a pipe-lead at the end half an ounce in weight, and 3 feet of hemp snooding with a strong Mackerel or Pollack hook. A knot or stop must be made on each side of the bung cork, leaving sufficient play for it to revolve freely on the line; to ensure this be careful to bore the hole both large and smooth with a hot wire.

At every fourth cork make fast a piece of line three or four fathoms in length, with a large stone to hinder the main line rising too high off the bottom, as well as to keep it from forming too much of a bow by the stream of tide, and moor the line at the ends with two heavy stones with buoy lines. Bait with living or dead Sand-Eels, two or three large Rag-Worms, Lugs,

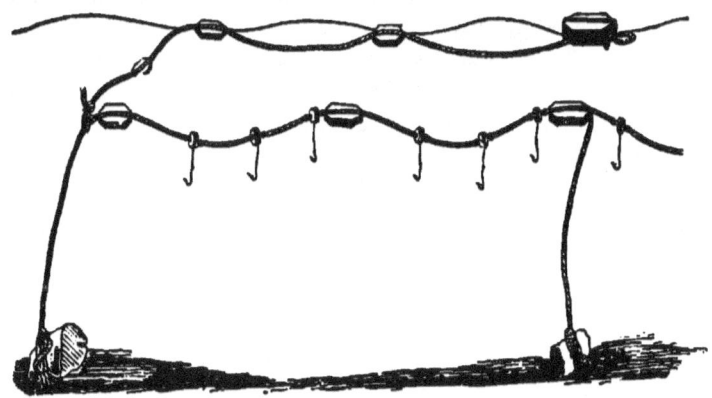

FIG. 35.—Floating Trot.

strips of Cuttle or Squid, or any bright fish, and shoot it on or near to rocky ground just before sunset, and raise it in the morning.

You will take large Pollack, Coal-fish, Bass, Bream, and Gar-fish, but for the last named use smaller hooks.

A line of thirty fathoms will be found a useful length.

At the commencement of the season in March, should fish be scarce, this gear wil. be found very effective, but may be used at any time. In common with other trots or bulters, this should be shot across the stream of tide, if not too strong. It may be used either on the surface, or at midwater, and should

H

be kept when not in use in a basket with a hook holder, shaped like a tuning-fork.

THE COAL-FISH.
(*Merlangus carbonarius.*)

This fish is of a much more rounded form of body than either the real or Silver Whiting, or the Whiting-Pollack or Lythe. It is found all round our own coasts as well as on that of the north of France, and Channel Islands, but is much more abundant on the east, north, and north-west of the kingdom, and amongst the Orkneys and Shetlands, than in the English Channel. The back is dark green, lateral line and belly white, but their hue varies much, many being quite blue on the back. It attains a larger size than the Pollack, reaching sometimes 30 lbs.; but the flesh is inferior. It has a number of provincial names, being in Devon and Cornwall known as the Race or Rauning Pollack, in Yorkshire as Parr and Billet, in the north of the kingdom as Saithe, Sillock, Coaley, Grey Lord, and Stedlock or Stenlock, and in Guernsey as Mutan. Its habits are very similar to the Pollack, being found on rocky ground, and it is taken with the same tackle and baits. I have caught numbers of 1 lb. or 2 lbs. weight in harbours with rod and line, and find them struggle harder than Pollack. Immense numbers of young Coal-fish are taken with rod and line in the Scotch lochs under the name of Cuddies, also on the Yorkshire coast, under the name of Parrs. Large fish keep more outside off headlands in strong streams of tide, and are also found on the smooth ground, where they are taken from the drift Herring boats, with hook and line, mingled with Cod. In the north of the kingdom and amongst the Scotch islands they are caught in large numbers by whiffing, with half a dozen rods stuck into a faggot or wisp of straw lashed to the thwart of the boat, with a very short line and a white feather fly (see p. 92), the boat being pulled very slowly along. A regular fishery is prosecuted in Norway, and they are also abundant on the North American coast.

THE WHITING-POUT.
(Morrhua lusca.)

Whiting-Pout, Rock-Whiting, Short-Whiting, Lady-Whiting, Blinds, Bib, and Blains, and the Ponchette of the Channel Islands, the 'universal Pout,' as I have heard it humorously styled, is one of the most common fish on the coasts, both on mixed and rocky ground; the larger ones afford good sport, and are very fair eating, if cleaned immediately after they are caught, as indeed all fish should be.

Those taken in harbours are, however, generally small. No. 3 line, p. 48, is the size recommended.

The best tackle for this fish is that described for Silver-Whiting, viz., the boat-shaped lead (fig. 7, p. 42), or the Kentish Rig (fig. 3, p. 38), which I consider about the best form of chopstick for ground-fishing. Weight of leads 1 lb. for the stern-lines, and 2 lbs. for the forward lines. Any of the gear, with moderate-sized leads, as described and illustrated for Whiting, will answer for Pout-fishing; but, as it is necessary to make a selection, I have chosen these two as the best, after having given all kinds of gear an impartial trial for some years. Supposing two hands in the boat, one line apiece is sufficient in ten fathoms of water and under, and even in deeper water one is ample for beginners, but in from fifteen to thirty fathoms, after having become expert in handling the lines, two may be worked by one as in Whiting-catching. In shallow water the line should be kept in hand and the fish struck sharply whenever a bite is felt; in deep water the fish are much bolder and more ravenous, and will generally hook themselves. This applies also to the true or Silver-Whiting. By using two lines you will not be losing time, as immediately you have taken your fish off one line, and fresh baited your hooks if requisite, you have only to throw the lead overboard and attend to the other. In putting your gear overboard, be careful to throw the hooks well away from the lead, which will ensure their going down clear, otherwise they will be very likely to foul. To set your line, act as in Whiting-catching—viz., first sound the bottom and then draw it up an arm's stretch or 5 feet (which should be the length of the line below the lead), then give it three or four turns round one of the thowl pins of your boat if using two lines. The Kentish Rig, and other kinds with short snoods, should be raised just clear of the ground. Using one line, rest the hand on the gunwale, raising the lead a few inches occasionally. For harbour fishing, or under ten fathoms, make the snoods of twisted or strong single gut; but in deeper water, foul bottom, and general outside work, fine cotton, hemp, or flax snooding will be found preferable, as you are constantly hooking large fish, and weeds and rocks, causing great wear and tear of gear.

Baits.—Much the same as for Silver-Whiting, Mussels of rather a smaller size, Lug-Worms, the largest Rag-Worms, a very large description of Flat Worm, sometimes near 18 inches in length and ½ an inch wide, which is found by digging under large stones at low tides, and sometimes also of a smaller size in the sand; this I have only met with in the Channel Islands, but I imagine it may be found in other parts of the kingdom, where the shore is composed of granite rocks and pebbles, with its accompanying sand and yellow clay. (*Note.*—A boat's bailer is always required when searching for this worm to turn out the water, which otherwise floods the pit immediately it is formed.) It is named Varm, see p. 190.

The other baits are a piece of any fresh fish, such as Herring or Pilchard, the soft part of a Limpet, or, when nothing better can be procured, garden snails, or the hard part of a Limpet, and the tail part of the Hermit Crab or Soldier, alias Crab-walk and Gann.

This fish being unprovided by nature with any shell for the lower part of his body, supplies the want of it by inhabiting that of a dead Whelk, and changes his abode as often as he finds his quarters too confined for his increasing size; these may be procured from the fishermen, as numbers of them, as well as Whelks, get into the Crab-pots, and are also taken in dredging and trawling.

I have found it an excellent plan to put out a drift-line at the stern when Pout-fishing, baiting with a large Mussel or a slip of any shining fish; use a strong Mackerel hook, without lead, if the tide be moderate; if in much stream a lightly leaded horse-hair line, as recommended for Pollack. I have taken many large Pollack and Blue Bream in this manner, and therefore advise my readers to make this a rule whilst Pout-fishing.

Nos. 12 and 13 will be found useful sizes for general Pout-fishing, but in harbours, and where they run small, use 14 in the cut of Hooks, fig. 63, p. 211.

THE POWER OR POOR-COD.
(*Morrhua minuta.*)

Among the Pout, you will generally take a fish of a longer form of body, known as the 'Poor' in Cornwall, Pouting at Plymouth, 'Gilligant' and 'White-eyes' at other localities. It is caught from 5 to 10 inches in length, and is a sweet-eating little fish, when fried of a delicate brown, with as little fat or butter as possible, soon after capture; but it becomes quite tasteless the following day, although it may be untainted by decomposition. The Rag-Worm is the favourite bait, and No. 7 Kirby hook tied on single gut. From six to twelve dozen are taken off the Hoe, Plymouth, in three or four hours, in an afternoon or evening when the tide suits; I have taken this little fish everywhere when after Whiting-Pout. It is the smallest of the Cod family.

From the beginning of July until nearly Christmas you will also take numbers of Chad, which are the young of the common Sea-Bream; in unhooking these be careful of their prickly back-fins. They make good bait for Conger.

A DAY'S GROUND-FISHING.

'We are going to have another day at Billingsgate,' was the greeting of a friend as I sauntered one morning down the ruddy cliffs impending o'er the pebbly beach at Budleigh Salterton, on which the wavelets, impelled by a light southerly breeze, were gently breaking, the precursors of a lop, or more agitated state of things, which followed later in the day.

The connection between Billingsgate and Budleigh Salterton—taking Billingsgate in its ordinary sense—cannot be very evident, seeing that the distance of near 200 miles intervenes between the two localities. I must therefore explain that the word is here used to denote a certain fishing-ground about two and a half miles off this little Devonshire watering-place, and although it cannot, as a matter of course, afford the variety attainable at the great metropolitan emporium, it yet offers to the fisherman considerable sport in hand-line ground-fishing, of which we had availed ourselves on an occasion not many days previous.

'What do you think of the weather?'

'A breeze, probably, but nothing to hurt, for I think the wind will be off the land again in the evening, although it will certainly not be as smooth as on our last visit to Billingsgate, when the sea was as little agitated as an unruffled lake.'

'What does Rogers say?' Rogers was the boatman, and, on being interrogated, took the same view of the weather as myself, observing, however, that we must pull out, for the wind was dead on end, and the sail 'would not be a ha'porth of good' until returning, when we should probably make our passage in a third of the time if the wind remained in the same quarter; 'but there's no hurry,' he observed, 'for if we get out just at the slackening away of the ebb we shall have three or four hours there, and that will be quite enough time to make a catch of fish, if there's any upon the ground.'

'How are we off for bait?'

'There's a peck of Mushels (Mussels), sir, and I've brought a dozen o' Pilchards out of a small lot we caught this morning in the little seine, so that we can try for a Conger as well, since you lost one when you were last upon the ground.' This occurred through the shortness of the ganging or hook link, beyond which, as the Conger had swallowed the hook, it gnawed off the line with little difficulty.

On this former occasion it had not been our intention to try for a Conger, and accordingly we had not provided ourselves with hooks specially fitted for the purpose; but on the present trip we were determined to have something proof against the effects of the teeth of his Congership, and I had fitted three or four hooks with a very strong gimp in the following manner, as is the custom also for Hake-fishing at Plymouth and elsewhere.

I cut off a piece of line sixteen inches in length, and a trifle less stout than the fishing-line with which it is to be used, and, splicing an eye at one end, unlay an inch and a half of the other and make a knot on the end, leaving the unlaid part about an inch long below the knot, passing through the knot the end of a coil of brass wire, by aid of which I convert my piece of line into a coarse hook-link of gimp, as will be seen presently. Bringing the knot about three-quarters of an inch below the flattened top of the hook on the inside, I bind the hook, the piece of line, and the wire firmly together with well-waxed thread, and fasten off with the invisible knot in common use ; and as it is necessary, for the successful conversion of the piece of line into gimp, that it should be as rigid as possible, I hitch the hook over a nail, the handle of a door, or other point of attachment *convanient*, and get the piece of line on the stretch by making it fast to some other firm object at the same height—it being essential that it should be in the horizontal position, that the wire may be wound regularly round it. The line and hook being conveniently placed as here described, they are to be bound as tightly together as possible with the wire, as far as the flattened top of the hook, when the wire is to be wound round the piece of line alone until the line is covered to about ten inches above the top of the hook, when the wire may be fastened off by interlacing it three or four times between the strands of the piece of line. The hook is now, to all intents and purposes, snooded or ganged with a piece of very stout gimp, and it must be an extraordinary Conger indeed to make any impression on it. Hooks fitted in this manner are procurable from Mr. Hearder, of Union Street, Plymouth ; and a dozen or two would be found very useful to all who may occasionally fish for Conger &c., as well as to others who visit or reside on the distant shores of our colonial possessions, where fish of such a size are met with that no ordinary gear will hold them, as testified by a correspondent from the Cape Colony, who mentions the capture of fish of 120 lbs. weight, with the escape of other monsters of unknown magnitude. This digression may, I trust, be pardoned, on account of its utility to would-be fishermen for Hake, Conger, and other huge denizens of old ocean's depths, whose sharp teeth and vice-like power of jaw make such short work of undefended snoodings.

My friend, his wife (for ladies go sea-fishing on the coast of Devon), and a relative, with myself, now proceed to the beach and bestow ourselves in a boat as desired—namely, two on the after thwart and one on the next forward, whilst the writer, as an old hand, is intrusted with an oar a little in advance of the last mentioned, and stands ready to assist in shoving off as soon as the word is given. A good smooth, as they term it, affording a favourable opportunity, with a shout of 'Now, now, now!' we glide into the swell, your humble servant assisting the efforts of the fishermen as they leap in over the bows, and continuing so to do until the boat is clear of the landwash, when my friend, his wife, and relative are desired to take their seats in the stern-sheets, and we proceed to trim the boat for our two and a half miles row to windward.

And now a word or two, *en passant*, on going afloat from an open beach.

It is a great matter to make a clean launch, especially if accompanied by ladies, and with ordinary precaution it is easily done without shipping water, which, especially at starting for a day's pleasure, is of all things to be avoided.

If the beach is steep it is the custom to launch stern foremost, the boat resting on a way, or skid, of holly or other hard wood, two or three feet from the stern. This keeps the keel out of the shingle, thereby preventing any check to her downward progress until the stern is water-borne, which object is facilitated by allowing no one to sit further aft than the after thwart, as mentioned on the present occasion. The stern is thus allowed to rise immediately it feels the swell, which it will not do if the passengers crowd aft, as the boat is thereby deprived of its buoyancy where and when it is most needed, and the stern caused to dip under the swell, deluging all the after part of the boat with water, to the discomfort and annoyance of all on board. On a flat beach, however, or one of a less steep gradient, a different plan is followed, for it is there the custom to launch bow foremost, because a boat has generally to pass a succession of breakers, which the bow is, of course, better calculated to deal with than the stern, whereas on a steep beach it is only the last wave which breaks, and even that is commonly avoided by watching a favourable opportunity.

On the present occasion, having made a clean launch, the rudder was shipped, and as the breeze was freshening, I pulled with the fisherman and his son, two sculls, or as we here term them, paddles, being pulled against an oar, any difference in strength of stroke being compensated by the helm in the hands of my friend. We pulled seaward some short distance, but finding the boat row heavily, we brought our killick stones (rather more than a hundred pounds in weight) about three feet further aft, a great improvement, the boat rising in a more lively manner, plunging less heavily as her bow fell down into the trough of the sea, against the waves of which, accelerated by the increasing breeze, we were advancing.

'Keep her a bit more west, if you please, sir,' says Rogers, 'for the tide's agoing east yet, and the wind's a westerning a little,' his object of course being not to drop to leeward of our required position and down the tide, which would cause delay in pulling up to windward again.

'What mark have you?' asked my friend, looking back over his shoulder at the land.

'Well, sir,' says Rogers, 'd'ye see the white tower, right astarn? That's the Belvedere at Bicton ; and you keeps the Belvedere between the wheel and the shed with the red-tiled roof (a little bit more west, please, sir) until the other marks comes on.'

'What wheel?' asks my friend.

'The whin (or whim) as they winds up the coal and stone with at the limekilns,' says Rogers.

'Exactly,' replies my friend, and steers accordingly.

A few minutes' more steady pulling brought us nearly up to the ground, when Rogers observed the tiller might be unshipped, and he would put the boat on the marks with the oars alone, 'for,' said he, 'you see, sir, this is a very particular place, and if you bayn't right to a boat's length, as you may say, you can't expect to have fish. Pull, Jim' (to his son); 'Back, if you please, sir' (to your humble servant). Squint number one at the north and south marks; squint number two at the east and west marks; and with a 'There, that'll do,' plunge goes the heavy killick stone towards the lower depths, carrying rapidly with it in its course the thin line used as a cable, which Rogers paid out until we were in our required position, when he made all fast, and we lay riding head to the sea. The Downend, forming the north termination of Little Haldon, was now visible in a certain gateway of the high land of Orcombe, near Exmouth, and the Bicton Belvedere, between the red-roofed shed and the limekiln wheel previously mentioned, forming two lines of sight intersecting each other in our position, at as near an approach to a right angle as could be found by selecting from among the objects of the neighbouring coast those four which would best supply the desideratum.

A supply of Mussels was soon opened, and being larger than were required for a single bait we cut them in half, the tongue of the Mussel in one half, and the circular gristle (by the expansion or contraction of which the shell-fish opens or shuts its valve at pleasure) in the other, thus leaving in each a sufficient hold for the hook. The hooks were baited and the leads thrown over without loss of time, you may be sure, as the wind still continued to freshen, and it seemed questionable whether or not we should be able to remain long enough to make a catch. The bottom was hardly sounded, however, and the lines set by raising the leads about four feet above the ground, before the fish struck every line, frequently a pair at once; then a pause, again a rapid run of sport, until our baskets began to make a fair show, and it was evident we should obtain, for the kind of fishing, a very good day's work.

I say for the kind of fishing, for we were not so much after large fish as for Pout (*Morrhua lusca*), here known as Blains, in Cornwall as Bib, in other parts as Short or Rock-Whiting, and in the Channel Islands by the appellation of 'Ponchette.' We had, of course, no objection to Pollack or other large fish which might give us the meeting, and which do so occasionally, as I took one on a former visit, not many days previous; but we look on these as chance fish, and the more to be welcomed when ground-fishing, because not to be reckoned on with certainty.

'But what have I here?' exclaims one of our little party, handling his line most gingerly, a course rendered necessary by the successive sharp tugs of a fish evidently of a different kind from any yet taken, and which, from its action, was more likely to part the snood or break its hold than any previously hooked.

'I reckon 'tis a Curner,' observes Rogers.

'Or a Baker,' I suggested; the former the local name for the Wrasse, or Rock-fish in its different varieties (Fam. *Labridæ*), the latter for the Braize or Bekker, a Blue-Bream (Fam. *Sparidæ*), which is here called 'Baker,' being commonly cooked in an oven, after a nicely seasoned veal stuffing has been introduced into his *penetralia*, supplemented by a modicum of beef gravy in the dish.

Whether Curner or Bekker, however, both gaff and hand-net were ready to aid in getting him on board, which, being safely accomplished, my conjecture proved the correct one, for a Bekker or Braize it turned out to be. This fish varies from $1\frac{1}{2}$ lb. to 4 lbs. in weight, and I do not recollect to have met with any small ones; in colour they are blue on the back, the sides silvery, but change their colour entirely after capture and become almost black, whence their name of Black Bream. The back is very much hogged and carinated, and the dorsal fin, when not erected, is received into a deep *sulcus*, or furrow. The teeth are large for the size of the fish, and always strike the attention of the observer, as the lips shrink back and leave them prominently bare. They are by no means a plentiful fish either here or on the coasts of the Channel Islands, and are never specially fished for like the Sea-Bream.

'Well, Rogers,' exclaimed one of our party, 'when do you intend to haul up that Conger?' addressing the fisherman, who, leaving us to the capture of such small deer as Pout &c., was endeavouring to coax one of those slimy monsters from his lurking-place amongst the rocky ground over which we were moored. In this he had so far succeeded as to feel several bites from one of the slippery individuals, probably the most crafty of any which swim or crawl at the bottom of the briny. 'But,' as he observed in answer to the question, 'he don't bite home, sir,' that is to say, he did not take the hook; for to feel a Conger is one thing, whilst to catch him often quite another affair. The fish in question had robbed him of no less than six Pilchards in succession, and had yet managed to escape the hook, showing clearly that he was a fishy incarnation of craftiness, requiring something even more *appétissant* as a bait than that oily member of the family Clupeidæ, the Pilchard, freshly caught that very morning as these fish were, and luscious enough apparently to satisfy the tastes of either Conger or— Cornishmen.

It is considered that scarcely a more enticing bait can be put into the water for a Conger than a Pilchard: but it has one great drawback—namely, if the fish do not feed heartily, but are in a picking humour, they easily rob the hook; consequently, Mackerel is preferred when attainable, as it possesses a tough skin compared with that of the Pilchard, which is more tender than any other fish of the same size. We had no Mackerel, however, on board, although we towed a couple of lines on our way out, in hopes of taking one or two, which would have been very useful to us on the present occasion, for the appetite of the fish having been whetted by the Pilchards, he would have been pretty certain to gorge a piece of Mackerel, 'Pilchards being a gathering bait and Mackerel a killing bait' —an old and very truthful adage of the Cornish fishermen. As his Congership, however, evidently required 'cream upon Pilchards'—as something out of the common, and usually considered unnecessarily rich, is termed in the west—we found an efficient substitute in a freshly-caught fish, which, placed on the hook in a particular manner, induced the Conger to 'bite home.' Selecting a 'Gilligant' and scraping off the scales, the fisherman entered his knife at the tail, and cut up towards the head to within an inch of it; then, turning the fish over, he served the other side in the same manner, and removed the backbone; and entering the point of the hook down the throat, he brought it out below the gills at the end of the incision, an inch from the head, and twisting one of the sides of the bait round, so placed it on the point of the hook that the outside was turned inside, and the *inside* would be first offered to the fish. Provincial names are generally an effectual disguise, and as my readers may probably be puzzled to know what fish is meant by this appellation, I had better perhaps inform them that it is nothing more nor less than the Power or Poor-Cod (*Morrhua minuta*), so frequent round our coasts, and known as the King-fish in Scotland (p. 101). This freshly-cut bait, although not in reality as rich as Pilchard, evidently possessed an irresistible attraction for our slippery friend, as he seized it without his previous caution, and that 'music on the water' which the fisherman loves to hear, technically termed 'sawing timber' (as the tightened line cuts into the gunwale), told an unmistakable tale of weight below.

The water being only twelve fathoms deep, the fish was soon alongside; one splash, a struggle, and taking the snood short in his hand, Rogers lifted him into the boat, and only just in time, for he had scarcely so done when the hook lost its hold, and the fish fell safely into the bottom of the boat, apparently uninjured, lashing its tail rapidly to and fro, and taking entire charge of that part of the boat in which I was stationed between the two after thwarts. There he lay, lead-coloured in hue, 'rather pale about the gills, and certainly doosed fishy about the eyes,' staring as the whole fraternity of Congers has been remarked to stare from time immemorial—whose fixity of gaze has thence passed into a proverb—sometimes remaining quiet and apparently at ease, then again gliding slowly over the bottom of the boat in changing S-curves, and threatening to invade the *sanctum sanctorum* of the stern-sheets with his unwelcome presence.

Although he took such a deal of catching, he was not very large after all, but a well-conditioned brute about 14 lbs. or 15 lbs., and 'worth eighteenpence,' the fisherman observed, contemplating his slimyship with an eye to business, for we had so much other fish in the boat that we did not regard the Conger except in the light of an accession to our catch, from which, having selected what we feel inclined, we leave the rest for the fisherman's private advantage. Not that a piece of Conger is unpalatable, if nicely dressed, particularly the head and six inches behind it, baked in a dish with veal stuffing, as I have previously recommended in the case of the Blue-Bream, Bekker, or Braize. This part of the Conger, the head especially, must be exceedingly nourishing from its very glutinous nature.

We tried with much perseverance to take more Congers, but none other gave us the meeting : they were therefore either scarce or shy on account of the daylight, for the Conger is generally a night-feeder, especially in water under ten fathoms in depth. I say generally, because I have met with occasional exceptions to this rule. When the water fires, or, as the fishermen term it, 'brimes,' Congers rarely feed well, for the line becomes apparently a cord of fire, which doubtless frightens the fish ; as is also the case with Herrings and Pilchards, which shun the nets when they become phosphorescent.

Much has been said of the ferocity of the Conger after capture ; but after having taken these fish more or less on various occasions for over thirty years, I can see no ground whatever for this assumption, having stood amongst a floor of Conger, so to speak, without any evidence of such ferocity. If a person is awkward or foolish enough to put his finger into a Conger's mouth the fish will very probably nip him fast ; and I have known them to catch the hook in this manner between their teeth, and so firmly will they hold it that I have been obliged to drive a thowl-pin into the mouth of the fish before I could succeed in dislodging it.

The increasing sea did not allow us to ride very quietly in our exposed position, and our inquiring gaze often scanned the horizon, watching for any indication of approaching bad weather which might render it prudent to abandon our sport and make for the shore. This, however, proved unnecessary, for although a larger wave than ordinary would occasionally break against the bow of our boat and sprinkle us with spray, the sea had not attained to such a state of agitation as to render safe landing on our open beach a matter of difficulty or doubt, and we therefore continued to fish on, keeping the weather eye open, prepared to act according to circumstances. As to standing up in the boat to haul the lines, it was perfectly out of the question for any of us who, from years of previous experience, had not shipped our sea legs, and, in fact, it had been so from the time we let go the killick stone. Still we put up with the kicking and jumping of the boat on account of the sport, for the intervals were very short in which the fish did not come on board from one or other of the five lines at work.

My three friends in the stern-sheets hauled away with the most praiseworthy perseverance, and the lines were kept going up and down without intermission. Our lady friend also, who entered thoroughly into the sport, hauled up the fish pair and pair constantly, unhooking her fish and fresh-baiting her hooks, and generally managing her own line entirely, in all but exceptional cases, when the fish in their eagerness would gorge the hooks, and some additional aid be required in their extraction; or when, by reason of the uneasy motion of the boat, or the attempts of a fish to escape, one line might be carried round another, causing a foul, and some assistance to be needful 'to undo the knotted hank.'

Rather suddenly, however, although not altogether unexpectedly, our sport slackened, and at length almost entirely ceased, for looking at the east and west marks, 'Drat it,' says Rogers, 'if us bayn't a drivin'! Bear a hand here, Jim, and get up the sling-stone; us 'ull bend on two instead of one, and let go again a little vurder out.' The sling-stone, or killick, having been brought on board, I took the oars with Jim, whilst Rogers bent on the second stone about a fathom behind the first. This is a much better plan than placing them close together, for one stone can thus be lifted on board at a time in hauling up, and, in addition, the space between the stones gives the chance of catching against two projections of the rock instead of one—a matter worthy of some attention. As soon as we had pulled far enough to the southward to bring the east and west marks a little open, one stone was carefully lowered over the side, and the second being cast after it, scope was given until the marks came properly on, and we were at liberty to go to work again without fear of driving. 'There,' exclaimed Rogers, 'that 'ull hold us till all's blue, and the cows comes home in the morning'—an old saying amongst seamen, sufficiently intelligible as regards the hue of the ocean, but not exactly clear respecting the holding power of killick stones or anchors in connection with any species of the bovine genus. It would indeed be a curious subject of study to trace out the origin of the sayings of seamen, many of which are doubtless of a remote antiquity, quaint, terse, expressive, and as much *sui generis* as might be expected from a class of men forming, so to speak, a nationality of their own.

But to proceed. A number of heavy intermittent tugs were felt on one of the lines, and much curiosity was expressed as to the cause. A merry laugh went round the boat, as a pair of dirty-looking speckled Dog-fish came on board with snake-like contortions, and were dropped into the bottom of the boat, only to be immolated on the altar of enraged fishermen, who in most cases, as in the present, setting one foot on the head and another on the tail, divide the vertebræ at the neck, and then throw them overboard to be devoured by either the finny or crustacean inhabitants of the deep.

We took also some three or four specimens of the Wrasse, or Rock-fish, one of the common kind (*Labrus rupestris*), and one or two of the variegated sorts having, from the brilliancy of their colours, the appearance of painted fish, namely the three spotted (*Labrus trimaculatus*), and the Cook or Cuckoo-Wrasse, of which the blue marks are very beautiful in their hue. When fishing in Guernsey in this manner we often take the Sea-Perch (*Serranus*), but it is not so frequent here as in the Channel Islands and off the Cornish Coast; it is not unlike the Pope or Ruff in appearance, but runs up to 8 inches in length, although commonly smaller. The flesh of the Sea-Perch is eatable, but that of all the varieties of Wrasse is wretched in the extreme, although much eaten in some parts of Devon and Cornwall, and looked upon as a regularly marketable fish in the Channel Islands, in which latter district it is customary to stew them with onions. I have seen them 7 lbs. or 8 lbs. in weight, but never took one above 6 lbs. myself. They vary more in size than any fish I have been in the habit of catching, and I have taken them of all sizes, from a quarter of a pound to 6 lbs.

'How do the Mussels hold out?' asked my friend. 'I hope we shall not fall short of bait, eh?'

'They be all opened,' said Rogers: 'and I think as many as we shall want, sir, for as soon as the ebb-tide makes away strong we shall have a nasty lop; and, as the tide makes alongshore before it does out here, we mustn't wait too long or we shall find bad landing.'

We continued, therefore, to fish as long as our bait lasted, and, having used the whole of our peck of Mussels, we decided on running home, reeled up our ground-lines, and made things snug in readiness for setting our canvas.

'Up killick' is the word, and a couple of hands tailing on to the cable, the stones are brought to the surface, and one after the other carefully lifted on board and stowed in the bottom, so as not to shift. The mast is now stepped, but, whilst the sail is being unfurled for hoisting, the boat has drifted broadside to the sea, and Old Father Neptune, more pressing than polite in his attentions, falls on board us with one of his little white caps on the weather quarter, as if to say, 'Now, my lads, it's time to clear out of this.' Requiring no second reminder, I put a paddle out on the weather side, and getting the boat's bow round towards the land, she immediately gathers way before the sea, and the lug being hoisted and sheeted home, urged by both wind and wave, we run rapidly in for the beach.

A couple of Mackerel lines, with leads of a pound weight, were put out on the principle of making the most of the occasion, but to little purpose, for the speed of the boat was such as to bring the leads nearly to the surface, and the only fish taken was a Gar-fish or Long-Nose (*Belone vulgaris*), which was caught on a spinner about two hundred yards from the land.

When we had arrived to within about a hundred yards from the surf (breaking in a line of rolling foam on the shingle), we lowered our sail, struck the mast, and put the boat under a pair of oars, intending to wait for a smooth and get on shore as dry as possible. My friends in the stern-sheets were requested to sit on the after-thwart, or as near to it as convenient, so that in case of a sea breaking over the stern they might meet with as little of it as possible. In other respects we kept all the remaining weight as far aft as we could, for if a boat is down by the head, and takes a heavy sea in the land-wash, her forefoot may catch on the shingle, and if she does not instantly free herself, the chances are that she capsizes. Not that on the present occasion there was any prospect of such an eventuality; but it is well to acquire habits of caution, if only to come on shore dry and comfortable. This we managed, with the exception of a slight sprinkling, of which no account is taken in beach-boating. Being now close in with the surf, and as fair a smooth presenting itself as we could expect in the agitated state of things, we seized the favourable opportunity, and with a few sturdy strokes sent the boat on to the shingle. Jim, who was ready with the painter, leaped over the bows, which being firmly grasped by half a dozen willing hands, the boat could not haul off again by the reflux of the wave, and we were all on shore in a twinkling, excepting our lady friend, who sat fast on the thwart, holding on by the gunwale, until the boat, by six-man power on the painter, was hauled nearly high and dry, and she could step conveniently out on the shingle. The fishing-lines and killick stones being taken out of the boat, the painter is passed through a hole made for the purpose in the forefoot, thus enabling us to lift and haul simultaneously; and timing our efforts with a succession of deep-throated 'Up-ho's!' from the stentorian lungs of one of our own crew, the boat springs four feet ahead at once over the greased ways placed in her track, and, topping the copp or high-water ridge of the slope of the beach, is deposited on the flat beyond the reach of danger.

The fish and line baskets are now carried up the path to the 'cottage on the cliff,' and whilst the contents of the latter are so disposed as to dry with all convenient speed, those of the former are emptied on the lawn and counted out; the result being twenty-two dozen Pout, three Bekker or Braize, the Wrasse already mentioned, and the Conger left in the boat.

My friend selected as many fish as he required for his own

use, a fry for myself, and another lot for distribution amongst his neighbours; the fisherman being desired to take the greater portion for his own advantage, whilst the Wrasse or Rock fish are devoted to the Prawn nets which are much used during autumn on the South Devon coast.

Thus ended our 'day's ground-fishing,' which proved a very fair day's sport for the kind of fish of which we went in quest.

The regular charge for an evening's fishing at Budleigh Salterton is 2*s.* 6*d.*, the man finding everything. If by the day, 1*s.* 6*d.* the first hour and 1*s.* after. Apply to James Rogers, or John Middleton.

THE DAB.

(*Platessa limanda.*)

The Dab is an excellent fish, and when in perfection, which is in the early spring, has a very delicate flavour.

This is very frequently confounded with the Flounder or Fluke (*Platessa flesus*), as it is of a similar size and shape, but much superior in quality; it may be easily distinguished therefrom, as the Dab is rough on the back and nearly transparent, whilst the Flounder is smooth and opaque.

It is a very nice looking fish when first taken out of the water, being of a delicate brown hue on the back, mottled with crescent-shaped spots of a bright orange, but these hues fade in death, and the spots become nearly invisible, so that the fish loses the attractive appearance it possessed on being first taken from its native element.

They are found on all the sandy and oozy shores of the British Isles, and are taken both with the trawl net and hook and line, of which six may be used, that is to say, three pairs of leads of the respective weights of $1\frac{1}{2}$, 1, and $\frac{1}{2}$ a pound, which will generally be found sufficiently heavy in any depth under ten fathoms; but if leads of $\frac{1}{2}$ a pound should be found too light, let those of 1 pound be used instead, and leads of 3 pounds take the places of those of $1\frac{1}{2}$ pound weight forward.

In some quiet bays with very little tide, much lighter leads may be used, such as of 4, 8, and 12 ounces' weight, and finer

THE DAB.

lines also, but all must depend on the depth of water and the strength of the tide. Provided the leads will keep the bottom, they are sufficiently heavy. The three pairs of the above-mentioned size are a useful average weight.

These fish mostly remain at the bottom, consequently plenty of line should be given in order that the leads and baits may keep the ground.

The boat-shaped leads (fig. 7, p. 42) are the best adapted to this fishing, and the snood with two hooks only (No. 13, fig. 63, p. 211) should not altogether exceed 5 feet in length from the lead. The baits should be fresh Mussels or Lug-Worms, of which Mussels are decidedly the better for Dabs, but for Plaice Lugs are preferable. The Mussels become more firm if opened the day previous to use. For baiting with Mussels, see under 'Baits,' pp. 50, 187 ; as to Lugs, it is only necessary to pass the hook through them three or four times. Fine white snooding or yellow silk is best for this fishing, unless you use stout gut or ordinary gut twisted double.

Your lines should be nearly twenty fathoms in length each, supposing the water to be from seven to ten fathoms in depth, as it is necessary to pay out much more line than the depth of water when there is any tide running, which is the best time to fish, as Dabs and other flat-fish—and in fact all fish—are then actively on the look out for food.

The finer kind of Whiting lines are strong enough for this sport. See No. 3, fig. 11, p. 48.

Having moored your boat and baited the two hooks, take the line of the heaviest lead in your right hand about eighteen inches above it, and the snood in your left hand, and swinging it two or three times to and fro, cast it from you on one side of the boat as far as possible (taking care not to hook your fingers whilst so doing), and thus proceed with the whole six ; the lines will then radiate from the boat like the spokes of a wheel (fig. 36, p. 117), and reach further as well as keep clearer than if merely lowered alongside ; they will also thus collect the fish together, as the lines, by the force of the stream, will be drawn towards a common centre. When fish are very plentiful, two lines fully occupy one person.

THE DAB.

These fish are from 10 to 15 inches in length, and 8 to 12 in breadth, and a man will, single handed, take from three to ten dozen in four or five hours when they are abundant.

Mussels of which the shells do not exceed 1½ inch in length are the best size for Dabs, small Whiting, or Pout; if larger they should be cut in two, leaving the tongue in one part and the round gristle in the other to hold the hook.

I have given here an illustration of the mode of spreading the lines round the boat in Dab or Flounder-fishing.

Dabs and Flounders constantly gorge the bait: to extract

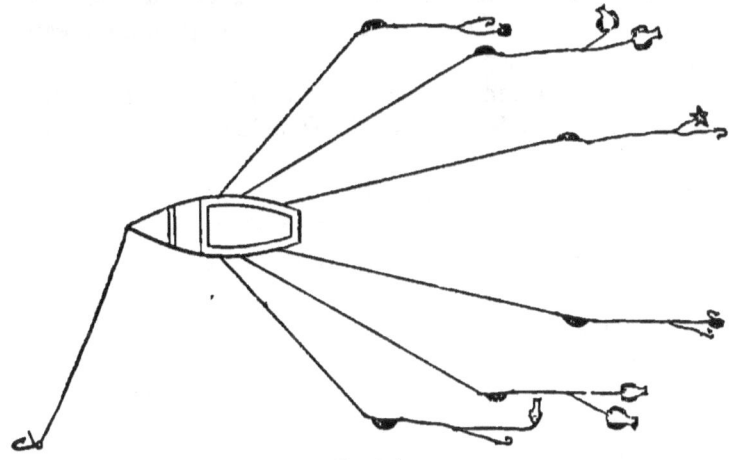

FIG. 36.
Mode of spreading lines round the boat in Dab or Flounder fishing.

the hook, insert your thumb in the gills and split them open, or use a disgorger. (See p. 53.)

Star-fish, Hermit-Crabs &c. are often troublesome, robbing the hooks, which must be frequently examined in turn. Great numbers of Dabs and also other flat-fish are taken with trots or spillers, which are long lines having a hook at every nine feet or two fathoms, tied to fine snooding three feet in length, and moored securely by stones, with a buoy line at each end; these lines have from one hundred to five hundred hooks, and a stone of about two pounds in weight attached every four fathoms, which prevents the rolling, and thereby the twisting of the snooding; at both extremities of the line the stones should be from ten to twenty pounds in weight.

N.B.—Trots are mostly shot across the tide in order that the hooks may stream out clear from the main line. Fishing in this extensive manner is comparatively rarely attempted by amateurs, as it is very troublesome and requires too much time; a short trot of forty or fifty hooks, however, may sometimes be used with advantage if set over night, as it may be taken up next morning, and is fishing whilst the owner is sleeping. I recommend, as a rule, however, that amateurs stick to hand-line fishing, which affords the best sport, although in quiet weather the result in the quantity of fish is generally greater with the trot or spiller.

In order to prevent entanglement of the hooks of the trot, bulter, or spiller, a basket with hook holders should be provided, both which are described at p. 143. The Dab is by no means confined to the shores of Britain, but has a very wide range over the Northern seas; Captain Dixon met with large quantities on the NW. coast of America in 1787, off Port Mulgrave. Whilst Dab-fishing you will now and then catch a ' sordid Dragonet' (*Callionymus dracunculus*), an ugly-looking flat-headed fish, having a three-forked broad spine over the gills; to unhook this fish safely place your foot on it, as the spines are sharper than needles, and may make an ugly cut in your fingers if due caution be not used. The points are nearly covered with skin, and may escape your observation. As a rule, be on your guard in particular against every sea-fish having a head like the river Bull-head or Miller's Thumb. The Dab is particularly good when nicely fried either with or without eggs and bread-crumbs. Having more than you know how to dispose of fresh, cut off the heads and that portion of the belly which covers the intestines, sprinkle them thickly with salt, and lay them an hour or two in a pan to allow it to penetrate, then hang them up on a line or drying frame (as mentioned under 'Whiting,' pp. 54, 55), or on a fish-stick, taking care when suspended that they do not touch each other, and they will keep good a fortnight or more after being dried. If you desire to preserve them longer, they must receive rather more salt at first, be dried still more, and opportunity must be taken afterwards to spread them in the sun occasionally. Toasted or broiled on a gridiron, they will be found an excellent tea and breakfast fish ; to extract a portion of the salt, after toasting pour boiling water on them in a basin and let them soak three minutes, then take out, drain, and immediately spread over them a bit of butter whilst still hot. The Dab is a very slimy fish and must be well scraped in cleaning ; it should not, however, be allowed to soak in the water, but be dipped sufficiently to rinse off the impurities only.

THE FLOUNDER OR FLUKE.
(*Platessa flesus.*)

The Flounder or Fluke frequents large tidal rivers, and, although evidently a sea-fish, will wander far into perfectly fresh water, and there live and thrive.

It is very similar in shape to the Dab, but much inferior in quality, yet in the winter until the beginning of spring it is tolerably firm, and being at this season full of spawn will be found very palatable, if broiled over a clear fire or nicely fried. (*Note.*—In cooking flat-fish by either broiling or frying, remove the roes and dress them alone, as they are rarely sufficiently done if left in the fish.)

Use the same tackle as for the Dab ; for bait I have never found anything equal to the Soft Crab, which is sufficiently tough, and not so quickly taken off the hook by the Crabs, which abound so much in all our estuaries. This Crab, warned by an allwise Providence of its approaching defenceless state from the casting of its shell, seeks shelter and concealment under stones or in holes, until a new one is formed over its naked body, as during this period it is equally an object of pursuit to all the finny race as to its brethren who have not parted with their coats of armour. The condition in which they are suitable for bait is just previous to their casting the old shell. Lugs are good bait, but the Crabs rob the hooks too quickly.

Use four or six lines, and act in the same way in casting

and arranging the lines as in Dab-fishing. To bait the hooks, take a Crab, and having cracked the shell and pulled it off, cut the body with a sharp knife into three or four baits, and place a piece on each hook, by putting the hook twice through it; crack the shell of the large and small legs also, and they will form one or two more baits. Whilst fishing for Flounders, you will also occasionally take large Freshwater Eels and Bass, which are very fond of the Soft Crab.

If you reside near any harbour having muddy shores, it will be quite worth while to contrive a number of artificial shelters for the bait-crabs, which is to be effected by procuring a quantity of old earthenware pots, old saucepans, or frying pans, to the number of two or three hundred or more, and placing them on the shore between half tide and low water mark; turn them upside down, leaving a small opening for the Crabs to enter, which they will not fail to do; you will thus have a supply of bait always at hand.

Where the water is entirely fresh, or in the upper part of an estuary where the fresh water preponderates, Flounders will take Earth-Worms as freely as Eels; but where the water is entirely salt, or the salt preponderates in the mixed water, the Soft Crab or Mud-Worm is preferred.

The quantity of Eels to be taken with the Soft-Crab bait in some estuaries is very great: I have heard of as much as a hundred weight in a single day's fishing in Southampton Water. Strong flax or hemp snooding is chiefly used for Eels in these localities, but horse-hair loosely twisted makes excellent snoods.

In fishing from shore for Flounders or Eels, use from four to six leger lines of very stout snooding, with a quarter-pound lead or stone at the end, and two hooks, No. 13 (see cut of Hooks, p. 211), a foot apart, tied to twisted gut or fine gimp or snooding,

This tackle is also suitable for Pout-fishing off a pier or rocks, with the addition of a couple of revolving chopsticks, which may be made of whalebone or brass wire. (See 'Eels,' p. 185.) Whilst angling with the rod and the Pater-Noster line from piers and quays, you will often take Flounders if you bait the bottom hook with a boiled Shrimp, which, being carefully peeled, is to be placed on the hook by entering the point at the larger end of the bait and threading it on nearly to the tail. For this bait a No. 3 Limerick or 13 Exeter bend is a good size, and preferable to a Kirby, the lateral curve of which is likely to break the bait (fig. 63, p. 211.) The Shrimp cannot be recommended for throw-out lines, for which the Rag-Worm or Crab is better calculated, not falling off the hook.

Flounder Spearing.—There are two kinds of spears—the Fork (fig. 37) and the Fluking-Pick (fig. 38, p. 122).

FIG. 37.

The Fork should be two-pronged, 6 inches in length, of square iron, the edges a trifle jagged to prevent the fish falling off, and of the stoutness of a tenpenny nail, securely fastened into a light stiff ash pole eight feet in length. To use it, procure a small flat-bottomed boat or one of light draught, and sitting astride across the bow, having first placed a heavy stone in the stern, propel the boat slowly up stream by help of the spear in shallow branches of the river, and carefully scanning the bottom you will frequently discover the fish, by its eyes only protruding above the surface of the sand, when you will find no difficulty in spearing it. You will also take many Flounders in the same manner, without a boat, in the drains and watercourses of embanked lands, and even with your hands, for the fish will often seek shelter under your feet if wading; this latter method is termed 'Grabbling.'

Calm and quiet weather, and clear water with a fine sandy bottom, should be chosen for taking Flounders with the fork, but for 'Grabbling' or 'Fluking' it is not so important.

It is perhaps needless to advise the reader to avoid a stony or rocky bottom, as it must of course speedily ruin any spear.

In forking Flounders, if two hands are in the boat, one should propel the boat at the stern with another fork and spear a Flounder when occasion offers, whilst the bowman watches the fish scuttling away ahead, and, marking where they stop to bury themselves, signals the steersman by inclining his spear either to the right or left, as required.

The Fluking-Pick or Pike.

In the narrow parts of harbours and tidal rivers, large numbers of Flounders are taken by the Fluking-Pick (fig. 38) by continually striking the bottom therewith, as the boat drifts down the stream; and where the water is sufficiently shallow, it is the custom to wade instead, in the summer season.

The Fluking-Pick is thus made : A piece of tough ash 2½ feet long and 2 inches square has introduced into it, at distances of 3½ or 4 inches, seven or eight teeth or tines, 5 inches in length, the edges of which have been jagged, and thus

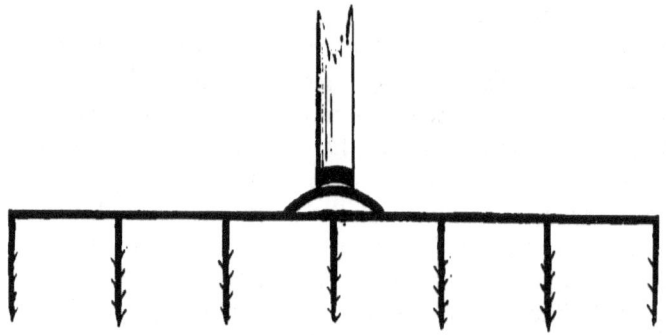

FIG. 38.—Fluking-Pick.

form barbs to prevent the fish falling off; a long spill or spike is set in this on the opposite side, and is securely driven into the end of a light fir pole from 10 to 14 feet in length. This is a cheap kind of pick, but a much superior one may be made for a shilling or two extra, by substituting iron for the wooden cross-head, and setting it edgeways, as this passes through the water with far less resistance.

The Halibut.

(*Pleuronectes hippoglossus.*)

This gigantic Flounder is found in all the Northern seas, and has been caught weighing from 400 to 500 lbs. Sometimes it is taken in the Channel, although it is not abundant, but becomes more numerous to the northward. On the banks of Newfoundland it is a very unwelcome visitor, and is often cut adrift by the fishermen, when it does not escape by its own strength.

It is common on the NW. American coast, and is there taken with Squid or sucker-bait, on a line floated with inflated bladders, a large trimmer in fact, after which it is speared. In the London market it often fetches 10*d*. or 1*s*. per lb.

The Mackerel.

(*Scomber scombrus.*)

This well-known fish is taken in large numbers by seine and drift-nets (see pp. 232, 247), and affords excellent sport with hook and line; it is taken by the latter in considerable quantities between April and September, but the best fishing is in July and August. They are usually caught from a sailing-boat during a fresh breeze, thence called a Mackerel-breeze, when the boat has good way through the water; but in calm weather a light rowing-boat is often had recourse to with much success, when the fish are abundant.

Sailing-boat for Mackerel-fishing.—A sailing boat for Mackerel-fishing should not draw more than 4 feet water, for long experience has shown the fishermen that those of moderate size take, during the season, more fish than the larger, possibly because they disturb the water to a less depth than others of a heavier draught, and consequently do not excite so much apprehension among the fish. On the other hand, a boat may be too small, and I would therefore recommend one of about 5 tons, 8 feet beam, and 21 or 22 feet over all, as the most advantageous size, figured and described as the Itchen river rig (fig. 83, p. 261). The minimum size for this fishing I would limit to 15 feet over all, because if smaller they will not make sufficient headway in disturbed water. On different parts of the coast boats of varied form and rig will be met with, each supposed to be adapted to the exigencies of its particular locality. Any fore-and-aft rigged boat is fit for Mackerel-fishing, supposing she is handy under sail, whether a sprit, a gaff, or working lug sail (see 'Boats and Boating'), but the ordinary or dipping-lug is not, as it must be lowered, dipped, and hoisted anew every time the boat tacks. In addition, it is a rig fraught with risk, unless in the most practised hands, for if taken aback, the portion of sail before the mast is so large, that the risk of capsizing is imminent, as the tack being fixed to the stem or weather bow, the boat cannot readily be freed from the pressure.

Lines.—For reeling or railing—that is to say, towing weighty plummets under sail—I do not find any kind of lines equal to those made of the double Bridport flax or cotton snooding, manufactured for Guernsey to the pattern furnished by the late Peter le Noury, which, consisting only of two strands, do not evince that tendency to snarl and become entangled so often characteristic of a threefold line (fig. 11, p. 48, No. 4).

Leads.—Much variety exists in the shape of Mackerel leads, but I find the most expert fishermen of both Great Britain and France have (in all probability without comparison of their

FIG. 39.—Plummet-lead and Revolving Chopstick.

experiences) adopted those of either a spherical or fig-shaped form (fig. 39), both of which will tow after the boat more quietly and steadily than those of a different shape; this latter property is an essential requirement, and the want of it will lead to continual fouling if using from four to six lines.

As a round lead, of course, evinces a disposition to roll, a slip of wood 2 or 3 feet long is commonly nailed along the seat or water-way of the boat to steady it, when you drop the lead to haul in a fish by the snood. Sailing boats of from 17 feet and upwards should be provided with a set of three pairs of lines, of

THE MACKEREL.

three, two, and one pound weight, with a pair of half-pounders for light weather; when the three pounds are in use, the halves may be kept on board, and *vice versâ*.

The length of line is, for the three pounders five fathoms, two ditto seven fathoms, one pound nine fathoms, and halves twelve fathoms; and the length of snood below the lead, one and a half fathom for the three pounders, two fathoms for the two pounders, and two and a half for the leads of one, and of a half-pound weight, to the ends of which two or three feet of the strongest single silkworm Salmon-gut is added, and to this the hook is bent. At Plymouth for some years gimp has been used for snood between the lead and the gut bottom. It is very pleasant to use, as it is weighty and does not readily blow on board when cast to windward, but must not be trusted beyond one season, as the salt-water causes verdigris, which rots the silk inside. The half-breadth sheer boat-shaped leads are now used for Mackerel-fishing under sail with success at Plymouth. See p. 45.

Six feet of very stout cord should be fastened to the inner end of each line, which will not chafe away so quickly as the line itself as it passes over the gunwale of the boat, and, at the other end of the line above the lead, a piece of whalebone or elder-wood, 3 inches in length, should revolve on a short leather sling to which the line is bent on, as shown in the cut.

The hole in this piece of wood or whalebone should be large enough to allow of its working freely, which will tend to prevent fouling.

The stoutest kip upper leather is preferred for the sling; cut a strip 3 inches long, the breadth equal to the thickness, make a half-inch slit near each end, pare off the edges, and round it by rolling it under foot on the floor. The best snooding is that of the double Bridport flax make, or double Shrewsbury thread, No. 18, laid up by a rope-maker. The lines with one pound and half-pound leads are also very useful in Pollack-fishing, in sailing or rowing, to save the wear of the horse-hair lines.

Boats above 6 tons often tow four-pound leads in front, and use besides a pair of threes and twos; this is chiefly in strong winds. With boats under 17 feet over all, leads of two, one, and half a pound are sufficiently heavy. Those of half a pound are weighty enough for a rowing-boat in a calm, but for quick pulling, one pound.

The best bait is a slice of the tail of the fish; and to cut it, take a Mackerel by the tail, and turning it on its right side, with the head between your knees, enter the knife on its left side 2½ inches above the tail, and cut down nearly to where the tail-fin joins the tail, then withdraw the knife, and passing it lightly across, take up the bait, and lay it on a piece of cork with the bright side downwards; now pass the hook through the smaller end, and all is complete (fig. 40).

Be careful not to cut below the red flesh, which is the first layer under the skin, or the bait will be too thick to fillip or flip in the water, upon which much of its attractiveness depends; the fish admits of the bait being cut thinner from the left than

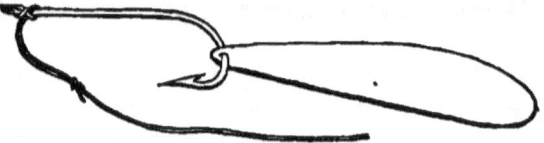

FIG. 40.—Baited Hook for Mackerel-railing.

from the right side, which is therefore preferred. This bait is termed a last, lask, float, or fion.

Having met with, or as it is called struck, the fish, you will probably take many without any check, but immediately you lose them, wear the boat round on her heel and run back over the same course, when you will probably strike them again.

Five or six hundred Mackerel, or even more, are frequently taken in a day by this method of fishing, under the favouring circumstances of a cloudy sky and a fresh breeze; but under oars a hundred fish is much above the average.

In the early part of the season, when the fish are shy, a dead Sand-Eel is used to bait the hook with much success. (See fig. 25, Freshwater Eel bait, dead, p. 82, for baiting the hook with it.)

Booms or Bobbers for Mackerel-fishing.—The Hampshire and Devonshire fishermen, and those of some other portions of the coast, use a stout rod or small spar rigged out over the gunwale of the boat, the length of which is about 15 or 16 feet; two of these rods, booms, or bobbers, are secured to the main rigging of the boat, so that they may not drag backwards by the strain of the lines, which hang from the tops of these bobbers, being made fast to a separate piece of line, one end of which is secured to the top of the rod, the other to the fishing line, the end of which with the reel is kept inboard. (See 'A Day with the Mackerel,' p. 131). In large boats exceeding 4 feet draught, they are useful to keep the lines out of the disturbed water, but in small boats their advantage is questionable. With the graduated lines here described, in Guernsey the fishermen are just as successful without them.

It is possible to go too fast for Mackerel : at four miles an hour the heavy leads will do good service ; carry therefore only sufficient sail to ensure good 'dray-way' through the water.

Artificial Baits consist of white and red feather, or other gaudy flies (see fig. 33, p. 89), or mother-of-pearl fish 2 inches long, and of late years artificial spinning baits have been introduced. They are especially useful for Mackerel-fishing, although other kinds of fish are taken with them. Every amateur should keep a few by him, as they are as cheap as useful. I prefer them with one hook for Mackerel-fishing. A bit of tobacco-pipe (fig. 41) is often used by fishermen with much success,

FIG. 41.— Tobacco-pipe bait for Mackerel.

and the best plan of fixing it is as here represented, which can, however, only be managed with a Limerick hook ; having no flattened top, this will enter the hole in the pipe easily. However, sea-fishermen often use it with a common sea-hook, and with success, but it is not nearly as snug a method of fitting it. The hook must be whipped on with fine silk, and the edge of the pipe-hole be scraped with a pointed pair of scissors, that it may not chafe off the snooding, which should also be lapped round with silk ½ an inch above the top of the pipe.

By rowing to windward of the shoal of the fish and casting the flies with a rod, as for Trout, Mackerel are also caught. Not unfrequently using two flies, you will take a Pollack on one and a Mackerel on the other.

Ground Mackerel-fishing.

It is the habit of Mackerel, in the latter part of the summer and through the autumn, to feed much at the bottom, or, as fishermen term it, 'to strike the ground.'

Ground-fishing for Mackerel is quite an institution in Plymouth Sound, and two men usually fish sitting down in each boat, one at the stern, the other at the bow, a line in each hand. It is followed from the middle of July until October. Fifty to eighty boats may be seen thus occupied, and many hundred dozens of Mackerel are thus caught in a day.

The ordinary tackle and bait, as described for whiffing, is not fitted for their capture under these circumstances, and a special arrangement is necessary. This ground-fishing for Mackerel is chiefly followed where there is very little or no tide, and leads of either the pipe or boat-shape may be used, the latter having a horse-hair loop or stiffener projecting from each end, not less than half the length of the lead itself. To this add a trace of four feet of gimp or triple-twisted gut, with two lengths of selected single gut for the hook link, and a hook size No. 10 (fig. 63, p. 211). The bait in general use for this ground-fishing is a piece of Pilchard, Pilchard gut, or Squid. To prepare a Pilchard for bait, remove the scales without damage to the skin, which requires some care, score it across diagonally with a very sharp knife or an old razor, at not quite ½-inch intervals, pass the knife underneath, and remove the pieces, which should be cut as thin as possible. Squid is often used together with Pilchard, and should be cut of a dagger form, and only $\frac{1}{8}$ of an inch in thickness. Hook on the squid first, the large end uppermost, and then the piece of pilchard, placing the silver side outermost. Sound the bottom, and raise the lead sufficiently to keep the bait just off the ground. Small mussels are quite as killing as pilchard for this method of fishing, but this is little known. Living Sand-Eels are also excellent bait. In a tide-way the light drift-lines with pipe leads answer with any of the above-named baits. It is best by jerking the line to keep the bait in constant motion, which is termed 'bobbing.' For want of a Squid or Cuttle, use a slip of parchment.

A Day with the Mackerel.

'There is not on sea or river (always excepting Salmon-fishing) any sport comparable to this delightful amusement. He who has experienced the glowing sensations of sailing on the Western Ocean, a bright autumnal sky above, a deep green lucid swell around, a steady breeze, and as much of it as the hooker can stand up to, will estimate the exquisite enjoyment our morning's Mackerel-fishing afforded.' Thus far the author of 'Wild Sports of the West.'

It was on a fine morning in the month of July that I determined on a day's hook and line Mackerel-fishing, or, as it is not unfrequently termed, reeling or trailing. The weather was fine, yet cloudy, whilst a pleasant breeze from the east, increasing as the sun approached the meridian, rippled the previously mirror-like surface of the summer sea.

The easterly wind has this frequent peculiarity, that it increases until about two o'clock in the day, or a little later, and sinks to rest with the declining sun. The locality is an open beach of pebbly shingle, terminated at its eastern extremity by a cliff, immediately under which there falls into the sea a small river, whose entrance is so barred with shingle that it is closed to even the smallest boat, ay, even to a Salmon, for nine hours out of twelve. Anything like a boat of tonnage cannot of course, under such circumstances, be made use of, as the harbour being so frequently inaccessible from the sea, and the sea from the harbour, the craft, although afloat in ten feet at low water inside the entrance, must there remain like a book in a glazed case, often looked at but very little used. I therefore owned a boat of such a size as could be beached at any time of tide, and hauled up by aid of a block or a small capstan at the top of the beach.

It chanced, however, that the day in question was one of those red-letter days, so few and far between, the weather being sufficiently settled to moor off or keep my little craft at anchor from the previous day. Yes, it was indeed, a *dies dierum*, for none of that carrying down of ballast was to be done, the ordinary accompaniment of boating from an open beach, with possibly the addition of a wetting to the waist in launching from the open shore, and the remaining afterwards wet during the trip, a most undesirable state of things, and certainly prejudicial to health in the long run if often repeated, notwithstanding many assertions I have heard to the contrary.

My boat was not a large one, for the reasons above stated, being scarce 15 feet over all, with a beam of 5 feet 8 inches, elm plank, and clench-built, with little rise of floor, and a keel not exceeding 5 inches deep, being as much as I could venture to put on her without crippling the garboard strakes in beaching—that is to say, the two first planks fitted and rabbeted into the keel. As our destination was about two miles to windward, and it is especially desirable in a beach-boat to take advantage of a favourable tide, we got under weigh an hour and a half after high water, so as to carry the remainder of the flood-tide with us for another hour and a half, by which time, if the wind held, we should fetch well into the bay to windward, and be ready for the fish at the breaking away of the ebb three hours after high water, until which time the eastern or flood-stream continues to run, notwithstanding the perpendicular fall of the water to half-tide level. We made a long board into the offing to keep the strength of the stream, and on going about found the western horn of the bay well under our lee, and that with the remainder of the flood under foot we should fetch well to windward of our fishing-ground.

As we were, therefore, quite at ease with regard to reaching the desired locality, I ordered the old pensioner by whom I was accompanied to put the lines in requisition, which were four in number and fitted with sugar-loaf leads, two with leads of 1 pound each for either quarter, and two others with leads of 2 pounds weight in front of them ; seven fathoms of line for the 2-pound leads, and nine for those of 1 pound, with two and a half fathoms of snood attached to a revolving chopstick for the two pounders, and three fathoms for the one pounders. Two-thirds of the snood of fine hemp, the remainder of fine silk line, or of the yellow silk known as barber's twist, used by shoe-makers in sewing upper leathers.

This is a corruption of 'Barbour,' the name of a manufacturing firm near Belfast, who must pardon the liberty taken by the fishermen with their patronymic.

We had not attained to the refinement of three lengths of Salmon-gut for the hook, or that of the double Bridport snooding for lines and snoods, but for lines used rather a fine three-stranded Whiting-line, or else a double twine line of home manufacture, spun up in six feet nossils (a nossil is a length of snood spun up by aid of the nossil-cock, a fisherman's spinning machine ; the length of snood determined by its height from the ground—see fig. 66, p. 216), and spliced together to give the requisite length according to the practice of my late old friend, Joseph Gibbs, of Budleigh Salterton, and other natives of the fishing village of Beer, in Devon. Notwithstanding this, the hook-link was fine, although not equal to gut, and excited the admiration of old John, who, as he unwound the snood, or sid as he called it, exclaimed, 'That there gare (*gear or tackle*), sir, is vine enough to catch any vish in the say (sea)'—f being commonly changed into v in the parlance of the SW. Coast. Fine enough I believe such snood to be ; but I prefer gut, as free from that tendency to foul characteristic of wet silk, especially when spun very fine.

Having handed over the baits, consisting of the usual thinly-cut Mackerel tail, they are affixed to the hook by placing them on a piece of soft flat cork, and forcing the point and barb through the smaller end, the two stern lines are speedily towing in our wake, being made fast to the after-thwart, whilst old John rigs out the bobbers or booms for the forward lines, which bobbers I had made fishing-rod fashion, of stout ash, in four pieces (the top hickory), for convenience of stowage. Twig, twig goes the line on the weather quarter—twig, twig is repeated by that on the lee, upon which I proceed to haul in one of the twain, desiring old John to bear a hand with the bobbers, as we were evidently getting well up with the body of the fish.

The reader will observe I hauled in only one of the twain, keeping the other towing astern until I should hook a successor, for the Mackerel, being eminently gregarious in its habits, will generally follow a hooked leader, until all or nearly all the shoal may be captured, provided the boat maintains an uniform rate of speed, so that the lines may not sink suddenly to a greater depth.

An additional minute or two allowed old John to get the bobbers in order, to which the lines were attached by the bobber strings, which being made fast to the bight of the lines, the reels were secured in-board, and the lines swung out clear of the sides of the boat, hanging by the bobber strings from the ends of the rods, which do not require removal to examine the lines, as the ends being made fast in-board, they can be overhauled at pleasure without it. Old John soon accompanied me in adding to the basket, and we continued for some minutes hauling fish, sometimes on the weather, sometimes on the lee lines, as fast as we were able to tend them in quick succession; until a sudden cessation took place, when we wore round and sailed back over the same ground again, overhauling the lines and renewing the baits if requisite, which should be occasionally done when they begin to lose their brilliancy, or do not play well enough in the water to have a lifelike appearance.

'There,' said old John, 'd'ye see, sir?' holding the lead in his hand, and jerking the bait as it towed in our wake, showing its silvery sheen and blood-red side by turns; 'seems to me, sir, as if 'twas jist the tail of a launce a-working'—in which view the Mackerel evidently concurred, as I hooked another the next moment, and exclaiming, 'Here we are again, John,' he dropped the lead like a hot potato overboard, and sprang to attend his bobber lines as if rejuvenated. The fish took thenceforward with great briskness, and I really believe we captured more than a hundred on one tack towards the offing, with little or no check, when they ceased suddenly, as is commonly the case, without assignable cause other than that we may suppose all the shoal to have been caught. We therefore wore again and stood in towards the land, but did not strike the fish for some time, of which I did not complain, for, to tell the truth, my back ached with continual stooping; and nature very plainly asserting that she abhorred a vacuum, we availed ourselves of the respite afforded to take some refreshment before renewing our labours.

In another quarter of an hour we again fell in with the fish, and both happening to be hauling at the same time, old John with the lee bobber and the writer with the weather-quarter line, by a piece of ill-luck or mal-adroitness, or both combined, the two fish contrived to sheer across the stern of the boat towards each other after the leads were on board, crossing not only each other, but also taking a turn round the lee-quarter line, thereby of course making a pretty considerable foul, increased every moment by the sudden movements of the fish— the personification of two animated weavers' shuttles, as they darted to and fro under the boat's quarter. This was, of course, particularly annoying, as we were clearly again in the midst of the fish, for the weather bobber plainly showed, by its jerking and bending, that a fish was fast to it; but we let the Mackerel remain to keep the shoal in tow, and getting one of the fish and the *foul* on board in a lump—for the other had escaped—we set ourselves to work, having put the fore-sheet to windward, 'to undo the knotted hank,' having first overhauled the remaining line, which yielded us a second fish.

Reader mine, were you ever similarly circumstanced? Have you been compelled to fish with snoodings of thread or barber's twist, and had the ill-luck to foul them when the fish were biting as fast as you could tend the lines?

If so, you will be able to appreciate the annoyance, for, *O miserie!* you may nearly as well endeavour to disentangle a wetted spider's web as to clear a foul of this description, and how it manages in so short a time to compass all the ins and outs is certainly a puzzle. However, there is, of course, only one thing to be done—that is, to draw largely on your stock of patience, and endeavour to clear away with all speed. The task was at length accomplished, and the lines again put out, with immediate success, and the unlucky Mackerel all this time towing on the bobber-line hauled on board in a semi-drowned state.

The basket was long since filled with the dead fish, and the midship part of the boat being likewise covered with Mackerel, they invaded the stern sheets, beating thereon in their struggles a lively tattoo with quivering tails. We had therefore taken a large number of fish, and it was now 4.30 P.M. The wind was failing fast, and as I had no inclination to pull back after the day's fishing, we boomed out our sails, and made the best of our way home, when we beached the little craft, hooked on the capstan chain, and hove her up, after a wash, preliminary to the regular scour she had next morning, which she certainly required after such a capture.

Old John had a large basket (as much as he could carry) for his share; and sending a servant for the garden wheelbarrow, it was piled full, after which another basket placed between its shafts received the rest, and being driven up the street, distribution was made right and left to all comers until the load was considerably lightened. The fish were counted into the baskets and the barrow, and the number amounted to 412. We took our first fish about 11.30 A.M. and ceased fishing at 4.30 P.M., and better sport with Mackerel I never had. My late old friend Joseph Gibbs was at it all day, and, delivering his first catch at noon, went out until evening, his day's work (assisted by his partner) yielding 920 Mackerel. I have known much larger numbers taken, even up to 1,400 or 1,500 a boat; but anything over a hundred may be looked on as a good day's fishing.

MACKEREL-FISHING AT ANCHOR.

In August, September, and October, these fish may be taken at anchor with the same pipe-lead hair-lines as are used for Whiting-Pollack, and the best bait is the purple, or, as it is generally called, the red Sand-Eel, to be used alive.

The Sand-Eels should be caught in a seine (fig. 70, p. 229), and towed behind the boat, as shown in fig. 20, p. 66.

To bait the hook, pass the point into the mouth and out at the gills, then turning it over, hook the fish across through the throat slightly ; this hurts the bait but little. (See fig. 16, p. 64.)

If the bait be more than six inches in length, cut it in two in the middle, after which, strange to say, it will move a considerable time. (See fig. 42.) In this fishing the bait should be kept just clear of the ground. Having baited the hook with the living fish, place it on the gunwale, and cutting it in two in a slanting direction, throw it into the water at the same instant, and pay out the line quickly, for the exudation from the freshly-cut fish renders it doubly enticing to the Mackerel.

If no Sand-Eels are procurable, get a quantity of the entrails of Pilchards, which are an excellent bait. Rag-Worms may also be used, but Sand-Eels and the entrails of Pilchards are standard baits. As large quantities of Pilchards are taken in summer and autumn on the western coasts of England, there is little difficulty in getting a quantity of offal almost daily.

If hair-lines with pipe-leads are not heavy enough in a tide-way, the one-pound leads with the hemp or flax line, used in sailing for Mackerel, will be found very useful, instead of the forward pair of hair-lines.

Whilst fishing at anchor for Mackerel, other fish are also frequently taken.

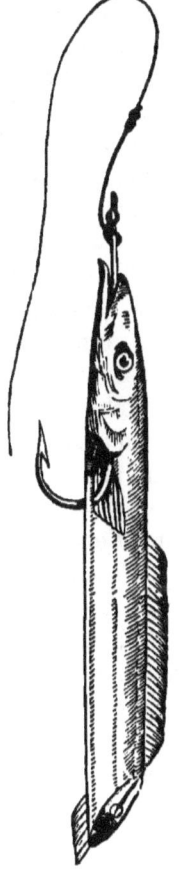

FIG. 42.—Red Sand-Eel, cut in two for Bait.

N.B.—This bait must not be cut before being placed on the hook, and then only just before it is cast overboard.

Some years since the fishermen of Guernsey discovered that numbers of large-sized Mackerel might be taken by this method at night, particularly during a bright moonlight. This has consequently come to be recognised as a regular means of capture in the autumn months, and I see no reason why it should not be adopted elsewhere in fitting localities. The favourite haunts of the large Mackerel in the locality referred to are certain spots on the edge of the bank of sand and gravel near St. Peter's Port, known as The Great Bank, and I cannot see why the plan should not answer as well on the coasts of Great Britain and Ireland in similar positions.

The kind of ground to try is a gravelly sand-bank, a mile or more detached from shore, having on it at low water from three to ten fathoms, with twenty fathoms or more in the immediate vicinity. These elevated banks are much frequented by both Mackerel, Gurnards, and the choicer flat fish; the Bahama Bank, for instance, off Ramsay Bay, in the Isle of Man, and many others. I feel satisfied the fishery might be much developed by attention to this one mode of capture. The Sand-Eels sometimes omit their visits to Guernsey in autumn in sufficient abundance for seining, in which case this fishery is not followed. These bank-ground-Mackerel are particularly fine.

When Pollack-fishing with drift-lines, in July and later, should the Mackerel make their appearance, cut the Sand-Eels as here shown. I have tried the slice from the Mackerel's tail frequently at anchor, but with little success; the Sand-Eel is superior in a tide-way to every other bait.

THE SCAD OR HORSE-MACKEREL.

(Scomber trachurus.)

This, although of the Mackerel family, is a coarse fish, and consequently not held in much estimation. It is taken indifferently, at anchor or under sail, when fishing for Mackerel, Pollack, Pout, or Whiting. The lateral line of this fish is curved and marked with a succession of pointed scales, besides two spines close to the anal fin, which it is well to remember whilst unhooking them. They are sometimes seen in enormous shoals entering coves and harbours, and on such occasions have been taken in cartloads by placing a net across the entrance. As bait for Prawns, &c., they are useful, but the flesh, being of a stronger taste than Mackerel, is in little request as food in general. It is often taken in the Mackerel seines, and not unfrequently with the angling rod when Pollack-fishing from rocks and piers. This fish is often salted for winter provision in Cornwall and the Scilly Isles.

THE BASS.
(Lupus labrax.)

The Bass, or Salmon-Bass, as it is frequently called, is of a dark blue on the back, which changes to a silvery white on the side and belly, and has a general resemblance in the eyes of a cursory observer to the Salmon, although of quite a different family—namely, the Perch. It is seen in large shoals in the mouths of rivers and harbours and off headlands of the English and French coasts, and attains to a considerable size, sometimes as much as 15 lbs.

These fish generally show themselves on the surface of the water, at the end of April or beginning of May, after which time they may be taken by fly-fishing, ground-fishing, and drift-line fishing.

Fly-fishing for Bass will naturally recommend itself to those who have been used to this method in fresh water, being identical in practice, except that it is usually from a boat. (See 'Flies,' pp. 89, 90.)

Any of the smaller kind of Salmon flies will answer for this sport, or very simple flies may be made by tying a pair of wings cut from a white goose feather on a white hook, with or without a body of red worsted. Two of these flies should be fastened, one to the end and the other 2 feet above, on a bottom or collar of the best Salmon gut, 5 feet in length, two-thirds double twisted and one-third single. White flies with silver bodies are also very killing, or a parchment fish, 2 inches long.

Two or three small mother-of-pearl fish, $1\frac{1}{2}$ inch in length, are also used, made out of card counters, but they fall more heavily on the water than flies, for which reason the flies are preferable; sometimes a small piece of the white skin of the belly of the Bass: the truth is that the fish, being of a ravenous and rapacious character, get so excited as to dash at any bright object in rapid motion.

At certain times of tides, frequently for days together, these fish rise to the surface in immense quantities, and may then be taken by rowing to windward of the shoal, and casting the flies or other baits amongst them, drawing the line quickly through the shoal of fish by short jerks, which will give a life-like appearance to the flies.

A gaff-hook or short-handled landing-net is frequently required to get your fish on board, as you sometimes hook them of more than 10 lbs. in weight: those who are not accustomed to the management of a boat will require a boatman, as they will find enough to do to handle the rod.

During a fresh breeze against the tide is the most favourable time for this sport, as it will render effectual aid in maintaining your position. Whilst waiting for the fish to show themselves, you may row up and down with the rod leaning over the side of the boat, and you will take Whiting-Pollack as well as Bass whilst so doing, also perchance a Mackerel, but immediately you perceive the fish on the surface, row to windward and cast the line as before.

It may be imagined by those unacquainted with the habits of this fish, that it would be just as effectual to row the boat through the shoal with the line trailing astern, as to cast the line with the rod to the fish; this is not the case, however, as they are of a shy nature, and immediately sink at the approach of the boat; whereas in the latter case they take the bait before the boat arrives sufficiently near to alarm them. You will generally find some gulls hovering over the shoal, which will serve to point out its locality, if you are not near enough to perceive the fish; and this fact of the birds accompanying the Bass arises from the fish pursuing the Sand-Eels to the surface, when the gulls hasten to share the repast.

For rod, winch, line, gaff, and landing or hand-net, see 'Fly-Fishing at Sea,' p. 86 and following pages.

GROUND-FISHING FOR BASS.

The Leger Line.—There are several methods of ground-fishing for Bass, of which the following is very useful, as it can be practised from on shore with very simple tackle :—

Procure a strong fine Whiting-line and a leaden plummet one pound in weight, which must be of a flattened form (p. 140) to prevent rolling, also a piece of an elder branch two inches in length and ¾ inch in thickness, bore out the pith, and make a groove round the middle. N.B. A small cotton reel with the rims at the ends cut off, will answer equally well. For size of line see fig. 11, p. 48, No. 3.

Take a Cod-hook, 1¼ inch from point to shank, and, having cut off a piece of line 15 inches in length, make a knot on the end and lay it on the inside of the hook, ½ an inch below the flattened top; now bind the hook and piece of line firmly together with some fine brass or copper wire, and hooking it over a nail or some firm object, continue to wind the wire round it close together over the whole piece of line, which will have the appearance of a harp or violoncello string, pass the end round the piece of elder wood in the groove, and bind it fast with some thread, waxed with shoemaker's wax ; make a knot on your line 18 inches from the end, and sliding up the piece of wood secure it by another knot below ; it will now revolve round the line by the wash of the water, and keep the hook clear, in which the wire, by stiffening the snooding, will much assist.

The chief baits I can recommend for this fishing are Squid and Cuttle-fish, which are very similar to each other and very tough, so that there is little fear of their falling off the hook ; they are to be procured from the fishermen. To bait, cut a piece three inches in length by one and a half in breadth, pass the hook through once; turn it over and pass the hook through a second time. Fig. 43 shows a leger-lead, trace, and baited hooks. Living or dead Sand-Eels are also good as bait, as well as the Soft Crab.

The lead is fastened securely to the end of the line by a

piece of leather, which will not chafe through on the bottom; or this part of the line may be served round with wire. To set the line, swing it to and fro, and cast it from you as far as possible. The most likely place to take Bass in this manner is in the mouth of a river or harbour, or near to it, off a pier-head or an open shingly beach, or from a point of rock whence you may cast your line so that it may rest on a gravelly or sandy bottom; the river's mouth is generally to be preferred. On a bottom of smooth sand it is a good plan to drive a 3-inch nail through the lead, thus converting it into a kind of anchor.

If two lines are used, let the lead of one be half a pound heavier than that of the other, and cast the heavier up and the lighter down the tide, by which means they will the better keep clear of each other.

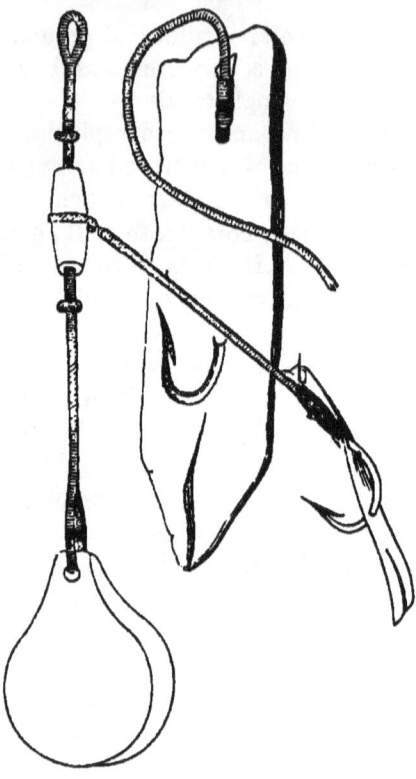

FIG. 43.
Leger-lead, Trace, and Baited Hooks.

In order that the bait may not slide down in a bunch on the bend of the hook, I frequently lash on a pin, bent at an angle of 45 degrees, at the back of the hook, and passing this pin through the top of the bait, all is kept snug in its required position. This is plainly figured in the smaller illustration. In throwing out these lines a stout stick (longer than the snood) is often used, having a notch in the end into which the line is placed just above the lead, which may then be thrown without risk of the hooks catching the fingers. A couple of joints of an old fishing-rod, with a forked wood plug at the end, make an excellent throwing-stick. This kind of throwing-stick is in frequent use at Lowestoft.

Every few minutes you should haul up and clear off any drifting weed which may have become attached to your lines.

The Bass feed best in rough weather and disturbed water, and will approach the shore when the waves are breaking six or seven feet high; in fact, in fine bright clear weather it is a loss of time to try for them, but when the weather is rough and squally, and it is not prudent to venture afloat, this sport may be followed with much success.

On open exposed shores it frequently happens that even in fine weather there is a considerable surf on the beach; if, therefore, there be any favourable point of rock or steep part of the beach from which you may cast your line clear of the breakers you will have a reasonable chance of sport.

Bass and other fish seek out the mouths of rivers along the coast; always therefore give these spots a trial. Spring tides generally answer best, and from half flood to half ebb. If there is much floating weed, from half an hour before to half an hour after high water will alone suit, but, if abundant, even in this short time many may be landed.

When fishing with two lines be careful to secure one whilst you hold the other, or you will very likely lose your tackle if you hook a large fish.

If you are likely to follow this kind of fishing much, I would recommend you to procure a piece of holly or beech wood, 18 inches long and an inch in diameter, shod with an iron point at the bottom, and having a ring or ferrule at the top; this you can drive into the sand or shingle, and wind the line on when you give up your sport. I do not recommend more than one hook for each line, unless you use a throwing-stick, in which case a second may be placed 18 inches above the first, which should be attached a foot only above the lead. Nos. 4 and 5, fig. 62, for ground-fishing with Squid or Cuttle bait; with smaller hooks, No. 8, fig. 63, p. 211 you may bait with a Sand-Eel and use half a dozen hooks, but the Squid and Cuttle are most to be depended on. Too much caution cannot be observed in throwing out, many painful accidents having occurred from want of it.

The Bulter, Trot, or Spiller (fig. 44).—Supposing you wish to put more hooks, you must make a trot, bulter, or spiller (for by all these names is it known), which may be set at low water, or with the help of a boat, the hooks not to be nearer than 6½ feet to each other, and tied to strong fishing line 3 feet in length; these are to be firmly fastened to the main line, which should be a strong cord or small rope, nearly half an inch in diameter, and to each end of this a cork line should be attached, if you desire to raise it in a boat.

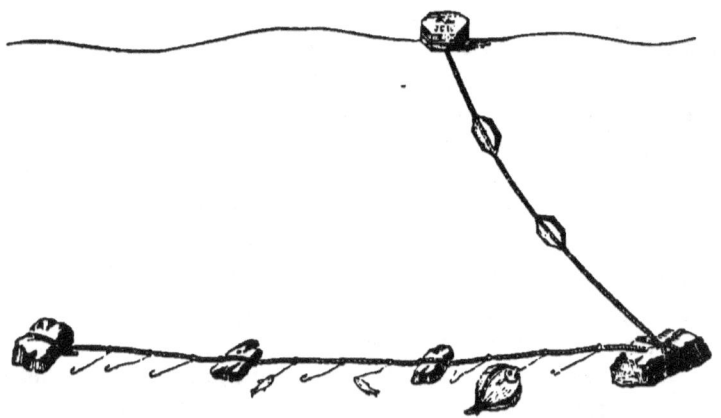

FIG. 44.—Bulter, Trot, or Spiller, for Cod, Conger, &c.

In positions where you have plenty of space, it is often more advantageous to place your hooks on the main or back line, as much as 12 feet apart, which will enable you to cover a far greater extent of ground; in a harbour's mouth, 9 feet may be a good distance. All must depend on the space you desire to cover. In shooting these lines, a common arrangement is to place the hooks all along the boat's gunwale, commencing close behind the after rowlock, and continuing backwards, the hooks not quite touching each other. Placing the first stone overboard, the hooks will follow each other as you pull along, and the line shoot itself.

In a sandy cove between the rocks, you may suspend it from the sides with advantage, as Bass frequent such places on a flowing tide. Very long lines of this kind are used by the fishermen for Cod, Conger, Turbot, Plaice, &c., sometimes with nearly a thousand hooks. Some boats carry twelve miles of these long lines, which are set in deep water at great distances from the shore.

On a flat sandy shore, a bulter may be set at low water and examined on the succeeding tide ; Sand-Eels or Mackerel are here the best baits, and hooks Nos. 8 and 9. The same line, No. 11 hooks, will catch Plaice with Lug bait. See fig. 63, p. 211.

The most killing method of taking Bass with which I am acquainted, is to bait a trot with living Sand-Eels as for Pollack, No. 10 hooks, not coiling down the trot, but dropping every hook into the water as soon as baited.

Trot Basket and **Hook Holder.**—These form a very useful combination for keeping a trot or spiller in good order. The line is dropped into the basket, and the hooks are passed over the holder, which should consist of a piece of tough ash a foot long, sawn down 9 inches, the two arms of which are to be rounded sufficiently to admit the hooks, which having been inserted until the holder is full, they are to be brought together with a clove hitch of the line to keep the hooks in their place. When one holder is full, place it on the line in the basket and fix a second in the wicker work at the side. A basket 15 inches wide and 10 deep is a convenient size for a small trot, but a bulter or Conger-trot, if of any considerable length, requires a larger basket. In Scotland and Ireland a basket of a scoop or coal-scuttle form is used, and some grass being laid flat on the fore part, the baited hooks are arranged thereon in regular rows, and the line coiled away behind it. Fasten a fig-drum in the centre of the basket for the baited hooks.

'**The Outhaul**' **Bulter.**—At low-water time put down an anchor or very heavy stone as far as convenient, or drive in a post if the ground will admit, and make fast a three-inch block, spliced to a piece of strong rope a yard or two in length. Let your line be of the size of $\frac{1}{3}$ of an inch thick, and double the length required to reach from just above high-water mark to the block; make fast sufficient Cod-hooks, in size $1\frac{1}{4}$ inch across from point to shank, to pieces of strong Whiting-line 4 feet long, and secure them to the main line at not less than 8 feet apart, and reeving the line through the block carry both ends up above high-water mark, and make fast to another post, which will enable you to haul in or out at pleasure: bait with pieces of Squid or Cuttle-fish, Pilchard or Mackerel, or Sand-Eels—I prefer the two former.

If you wish your tackle available at all times, make a separate bulter half the length of the outhaul, to which you must join it, and in that case you may leave down the outhaul continually. Five feet nearer to the block than the outermost hook lash on a stout piece of stick six inches long, to prevent the snooding of that hook choking the block.

This outhaul bulter must be used when the water is free from floating weed, which would probably choke the block; at such times fish only with the leger lines. If you have plenty of bait, you may of course use more hooks.

Should you leave this line down after dark, you will also take Conger, and if you do so use stout double snooding, which should be traced over with hemp, as shown in fig. 47, article 'Conger' (p. 178).

Harbour Ground-fishing.

(*From a boat.*)

Use a heavy boat-shaped lead and a three-fathom snooding of strong hemp, with two hooks, size No. 8 (fig. 63, p.211) tied to two feet each of stout gimp; bait with a living or dead Sand-Eel—the former to be preferred, or pieces of Squid or Cuttle-fish three inches long and one and a half wide; hook them by one end, and turning over pass the hook through the second time. A whole Soft Crab is also a good bait.

When a Sand-Eel, enter the hook at the mouth, and, bringing it out at the gills, hook it slightly across the throat.

A fifteen-fathom line without lead may also be used with advantage. Anchor the boat in a run of tide, and fish on the ground with the heavy leads ; with the Sand-Eel as bait, large Flounders are also taken.

In the north of Devon, at Instow, in the mouth of the estuary of the Taw and Torridge Rivers, they use for harbour ground-fishing a four-pound lead of the shape of half an orange, having a projection on the top through which a hole is bored to receive the line. This shape is chosen because the flattened base sucks into the oozy bottom, and anchors the line (so to speak) in the very strong tide which runs in that locality. A stout three or four inch revolving chopstick is placed between two knots close above the lead to receive the snood, similar to that of the Mackerel-line (fig. 39, p. 124). This method of fishing is by no means equal to the following, which is superior to all others.

Drift-line Fishing for Bass in Bar-harbours.

This differs in nothing from Pollack-fishing with the drift-lines, except that you would try at known resorts of Bass, such as in the mouth of a bar-harbour in a rippling tide at anchor, in which locality you will take both kinds of fish. In harbours also which have deep water entrances, there are often numerous Bass in summer and autumn, and the most certain places to find them are in the gullets or narrow throats of these harbours and rivers. It by no means follows that the actual mouth of a harbour is the best place for Bass, for the contraction of the channel, or the presence of a tongue of rock at the forking of two channels, may be excellent pitches for Bass-fishing, although miles up from the open sea. From half flood until high-water, is commonly the best time of tide, well up from the sea. In this method of fishing, where you are always moored, the living Sand-Eel is the only bait really to be depended on, second to which is the Mud-Worm. One line without lead should be always kept out, as the Bass is a fish which so frequently comes to the surface.

This method of taking Bass can be strongly recommended to visitors to the sea-side, who may object to open sea-fishing on account of suffering the *mal-de-mer*, which so completely prostrates many people that they dread even to get into a boat. It can be carried on in sheltered positions, where, under the ordinary conditions of wind and water, the violent undulations of the sea rarely enter, to the great satisfaction of amateurs of aquatics of either sex. As many sea-side visitors may not know in what a bar-harbour consists, a short description of its leading characteristics will not be amiss, as it may help many to obtain sport which they may not be aware can be developed in such localities. In brief, then, a bar-harbour is a harbour where, owing to the formation of the shore, the opposing currents from the river and sea cause an accumulation of sand or shingle across the entrance, greatly reducing the depth of water and causing the sea to break heavily when the wind is on shore. Inside the bar the depth of water generally increases rapidly.

The banks of sand which are found, either on both or one side only of the channel outside the throat of the entrance, are termed 'poles,' and are often fruitful of quantities of Sand-Eels, which bury themselves in these banks on the receding tide. Here they may be obtained by the well-known method of scraping. A bar-harbour has generally on one side of its entrance a nearly straight shore, with a point of sand and shingle running towards it at a right angle, past which the tide runs with great strength, and scours out the sand and gravel to a considerable depth by the force of the ebb stream. Just inside the entrance there is usually a deep pool, generally the deepest water of the harbour, which is caused by the strength of the current impinging on the straight beach during the ebb, whence it is reflected in a rapid eddy during spring tides. At the last quarter of the ebb fish accumulate in this pool, which gradually becomes tranquil as the current slackens, and at the last hour of the ebb they may be fished for with success, until the current of the flowing tide perceptibly runs upward through the pool. When this occurs the chance of sport may be said to cease, and it is then necessary to go more seaward and anchor outside the point. The precise distance cannot be given, except in the case of any particular harbour. The Sand-Eel baskets or courges constitute the most important feature in live bait fishing, either at sea or in harbour, for without them little or nothing can be done, simply because no efficient substitute has been found in which the Sand-Eels can be preserved alive. Many persons imagine they can keep seafish alive in a tin, or bucket, or bait kettle, without suspending it in the water, but there can be no greater delusion, for the oxygen therein contained soon becomes exhausted, causing the fish either to die at once, or to get so faint and weak that there is no vigour, upon which their attractiveness depends, left in them. The reader will find this basket illustrated and described at fig. 19, p. 66, together with the method of manufacture. The Sand-Eels are best and most lively when caught in a seine net (see fig. 70, p. 229); but if no such net can be met with, resort must be had to raking or scraping, as it is termed. They suffer, however, so much by this method of capture, that they are but little use unless fished with the same day as caught, for they have to be snatched up in the hand, which is decidedly injurious to them. The scraper is of the form of a hoe, and the iron plate should be 10 inches long by 7 inches wide, with a handle 2 feet 6 inches long made out of a mop or brush stick. The method to be observed, when scraping, is as follows, and should be strictly adhered to if you desire sport. Proceed to the sands just before low water, being provided with a scraper and Sand-Eel basket. If there is any loose gravelly sand near the edge of the water, dig there, first taking care not to trample on any portion of the sand you are about to dig, for if it be trampled on the Sand-Eels will be alarmed, and depart speedily from your vicinity. Place the basket in water four or five inches deep, and put the Sand-Eels into it as soon as captured. If you find it necessary to dig at twenty yards from the water's edge, which is sometimes the case, a pot or bailer should be taken for the purpose full of water, and when half a dozen baits have been procured, run over and empty them into the basket. An ordinary paint-pot is a convenient size for this purpose, supposing it has not been used, or an Australian 6-pound meat tin, with a bit of galvanized wire across the top for a handle. When scraping it will be found a good plan to dig a trench in the form of a circle, 10 to 15 feet in diameter, for by this means you will enclose an area of undisturbed sand. Work towards the centre round the inner edge of the trench; the Sand-Eels will retire before you, and in the last twenty or thirty strokes you may get a number of them. I have struck out of the sand as many as half a dozen at a stroke when they are plentiful. To all these details, whether it be regarding the actual fishing or in the procuring of bait, due attention must be paid, or sport cannot be expected. The following description of a few hours' fishing on September 2, 1872, at Teignmouth harbour, as it gives the whole *modus operandi*, will doubtless be welcome to our readers. We anchored in Shaldon Pool, just inside the harbour's mouth, at 10.30 A.M., wishing to commence fishing at three-quarters ebb tide. At and near the spring tides quantities of weed are in motion, and fishing is therefore impracticable until nearly low water; that is to say, until three-quarters of the ebb tide has run out, at which time the current has considerably slackened, and the drifting weed for the most part been carried out to sea. As the bottom of this pool is much of it very rocky, or, as sailors term it, foul, it is necessary to scow or bark the anchor, by which method you are generally enabled to lift it if it should drag into a rock. For this method, see p. 202. A stone is not, in a locality of this sort, so well adapted as a scowed anchor, as the anchor will hold in a gravel bottom, clear of the rocks, which a stone will not. Having moored the boat, we prepare for fishing in the following manner. In the first place we remove the basket from its position at the stern, and make fast its cord to the middle thwart of the boat, in order that it may be both readily accessible and also out of the way when playing fish, which sometimes sheer across the stern from side to side in their efforts to escape, and might probably get free if any obstacle were offered to the adroit handling of the line. We are provided with a bucket of a coarse open canvas known as cheesecloth, depth one foot, diameter 9 inches. It is kept open by two cane hoops at top and bottom, and a lead of couple of pounds weight, attached by a sling to the lower hoop, keeps it sufficiently submerged, that is to say, about a third of its depth. We sling this over the side, and pour into it out of the basket about a dozen and a half Sand-Eels at a time. Here they remain well alive, and a bait can be taken out of it readily without opening the basket, recourse to which is only had when nearly all the Sand-Eels in the bucket have been used. At the stern of the boat we use two drift-lines, and bait them as for Pollack, p. 64, fig. 16.

Dropping the bait instantly into the water as soon as securely attached to the hook, we pay out as much length as we can venture, according to the depth of water, here from 20 to 25 feet, and the strength of the current will allow. At one hour before low water, a considerable strength of current yet remains, consequently several leads will be required, and as I have previously explained that 12-feet intervals are left between the leads, the exact length veered away can be at once known by counting them. Each lead can be easily marked in Roman numerals, which was generally done by our late old friend Peter Le Noury, to avoid hauling in the line to count the leads, which is undoubtedly an excellent plan. From 20 to 25 feet of water will require about 6 or 7 leads, for they are small, barely an ounce weight each. As there is always some drifting weed, the hooks require examining every minute or two, for if any weed catches on the bait it renders it useless; it is the best plan too, if two are in the boat, for each to manage a line. As the tide slackens we shorten the lines to avoid fouling the bottom, and the fish beginning to feed we catch three Bass; when, the tide being entirely done, we cease to feel them. When the tide is running strong a heavy line with a single plummet is sometimes very killing, and I therefore put out a line ordinarily used for Mackerel, having a one-pound plummet. See p. 124. It is now dead slack, and the lines are perpendicular, and as the flood current will shortly set into the pool, we haul up our anchor and prepare to whiff; that is to say, we pull along slowly through the pool and past the point, with six leads out. We have two or three more bites and get another fish. By this time the flood current is well made, and pulling out about a hundred yards past the entrance of the harbour, we anchor in the middle of the channel, putting out four leads to begin with, increasing the number up to eight or ten, as the strength of the stream requires. The fish now come on faster, and we are very busy for some time, until the weed becomes very troublesome, and the fish consequently few and far between. On looking over our fish, we find we have thirteen Bass and one Pollack, varying from 1 lb. to 4 lbs. in weight. These larger fish fight well, sometimes requiring five or six minutes to kill them, and always causing the landing-net to be brought into use. Bass occasionally are taken up to ten or fifteen pounds, and even larger; but I find, after taking a great number and regularly fishing through several seasons, that 2 lbs. constitutes a good average weight. In drift-line fishing I have never saved a Bass above 8 lbs., but having frequently been broken by heavy fish, I doubt not they have been beyond that weight, as I have caught them in the same localities by ground-lines, or with a trammel net, of 12 and 14 lbs. I have heard of Bass of 28 lbs., but do not happen to have seen them, although this size is well authenticated. As regards the length of line to be used, it depends on the depth of water and strength of the current. The bait should not touch the bottom, because it will then for certain be spoiled by the green harbour crabs; if, therefore, it should be pinched by these little plagues, haul up a yard or two, and the bait will escape damage. After the tide has risen about 2½ hours, the current in many bar-harbours becomes too strong to allow the bait to be kept at the required depth. In this case anchor on the flat ground on either side of the entrance in not less than 2½ feet depth, and nearer or more distant from the throat of the entrance, according to the strength of the stream. Where the water runs over the edge of a bank will often be found good fishing. The greatest drawback to this method of fishing in bar-harbours is the frequent abundance of weed drifting along with the tide at all depths.

WHIFFING FOR BASS.

Good sport may also be often obtained by sailing in a tideway with a dead Sand-Eel or other small fish bait: to bait the hook, see page 82. Artificial baits—particularly the two spinning Sand-Eels (figs. 28, 29, p. 83.)—are successfully used in whiffing for Bass when good natural baits cannot easily be procured, but the latter are unquestionably superior. Use the Cornish whiffing line as for Pollack, p. 81. Bass caught in this manner are frequently very large. Use hooks No. 9, fig. 63, p. 211. Flies are also often successful, but less so than when thrown with the rod.

A brisk breeze blowing out against the flood current is the most favourable time, and during spring-tides. A horn of the sucker or octopus five inches long, baited as the Freshwater Eel (fig. 25, p. 82), will also take them. When Pollacking I have sometimes caught them with the Freshwater Eel bait, in common use for this fishing in many places. The Exmouth men are very expert at this method with dead Sand-Eel bait, and they use a pair of 12-feet poles to give the lines greater spread. The hook is brought out at the tail of the bait at Exmouth.

ANGLING FOR BASS.

Bottom angling for Bass may be practised with success from off a pier-head, with a strong rod sixteen to eighteen feet in length, and either a floated or a Pater-Noster line, as described for Whiting-Pollack, p. 95, only that your tackle should be somewhat stouter, and your hooks tied to triple gut or gimp; the hooks should be one inch across from the point to the shank, and the bait half a Soft Crab, a piece of Cuttle-fish, a living Sand-Eel, a dead one, or a piece of Pilchard; plumb the depth with a plummet, and fish just clear of the ground if you have a float, and if fishing off a point or headland, with a wash on the shore caused by a breeze from seaward, which is the most favourable time to try for them, and when a float is most useful.

You must be provided with a landing-net or gaff-hook, and when you get hold of a large fish play him round to the nearest convenient spot and dip him up or gaff him. See fig. 64, p. 213.

Whilst fishing with the rod, you may also throw in a ground-line for Bass, if a convenient place can be found to make fast the line. At the head of many coves, amongst the rocks, great accumulations of seaweed take place, which in the process of decomposition breed immense numbers of maggots. At spring tides, during a fresh breeze, this bank of weed being washed by the water, the contents are dispersed and the Bass come quite into the land-wash to feed, and are found in almost every cove and bay under these circumstances in the summer.

The Dory.
(Zeus faber.)

The Dory or John Dory is one of the most grotesque and at the same time one of the very best fish afforded by our seas. Head very large and very ugly, body deep and compressed, colour olive-brown with a golden yellowish tinge, a deep notch in front of the eyes, a black spot behind each gill-cover, mouth capable of great protrusion, and head having a very lantern-jawed appearance. It is very sluggish in its movements, floating or drifting along with the tide, but can exert itself when its prey is in sight. The Dory is taken amongst other fish in the trawl, seine, or trammel-net, but is at times caught with hook and line. It will swallow with avidity a small live fish already hooked, and I have now and then taken one when Pout-fishing, the Dory having gorged a hooked Pout. If you perceive a Dory hovering round the boat, and throw out a line baited with a small Pout, Pollack, or Bream, 'all alive and kicking,' you will often get the Dory, sometimes even without hooking it, for there are two projecting bones in the inside of the jaw over which the snood catches and often enables you to haul them to the surface, when a hand-net or gaff, or, in default of either, your fingers in the gills of the fish, may be used to get the Dory on board. They are not unfrequently caught when Pollack-fishing with the living Sand-Eel. The Plymouth men make fast a small Bream by the tail, without a hook, as a bait, and on a Dory gorging it, the dorsal and pectoral fins stick across the throat and thus do the work of a hook. The Dory frequently runs itself on shore; it grunts when taken out of the water.

The Grey Mullet.
(*Mugil capito.*)

This, although a sea-fish, frequents brackish water even more than Bass, and is perhaps of all sea-fish the most capricious in taking the hook; it lives by routing up the bottom for worms, soft substances, insects, &c.

Large quantities are to be met with in the various docks, and about piers and harbour works, the gates of tide-mills, &c., and are to be taken with a rod and a pater-noster line of strong gut, fitted with four or five Limerick hooks of the sizes 7 or 8, which are preferable to larger, as they have but small mouths. In the article on the 'Smelt and Sand-Smelt' the reader will find a pater-noster with six hooks figured and described, fig. 45, p. 156.

The baits are the Red Rag-Worm found in the mud or sand of most of our harbours, or part of a raw Shrimp taken out of its shell, either of which should be carefully put on the hook, or they may suck it off without being perceived. A bit of pilchard-gut is also excellent for bait.

They occasionally become perfectly ravenous, and at such times may be taken as fast as you can throw in your line. Their feeding appears to be more in winter than in summer, and as far as my experience goes they rarely take a bait except from between the end of September to the end of March.

It is a good plan to collect them together by throwing into the water any kind of refuse fish pounded up soft with a few mealy potatoes, a part of the roe of a Cod-fish mixed with water, a few pounded Green Crabs, and a little chalk, as they will remain for hours together in the corner of a dock or other spot in which they find food. This method of baiting the water is termed ground-baiting in freshwater-fishing, and is only useful in still water, as the fish would immediately disperse in pursuit of it, when carried away by the current.

The Grey Mullet is of a very lively and sportive nature, and in fine weather delights in basking on the surface of the water in large shoals; it is said they may be taken with a fly at such times, but although I have frequently tried I have never been successful, yet I have seen them follow it with great eagerness.

From some of the large tidal rivers on various parts of the coast a considerable extent of land has been embanked, leaving here and there large ponds of water communicating by drains with the sea; in such places you will always find Grey Mullet, Freshwater Eels, a few Bass and Flounders, and occasionally Trout, with immense quantities of Shrimps and Green Crabs, on which the fish are supposed to feed; in these pools fish are easily taken by a casting-net, and the Eels by hook and line, of which the other fish seem very shy. It would be well worth while to introduce Perch, as an experiment, into such ponds. When many Mullet are enclosed in a seine or drag-net, numbers will escape by leaping over the cork-line in rapid succession, as sheep will follow each other over a fence; a trammel-net, however, is a very effectual means of capture, particularly if backed by a second at a distance of three or four feet.

The Greeks have an ingenious way of preventing their escape by extending a piece of net from the cork-line on canes on the surface of the water.

The Chervin, or Shrimp Ground-Bait.

The following information, communicated by a former resident in Jersey, is well worth notice, the practice described being most successfully carried out by two or three knowing old stagers in that island: with a very fine horse-hair net, they repair to the flat sands at low water and catch a large quantity of the young Shrimps in the small rills of water, salting them down in a jar until wanted, when they take a pint or so with them, and mixing some water with the Shrimps or Chervin to about the consistence of gruel, throw a little into the water beside the line with a spoon occasionally, which collects and keeps the Mullet about the place, and seems to whet their appetite for the bait—a piece of a good-sized Shrimp minus the shell. Off-shore winds and smooth water are preferred for this fishing.

The horse-hair net used is of a triangular form on a frame each side about 30 inches, and being crossed by a pole, it is pushed along on the sand like an ordinary strand Shrimping-net (fig. 75, p. 244). To the bottom of the net a pipe or hose about 8 inches long is sewed, made open at the bottom, but closed by a string when in use, which being untied the contents are at once emptied without loss. Before salting down, all weeds, &c., must be picked out, or it will become putrid.

THE SMELT AND SAND-SMELT.
(*Osmerus eperlanus.*) (*Atherina presbyter.*)

The Smelt, so called from the odour which it gives out when fresh caught, is a silvery transparent little fish from 6 to 9 inches in length, and of a very delicate flavour, abounding in many of the harbours on the eastern and western coasts, but rarely seen on the southern, where its place is supplied by the Atherine or Sand-Smelt, known under the name of the Rosselet in the Channel Islands, which, although a sweet-eating little fish, is not equal to the real smelt in quality. Some anglers use a floated line, but the Pater-Noster is the most killing, both for this and Mullet-fishing, the hooks a size smaller, say 9 or 10, Limerick make (fig. 63, p. 211), and the bait a piece of Rag-Worm half an inch in length, or a piece of Shrimp the size of a pea ; the hook need only be stuck through the bait with the point showing, which does not prevent these fish biting. Both Smelt and Atherine will take a bait cut from their own brethren or any small bit of brightly shining fish ; but no bait is more killing than a bit of Rag-Worm, and your hooks become unbaited when using it less than with any other, a matter of no small moment in fast fishing. The gut-line should be not less than 6 feet in length, and of stout Salmon or ordinary double gut, with 5 or 6 hooks on it, and a small dip-lead at the bottom about a quarter of an ounce in weight, or heavier if there be any current to require it ; the rod light, stiff, and about 14 feet in length ; the line the same length as the rod. Smelts are to be sought for where any drain, sewer, or stream of fresh water enters a harbour, at the gates of a tide-mill, or in rocky coves within any sheltered bay on the coast they frequent. Both Smelts and Atherine bite so gently that frequently you can scarcely perceive it, therefore you must not wait until you feel a bite, but every four or five seconds strike smartly but gently, and raise your rod, for by acting in this manner you will catch two to one to those who wait to feel the fish before they strike.

Be very particular that your hooks are sharp, or you will lose as many as you catch, and in fitting out your line take care that the gut to which your hooks are tied be not long enough to reach each other, or they will frequently foul; these pieces of gut should not be more than 5 inches in length each, and if you place them a little more than 10 inches apart they will then keep clear.

In the illustration of the Pater-Noster (fig. 45) the loops of the hook-links are shown untightened, that the method of attaching them may be more evident. The knots made in joining the gut will, if it be short, be found at convenient distances for looping on the hooks, but if long, extra knots must be made, or split shot placed on the line at necessary intervals. Both Smelts and Sand-Smelts may be taken from a boat if there is no convenient place to fish from shore, either with the rod, or with lines fitted out after the same form as Whiting-lines, for instance, like the boat-shaped or Kentish Rig (figs. 3 and 7, pp. 38, 42), the leads from two to four ounces in weight, according to the strength of tide. Although so small they afford good sport, as they take a bait very fast, and it is not at all uncommon to basket from fifteen to twenty dozen under favourable circumstances. The real Smelt are said to thrive in fresh water as well, and in the harbours of South America—viz., at Buenos Ayres, the Straits of Magellan, &c.—are taken as much as twenty inches in length. I am here to be understood as speaking of the real Smelt, which on those coasts weighs sometimes

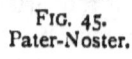

FIG. 45.
Pater-Noster.

nearly 3 lbs., and not of the Atherine or Sand-Smelt. The principal difference between the Smelt proper and the Sand-Smelt is that the former has the second dorsal fin rayless and fleshy, in this respect similar to that of the Salmon, whilst the Sand-Smelt has several rays in that fin, and the eye is smaller and the mouth opens more vertically than that of the real Smelt; in addition to this, the back is much more straight. Large numbers of both kinds are taken by seine or draught nets, and also by dip-nets laced on iron hoops, from 4 to 12 feet in diameter, which are lowered from quays or wharves by a crane or derrick, which is sometimes rigged out from a boat for the same purpose.

Pieces of fish may be fastened in the net as bait, or Green Crabs and chalk or oyster-shells may be pounded up and mixed together, and then thrown in over the net to attract the fish. Other fish may be occasionally taken by this method, such as Mullet, Eels, or Flounders, and particularly with Pilchard offal, of which all sea and estuary fish appear remarkably fond. In addition to seine or dip-nets, Atherine or Sand-Smelts are caught in large numbers by meshing-nets forty to fifty fathoms long, eight feet deep, and one-inch mesh, corked on the head and leaded on the foot; these nets are shot across the entrance of small coves; a portion of a harbour, or a semi-circular space may be enclosed where any small stream of water falls into the sea. The net is not to be hauled as a seine, but the boat being pulled to and fro within the enclosure twice or thrice, a second man at the stern splashes with another oar, and thus causes the frightened fish to mesh themselves. The net should be made of fine thread—No. 25 Shrewsbury is the favourite with the fishermen in Guernsey. So fine is the thread, that a bag three feet long and fourteen inches diameter will contain the net. Thousands of Sardine-sized Pilchards are also taken with it. It is a pity this method of fishing is not known and practised on the Devon and Cornish or other coasts, as much money might be earned by it by taking Sand-Smelts for sale. They are good bait for trots or long lines, for Turbot, Rays, or Conger. Whilst Smelt-fishing I have sometimes taken a Herring or two with rod and line. I find from a London fishmonger that the real Smelt is now chiefly supplied from Holland.

The Atherine or Sand-Smelt feeds well after sunset, and may in most harbours of the English Channel be taken from piers and quays or landing-steps under a lamp, which it is believed attracts them to the spot, particularly on the flowing tide. At the mouth of a drain or stream which is *below* the level of the high water I have been most successful when the current has commenced to show itself after the tide has ebbed a little and at spring tides; but where a perceptible in-pouring of water exists from a higher level, you may take them well at neap-tides in any depth above three feet. My best fishing has been from October to the end of March. In the Scilly Isles the Sand-Smelt often exceeds 8 inches in length.

THE COD.

(*Morrhua vulgaris.*)

The Cod-fish is so well known as hardly to need description, and although it does not come within the reach of amateur fishermen as often as other kinds, yet from its importance in a national point of view, and the amount of capital invested in vessels and gear employed in its capture, it must not be passed over unnoticed. Cod are caught in the seas all round Great Britain, but more particularly in the North Sea, where very large boats and decked vessels are employed, as this fishery is frequently followed at a great distance from the land. Numbers of the North Sea smacks are fitted with wells, by aid of which the Cod are brought into port alive. Grimsby is now the chief station. The same kind of tackle as that used for Silver-Whiting is adapted for this fishing, but the hooks and snoods are much larger and stronger, and only two on each line, the hooks not less than an inch and three-quarters to two inches from point to shank, and the snood strong fishing-line served round with wire as for Bass, or green hemp as for Conger, or else extra stout. Size of line, No. 2 (fig. 11, p. 48).

The best baits are fresh Squid or Cuttle-fish, fresh Mackerel, or Herrings, or Pilchard, and in fact when the fish are plentiful any kind of fresh fish; the method of fishing as for Silver-Whiting at anchor.

The Sand-Eel or Launce is also an excellent bait, and much used in the Newfoundland Fishery. Whelks are also employed, and form the chief bait of the cod smacks.

A great many Cod are taken by trots, or long lines, for which see under 'Bass' (p. 142). Size of line, fig. 11, No. 1, p. 48.

The supply of salted Cod-fish is procured, as is well known, from the banks of Newfoundland. When fishing for Silver-Whiting in the autumn, you will now and then take a Cod-fish. Small Cod are termed Codlings or Tom-Cod, and are to be fished for in every respect as for Silver-Whiting.

It will be seen on referring to the article 'Conger' that the gear is of the same fashion as for Whiting, but heavier and stronger, and with three swivels; this is well adapted for Cod-fishing. The banker's lead is much used in Newfoundland (fig. 6, p. 41), and the Sprool Rig in the North Sea (p. 37). The North Sea smacks catch a great quantity of fish drifting, laid to.

The vessels belonging to the North Sea Fishing Station of Lerwick in Shetland, when they leave for the fishing-grounds in April, are each provided with a square sail containing about one hundred and thirty yards of canvas, which when fishing is attached to warps and put overboard to act as a drag and check driftage.

This of course it does not entirely effect, but nevertheless checks the drift of the vessel very considerably, so that she does not travel over the ground nearly as fast as would otherwise be the case by the action of the wind or tide on her hull and rigging. The fishing-vessels return towards the end of August.

NEWFOUNDLAND COD-FISHERY.

'The Great Bank of Newfoundland is an extensive shoal lying to the south-east of the island, measuring upwards of three hundred and thirty miles in length, and about seventy-five in width, the water varying in depth from sixteen to sixty fathoms.

'Cod formerly here abounded in such countless numbers that it seemed impossible any diminution in the supply should ever arise; frequent complaints have, however, of late years been heard of the deterioration of this important fishery.

'The season for the fishery commences about May, and continues till August for exportation, but by the inhabitants is continued into September. Each vessel, as she arrives at the island, takes her station opposite any unoccupied part of the beach which may afford a convenient situation for the curing of the fish. The first proceeding is to unrig and take down the upper masts, &c., of the vessel, and to erect or prepare the stage on shore. This is a covered platform projecting over the water, strongly built, and guarded with piles to prevent injury from the boats. On the stage is a large firm table, on which all the processes to be hereafter described are performed.

'Near the shore the fishing is carried on in boats about eighteen feet keel, in the offing by large boats, and on the Great Bank itself in schooners, with a crew in proportion to the size of the craft.

'Each person manages two lines, and each line carries two hooks; so that if there are four men in the boat, which is usually the case, there are sixteen baits out. The bait consists of Herrings, Capelin, Squid, Mackerel, or Launce. Good hands will take from three to four hundred fish in a day; but it is severe labour, from the weight of the fish and the extreme cold felt in such an exposed situation during easterly winds; in fine weather, with the wind between S. and W., the temperature is agreeable. Two men have sometimes caught above 2,000 in a day.

'The boats take their station on the edge of the shoal, and the lines being baited are thrown out. When a sufficient load has been taken, it is carried to the stage; for if the fish were kept too long unopened they would be materially injured.

'Each fish is taken by a man standing on one side of the table, who cuts its throat with a knife. He then pushes it to a second on his right hand called the *header*; this person, taking the fish in his left hand, draws the liver out, which he throws through one hole into a cask under the table, and the intestines through another which is over the sea, into which they drop. He next separates the head, by placing the fish against the edge of the table, which is constructed curved and sharp at this part for the purpose; and pressing on the head with the left hand, he with a violent and sudden wrench detaches the body, which by the action is pushed to a third man opposite to him, the head falling through an opening in the stage into the water. The man who performs this feat sits in a chair with a stout back to enable him to use the necessary force, and his left hand is guarded with a strong leathern glove to give him a better hold.

'The *splitter* cuts the body open from the neck downwards with rapidity, but with a skill acquired by long practice; the value of the body depending on its being done in a particular way. The sound-bone is detached by the process, and is suffered to fall also into the sea, unless the sounds and tongues are intended to be saved for use, in which case the requisite number of them and of the heads are thrown aside, and removed, so as to offer no interruption to the main business.

'When the barrow into which the split bodies are thrown is full, it is removed to the *salter* at the further end of the stage, who piles the fish in layers, spreading on each as he takes it out a proper quantity of salt, which must be apportioned with accuracy and judgment, a deficiency or excess of it at this part of the process being detrimental to the proper curing.

'It is the custom in some places, or by some fishermen, to place the split fish in vats or oblong square troughs, instead of in open piles.

'After remaining from four to six days in salt, the fish is washed in sea-water, in large wooden troughs, seven or eight feet wide and three or four deep, a quantity of the bodies being put in at a time; each is taken up singly and carefully cleaned with a woollen cloth, and then laid in long rows on the stage to drain for a day or two. When a sufficient quantity is thus prepared it is spread to dry on stages, made either of wattles supported on poles or else of more substantial timber, the object being that the Cod should be thoroughly and equally exposed to a free circulation of air. Every evening the fish are gathered into heaps of three or more, placed one on top of the other, the backs being uppermost, to guard against rain or damp during the night. These piles are increased as the fish become more dry; but during the day-time they are spread out on the *flakes*, or stages, separately. On the fifth evening, the night piles consist of from forty to fifty fish each, laid regularly, with a few at top, disposed like thatch to throw off the rain; and when finally made up into stacks ready for shipment, tarpaulins and rind of trees kept down by stones are used for the same purpose. It is left in these stacks for a considerable time, being occasionally spread out again during fine weather; and as damp getting into the fish will spoil not only the one so wetted, but often the whole pile, great attention is paid to the weather while the fish are spread on the flakes; at the slightest signs of the approach of wet they are all turned back uppermost, and, as sudden showers are frequent during the summer season, the hurry and confusion of the time the fish is drying is indescribable. Even the Sabbath, during divine service, affords no respite if this source of danger is apprehended, for the whole fruits of a voyage may depend on the exertions of a few minutes: the flakes are on such occasions surrounded with men, women, and children, turning the fish or piling them up, to shelter them from the coming rain.

'The whole coasts of Labrador and Nova Scotia, as well as Newfoundland, are the scene of these fisheries. Twenty thousand British subjects are annually employed, with from two to three hundred schooners, on the Labrador stations. About four-fifths of what we prepare is afterwards exported to the Catholic countries of Europe. A great quantity of Cod is imported *green*, that is, it is split and salted, but has not been dried at the stations.

'Cod are also taken with large nets, called *Cod-seines*, from 80 to 100 feet deep.

'When the fishing-stations are at a considerable distance from the shore, so that too long a period would elapse before the cargo could be salted in the regular manner, it is usual to perform this process on board, and boats laden with the fish thus partly prepared are continually being despatched to the mainland, for them to undergo the subsequent processes of drying. These boats, as they arrive, are moored to an oblong square vessel made of planks, put loosely together, so that a current of sea-water is always flowing through them. This vessel, called a 'Ram's Horn' (supposed to be a corruption of the French term *rincer* or *rinçoir*), is fixed at the head of the stage. Three or four men stand in it to wash and scour the fish with mops as they are thrown singly out of the boat into the vessel; as fast as they are cleaned one of the men throws the fish up on a scaffold half the height of the stage, and from thence others throw them on the stage itself, where they are received into barrows, and removed to the flakes to dry.

'The livers of the fish, it has been mentioned, are collected in casks, placed for the purpose under the table; these tubs are emptied as fast as they are filled into larger puncheons, which receive the full action of the sun's rays; in about a week the livers resolve into oil, which is drawn off by a tap at about half-way between the top and bottom of the puncheon, so as to leave all the solid and dirty parts behind; the oil thus separated is again further purified by a similar process, and being put into clean hogsheads, is exported as *train-oil*, a name given to it on the spot to distinguish it from whale, or seal oil, which is called *fat-oil*. The refuse in the first puncheon, consisting of blood and dirt, is let out, and boiled in copper cauldrons, by which a further portion of inferior oil is obtained. The Cod-oil is employed in dressing leather, and the better quality as a medicine.

'Besides Cod, Newfoundland and the adjacent coasts and rivers furnish Salmon, Herrings, Capelins, Plaice, Soles, Haddock, Mackerel, Halibut, &c. The Capelin is a small species of the Salmon genus, and is an excellent fish; it resorts to Labrador and Newfoundland in shoals rivalling in magnitude those of the Herring; these generally arrive about the middle of June, and the fishery is carried on by two persons in each boat, which they easily fill in a couple of hours. They employ a cylindrical net, open at both ends, one of which is loaded with lead to sink it, and the other is gathered in by a running rope. The fisherman holds the rope in one hand and the top of the net in his teeth, and spreading out the lower end with both hands, he drops it over a shoal of the fish ; the net is then quickly pulled in by both men, and being emptied of its contents, it is again cast : a load is thus frequently obtained without the necessity of moving from the spot.

'As the Capelin, independent of its being an excellent article of food, is extensively used as a bait in the more important Cod-fishery, immense numbers are annually taken.'

Dried Capelin are imported to some amount, and may be procured of fishmongers and grocers. They are commonly eaten as a tea or breakfast fish. The Capelin is a variety of the real Smelt, and belongs to the Salmon family.

This notice of the Newfoundland fishery was extracted from the *Saturday Magazine*, and has been revised with the assistance of Newfoundland fishermen residing in Guernsey and Teignmouth. A good Cod is thick down to the tail, has its sides ribbed, and a deep *sulcus* or furrow in the nape of the neck. Channel Cod-fish are often indifferent in quality.

THE LING.

(Lota molva.)

The Ling, as its name implies, has a long form of body, and is a very voracious fish. As an instance of its voracity I may mention I knew of one being taken through swallowing a fine Red Mullet and a portion of the trammel-net in which it had become entangled, as the net became hooked over the large teeth of the Ling, which was consequently unable to free itself. They are chiefly caught amongst the Cod, as they will take the same baits, and are found on the same fishing-grounds in the North Sea, off the Irish coasts, &c. Numbers are also taken on certain grounds in the vicinity of the Scilly Isles, and are split, salted, and dried for winter consumption, for which they are well calculated, being fine thick fish and taking salt well. They are sometimes caught on a Whiting-line, attain nearly 5 feet in length, and in common with Cod are often found in a puff-blown or distended state, floating, either dead or dying. The liver of a good Ling is large and white, and much oil may be melted out of it over a slow fire; when in poor condition the liver is red and affords no oil. The fish should carry its thickness well down to the tail.

THE SEA-LOACH OR ROCK-LING
(*Mustela vulgaris*)

is a small fish from 6 to 18 inches long, mottled over with spots. There are three kinds, Grey, Red, and Black. The smaller frequent rock-pools, and may be taken with rod, line, and a small hook baited with the soft part of a Limpet; the larger are sometimes caught when ground-fishing. They are best fried.

THE HADDOCK.
(*Morrhua æglefinus.*)

The average size of the Haddock is not more than two or three pounds, although they occasionally reach eight or ten. They are very numerous along the eastern coast of Great Britain, but some are found in all the British waters. Scarborough, with other fishing towns on the east coast, sends numbers to the London market, and the coast of Scotland is even more prolific of this fish, where they are salted and smoked for exportation, and have of late years obtained a high repute, a 'Finnan Haddie' (so called from a village of that name) being considered quite a delicacy. Use the same tackle and baits as for Silver-Whiting, such as Mussels, Lugs, Mackerel, fresh Herring or Pilchard: in the west of England, off Plymouth, &c., a piece of Squid or Cuttle-fish, with a large Mussel on the point of the hook, is considered the best bait for Haddock. Many are caught amongst other fish in the trawl, and also with spillers or long lines. Size of hooks Nos. 8 and 9, fig. 63, p. 211.

THE HAKE.
(*Merlucius vulgaris.*)

The Hake is a fish which of late years has risen into repute, and large quantities are sent by railway from Plymouth and Brixham to the London, Bath, Bristol, and other markets; it is taken both by the trawl-net and by hook and line.

The drift-fishing for Pilchards and hooking for Hake

usually proceed at the same time, and are managed in the following manner : the Pilchard nets having been shot out in a straight line, the boat is made fast to the last net and drifts along with the tide ; the lines, four or five in number, are then put out baited with fresh Pilchards. Hake vary in size from five or six to twelve pounds in weight, and large hooks and strong tackle are therefore necessary ; the hooks are not less than two inches from point to shank, with 8-inch shanks having an eye at the top. Their sharp and huge teeth would soon cut through the snood of a short-shanked hook. Pilchard is considered the best bait, but they will also take a piece of Whiting, Mackerel, or other fresh fish. Before the Pilchard season commences it is the custom to fish for them at anchor, but the success is not usually so great as when drifting with the Pilchard nets.

Haking, as it is called, is always practised at night, for they do not feed well during the day. I have known as many as 50 dozen Hake and 5,000 Pilchards taken in one night by one boat, with a crew of three or four men and a boy.

To bait with a Pilchard, enter the hook close to the tail and bring it out the other side, then pass the hook through the second time, as shown in the frontispiece. Half a small Mackerel or Whiting is also often used, for which see under 'Conger,' p. 179. For Hake or Conger-fishing a stout truncheon like a policeman's staff, should be provided, made out of a piece of heavy wood, to stun the fish with a smart blow or two behind the head, without which it is difficult to hold them. This will also serve as a disgorger (see p. 53).

Use the boat-shaped sinkers as for Whiting, half a pound leads aft, one and a half pound amidships, and three pounds forward if you use six lines, but four are mostly sufficient when drifting. If at anchor, heavier leads may be required. For size of line, see fig. 11, No. 2, p. 48.

You try first at the bottom, then a few fathoms higher, then shallower still, until you meet with the fish, as they are not always at the bottom, but swim at various depths, following the shoals of Pilchards, Sprats, &c., on which they feed.

The length of line from the lead to the hook is usually

three fathoms, and the Hake will sometimes come so near to the surface of the water that they are taken whilst holding the lead in the hand. The Hake is said to abound on the coast of Ireland, particularly off Waterford and Galway. This is so good a fish that it is scarce possible to cook it amiss : it is excellent in steaks or baked upon potatoes ; the back-bone having been removed, and the fish larded with strips of fat bacon, with plain salt and pepper seasoning.

The ham and bacon factors of Plymouth smoke a great many Hake, and although not equal to Salmon thus prepared, they are a very palatable breakfast fish.

Very large Pollack are caught whilst Haking.

Hake-fishing has very much deteriorated on the SW. coast of England during the last 20 years.

THE SOLE.

(Solea vulgaris.)

This excellent fish is generally taken with the trawl-net, but is sometimes caught with the hook. A trot or spiller, as recommended for Dabs, is suitable for this work, and the baits Lug or Rag-Worms ; they feed best in the night.

Soles are also often taken in a trammel net.

The Sole, in common with other flat fish, frequents the sandy and oozy bottoms of our coasts, and is taken also in the various tidal rivers whilst of a very small size, especially in the lower part of the river Exe, by seine-nets. I have seen Soles thus caught not more than 6 inches in length, which from their small size are locally called *tongues*, and are sold in quantities at times by the fish-hawkers. At night in the open sea the fish frequent the shore, and, when the wind is strong enough off the land for their purpose, trawling vessels scrape the coast line, to the prejudice of the fishery in general ; the laws being a dead letter, because no one is charged to enforce the same. It has always been considered that the bays and shores are the nurseries of the smaller fish, which therefore should remain undisturbed by trawl nets, whose proper sphere of operations is the offing, but who continually work the shores whenever favourable circumstances render it worth their while so to do. The greater part of the fish supply of Plymouth having been caught in sight of my residence on the confines of Devon and Cornwall, I have had constant opportunities of seeing, from the cliffs as well as from the sea, these operations in progress amongst the passing vessels. It is a great pity that our legislators are not more qualified by their own experience to deal with our fisheries, as important evidence respecting them is frequently suppressed, and that given before Commissions so cooked as not to afford any means of arriving at the truth. That the shallows of the shore are the nurseries of the small fish is constantly proved by the quantity of small Plaice, Soles, Turbot, &c., caught when shrimping in the sand-pools on every strand. I have taken numbers myself in these situations, allowing them to escape by inverting the net. In a trawl this is impossible, as it is dragged long distances, and the fish are killed, the whole mass not unfrequently being churned up, so to speak, into a kind of paste. The destruction permitted in this way is something frightful to contemplate, and much is caught and sold as manure which, if allowed to live and grow, would tend to make up for the destruction of fish-life always in progress.

The destruction in river-seining has also been great, as it is the habit to capsize the bunt of the net, and leave quantities of young fish to die in the sun. It is expected that reservations of portions of bays will be made shortly, as nurseries or breeding grounds.

THE PLAICE.

(*Platessa vulgaris.*)

Large quantities of Plaice are taken with the trawl, also with the trot or spiller, consisting of a Whiting-line and hooks fig. 63, No. 10, p. 211, on a double thread snood of fine twine laid up with twisting machine; single twine will not answer as well, being more liable to foul. Occasionally they are caught with hand-lines, and where they are very plentiful may be fished for in a similar manner to that employed for Dabs; best baits, Lugs and large Rag-Worms. Small Plaice, Dabs, Flounders, and Freshwater Eels may be taken from off most piers on the coast by the rod and a Pater-Noster line, hooks No. 4 size, Kirby or Limerick, and the bait a boiled Shrimp peeled, Rag-Worms, and Lugs. For harbour-trot, use No. 13 hooks (fig. 63, p. 211, size of line, fig. 11, No. 2, p. 48).

Many Plaice and sometimes Turbot are taken by spearing on the sand flats between the Scilly Islands, also on the west coast of Scotland.

THE TURBOT.

(Rhombus maximus.)

This highly-prized fish frequents sand-banks in all parts of the British seas, and is taken by the trawl and long lines or bulters. The size of the hook one inch from the point to the shank (p. 210), fastened to strong snooding three feet in length, and from 1 to 2 fathoms apart, on a stout line of about the thickness of window-sash line, coiled in a box, tray, or basket, the hooks baited with half a Smelt, or Atherine, large Sand-Eels, a piece of Herring, Mackerel, Long-Nose, otherwise called Gar-fish, or other fresh fish. The Lampern is also an excellent bait and is much used by Dutch fishermen for their long lines. A fishery for Lamperns has existed at Teddington on the Thames for ages.

This line is secured by stones at certain distances, or anchors, and supplied with buoy-lines to raise it when necessary.

From two to three thousand hooks are sometimes attached to one line, extending a mile or more in length, and shot across the tide. Large boats in the North Sea carry frequently as much as ten miles of long lines or trots.

THE DRIFT-TROT.

In Guernsey, whilst at anchor on the bank waiting to raise the trot, the fishermen use a ' Ligne Longue,' as they term it, consisting of a 50-hook trot bent on to a ground-line, and which drifts out with the tide. Turbot are occasionally caught with ordinary hand-lines whilst waiting for slack tide to raise the long-lines; the line such as used for Silver-Whiting. Turbot vary in size from one or two to thirty pounds in weight.

The Brill.
(*Rhombus vulgaris.*)

The Brill, like the Turbot, is a bank-frequenting fish, and is commonly caught in the trawl-net, occasionally on Turbot trots, but rarely on a hand-line. Although a good fish, it is not held in the same estimation as the Turbot, the flesh not being so firm.

The Wrasse or Rock-Fish.
(*Labrus.*)

This fish haunts rocky shores and weedy grounds, and it is very widely spread over the world, not being confined to Europe. It has a remarkable set of both cutting and what may be termed grinding teeth in the throat; with the first it tears off the young Mussels from the rocks, with the last it comminutes their shells and those of small Crabs, &c., of which it is particularly fond. The protruding and flexible lips have given the name (*Labridæ*) to the family, of which there are several members, some beautifully coloured. The best tackle for this fish is that described as the Kentish Rig, fig. 3, p. 38; the hooks should be strong, fig. 63, No. 12, and the snoods about fifteen inches long. It grows to 8 or 10 lbs. weight, is a poor fish for the table, but the larger are eaten in Guernsey stewed with onions. They are constantly caught when Pout-fishing. Fresh it is a good bait for Crab-pots, and when stale for Prawn-pots or the dip-nets (fig. 76, p. 246).

Varm or Sea Tape-Worm, Mussels, and Lugs are the chief baits, with large Rag-Worms. They also take well both a piece of soft crab and small hard crabs the size of a shilling and under. An India-rubber band round a small crab secures it to the hook without injury. This bait may be used with success for Bass when angling with rod and line from piers or steep rocks.

The Sea-Bream.
(*Pagellus centrodontus.*)

The Sea-Bream, Brim, Chad-Brain, or Red Gilt-Head, for by all these names it is known, is very numerous on the coasts of England, &c., and is frequently taken whilst fishing for Whiting.

The young of this fish perfectly swarm, both off the open shore and in the deep harbours of Devonshire and Cornwall, from July until the end of October; they are there called Chad. The large fish reach the size of four or five pounds, and struggle lustily when they feel the hook, which should be baited with Rag or Lug-Worms, a living or a piece of dead Sand-Eel, a piece of Mackerel, Pilchard, Herring, or a Mussel; they will also take freely the soft part of a Limpet, or when well on the feed a garden Snail, of course minus its shell, and even the hard part of the Limpet. To bait with Lugs run the hook down from the head about an inch and pass it through three times. Rag-Worms for Bream should be large and the hook passed four times through them. Of Mackerel use a piece an inch long and half an inch wide. Mussels as for Whiting, if large use half only. No. 11 hook (fig. 63, p. 211); for Chad or Small Bream, No. 14, when harbour-fishing or close to shore.

The soft part of a Limpet is such an excellent bait for Bream, and is in general so easily procured, that I feel it quite worthy of a special description as regards its use. Procure about fifty or more of the largest Limpets, and prepare them for bait in the following manner—the soft part of the Limpet is the more attractive, but as this is too soft to hold well on the hook alone, it is necessary a small portion of the hard part should be included in the bait. Take the Limpet out of the shell with a round-topped knife, and passing the knife between the hard and soft parts, cut off with the soft that portion of the hard also in which the horns are situated, about the size of a silver threepenny piece. Passing the hook through the small hard piece first and then through the soft, if the fish knocks off the soft at the first nibble, it is often caught with the hard. This bait is much the better for being prepared and placed in the sun an hour previous to use. The Sea-Bream feeds at all depths, and although often caught on ground-lines, may be taken in far greater quantities with the lighter kinds of drift-lines used for Pollack, namely the horse-hair lines, one pair with half-, the other with quarter-ounce leads. In addition, I put out frequently a line without lead, often very effective.

The Chad or small Bream, as previously observed, come into the harbours and close alongshore during the summer and autumn months, but the full-grown Bream can rarely be taken in any quantity in less than the depth of ten fathoms, and they will not often bite in water of so small a depth as this until the sun is setting or twilight commences. Rocky ground, or a sandy spot close thereto, is the locality suitable for Bream-fishing, many of which are to be found on the coasts of South Devon and Cornwall, Ireland, &c. Let go your slingstone or killick on the rocky ground, and pay out the cable until you are over the end of the rocky ground, and make all fast.

Having ascertained the depth with a ground-line, bait your light drift-line as before directed, and pay out as much as the depth of the water, if the tide be moderate; if dead slack a fathom less; if rather strong, half as much again, and drop the lead inside the gunwale; if too much stream, and the lead drags overboard, make fast a loose thole-pin, lay it down in the boat, not forgetting to secure the end, with a couple of fathoms of slack between the reel and the thole-pin, in order that the fish may not bring himself up short, and carry away the gear before you have time to give line.

Before going to sea procure, say, half a hundred shore Green Crabs, and pound them up in the boat's bailer with the Limpet shells and hard parts of the Limpets chopped small.

This ground-baiting kind of mixture is termed 'burley' in Australia; and on the coast of North America is much used in the Mackerel fishery and elsewhere, although the ingredients vary much with the locality.

When the tide is moderate or quite slack, throw in a little of this 'hurly-burley,' and you will find it very effective in collecting the fish and keeping them about the spot.

The most agreeable weather for this fishing is a bright moonlight evening, but they will bite when it is so pitchy dark that you cannot see to bait your hook; a lantern, therefore, is a desirable appendage to your apparatus, and, besides, will render effectual aid in clearing any entanglement of the lines, which is sometimes caused by a large fish sweeping round them all before he can be taken on board. To take large Pollack whilst Bream-fishing scale a small Bream, and cutting off one of its sides divide it in two lengthways, and put half of it out on a stern drift-line without lead, with a No. 6 hook (fig. 62, p. 210).

A short-handled net is very convenient to get the large Bream on board.

The back fin of this fish is studded with sharp spines or prickles, which renders great care necessary in taking it off the hook. In Bream-fishing the oilskin petticoats worn by seamen are very useful, and, clad in these, you may venture to clip the Bream between your knees, then, putting finger and thumb into the eyes, you will be able to hold them whilst withdrawing the hook.

The Sea-Bream is not usually held in great estimation, but if stuffed with veal stuffing and baked, and occasionally basted with butter, will be found far from despicable. It is getting into greater request than formerly.

THE BRAIZE, OR BECKER.
(Pagrus vulgaris.)

This fish is blue on the back, the sides are bright and silvery when first taken out of the water, but they turn almost black in death. It takes Mussels, Rag-Worms, and Lugs, and is often caught when Pout-fishing on the ground lines, but never in very considerable numbers, as it is not a plentiful fish on our coasts.

I have frequently taken them when at the Pouting ground by throwing out a drift-line without any lead, and baiting with a fresh Mussel or slip of any fresh fish.

If hooked on fine gear, the line must be carefully handled, as it is a very strong fish, and struggles to escape with a succession of sharp jerks. Its flesh is better flavoured than that of the ordinary Sea-Bream, and if stuffed, baked, and basted with a little butter or beef gravy, will be found really excellent. There are other varieties of the *Sparidæ*, or Bream family, but which have no special means of capture in our seas. They will be found in Mr. Couch's voluminous work on British Fishes.

The Gar-Fish, or Long-Nose.
(Belone vulgaris.)

The Gar-fish, Long-Nose, Snipe-Eel, or Sea-Needle, has a very long slender body, and a bill like a snipe or woodcock, studded with small sharp teeth ; it varies from two to three feet in length.

This fish is usually taken in nets, and frequently accompanies the shoals of Mackerel, but it may be taken by a light drift-line in a tide-way or under sail, with a small-sized Mackerel-hook, and a slip of any shining fish as bait, with or without a Rag-Worm on the point of the hook ; also with the half Sand-Eel as used for Mackerel (fig. 42, p. 135), and on the Floating Trot (fig. 35, p. 97).

The best lines for this sport are the horse-hair lines without lead, which, to take Gar-Fish, must be handled in a peculiar manner. From their beak-like form of mouth they cannot take a bait quickly like other fish ; when, therefore, you perceive a bite, give two or three yards of line, to enable the fish to get the hook well within its mouth, then strike with a smart jerk, and you will generally be successful. It has somewhat of a Mackerel flavour, but is of a much stronger taste, yet, if cut in four-inch lengths, nicely fried, and the green bones removed before being sent to table, is by no means an unpalatable fish. It is useful as bait for Whiting, Turbot, or Conger.

They are of a very sportive nature, and may often be seen leaping out of the water in rapid succession ; in fact, so fond are they of this, that the fishermen frequently throw into the sea a small spar or sprit, for, if there are any in the neighbourhood, they are almost certain to be attracted, and commence throwing themselves over it, thus betraying their locality. In France and the Channel Islands it is known as the Orphie.

The Red Mullet.
(Mullus surmuletus.)

These fish have always been held in great repute both by the ancients and moderns, and are taken both in the trawl, seine, and trammel net, but more particularly in the last mentioned; they are sent generally to the London and other distant markets, as they commonly fetch too high a price for ordinary consumption.

A few are occasionally caught by hook and line or a trot, the bait Rag-Worms or a piece of the large Flat Worm found by digging under rocks at very low tides, commonly known where it is found as Varm or Sea Tape-Worm. The Red Mullet appears to attain a larger size on the French than on the English side of the Channel, which is also the case with the Atherine or Sand-Smelt.

They are in better season and greater abundance from July until the end of November than at other times of the year, and as soon in the autumn as the weather becomes sufficiently cool, numbers are sent from Guernsey to Billingsgate. With hook and line they can occasionally be taken from a pier, with a revolving chopstick line, having a lead plummet at the end from $\frac{1}{4}$ lb. to 1 lb. in weight, according to the tide. Hook No. 3, Kirby or Limerick, or 14 round bend sea-pattern (fig. 63, p. 211.) They should be tied on 7 inches double or best single Salmon gut. Three wire or whalebone chopsticks $4\frac{1}{2}$ inches long—in fact, the improved Portsmouth Rig. During the year 1867, 58 were taken by one individual from Guernsey Pier, five being the greatest number in one day.

Whilst Atherine-fishing in any of the harbours of the west of England you may chance to catch a Red Mullet. Many yachts take a trammel net as part of their outfit, principally for the sake of this excellent fish. It is, however, a net to which nothing comes amiss, and will catch Crabs and Lobsters, Crayfish, Soles, Plaice, or Whiting-Pollack, &c., in fact, everything with equal facility. (See 'Trammel,' p. 222.)

THE GURNARD OR GURNET.
(*Trigla cuculus.*)

The Gurnards, so called from the grunting sound which they emit when taken out of the water, are inconceivably numerous round the coast of Great Britain, and in the Bristol Channel the water is sometimes quite alive with them.

. Great numbers are taken with the trawl-net as well as by hook and line. When fishing for Whiting they are frequently hooked, and occasionally also the Sapphirine Gurnard (*Trigla hirundo*), provincially known as the Piper or Tub, which reaches to the length of two feet and to four or five pounds, or an even greater weight. This large variety makes an excellent dish when stuffed with veal stuffing, and baked with gravy or butter. The Grey Gurnard (*Gurnardus griseus*) is usually smaller than the Red, and is often caught when sailing for Mackerel, particularly when the wind falls light and the lines sink more deeply in the water. The two smaller varieties of Gurnards are much used by the fishermen as baits for Crabs, Lobsters, Cray-fish, and Prawns.

All the varieties of Gurnards have a square form of head, the projections of which are of sharp ragged bone; these, together with the back fin, must be guarded against in unhooking them.

The Gurnards take well all the baits used for Whiting and Mackerel, including flies and spinners.

THE CONGER.
(*Conger vulgaris.*)

The Conger is the largest of the Eel tribe, sometimes reaching 100 lbs. in weight; it is found in all the British Seas, and taken both by hand-lines and trots or bulters. It is to be searched for on or close to rocky ground, although not exclusively found there, as it roams far and wide in search of food. Any of the methods of fitting ground lines for Whiting with long snoods will, as far as the form of gear is concerned, answer for Congering, but that represented in the cut (fig. 46, p. 126) fitted with three copper swivels, is highly recommended for Conger and Cod-fishing. The strength of the Conger is very great, and when hauled into the boat it has the habit, as you hold its head up for the purpose of unhooking, of screwing itself round with force and rapidity, and unless your line is provided with swivels, everything is likely to get twisted into a foul.

THE CONGER.

A very necessary addition to your gear is a short truncheon wherewith to give the *quietus* to either Hake or Conger; it is usually termed a Hake or Conger Bat, and a useful length will be 18 inches. This stick, if tapered at the small end, may be usefully employed as a disgorger. (See p. 53.) Any heavy wood will answer, and it should be about 2 inches in diameter at the club end. Give the fish a heavy blow over the head and another or two at the termination of the fin on the abdomen before you attempt to withdraw the hook. Lines of course must be stout (fig. 11, p. 48, No. 2). The boat-shaped lead, from 3 to 10 pounds weight, may be used, with three copper swivels, as

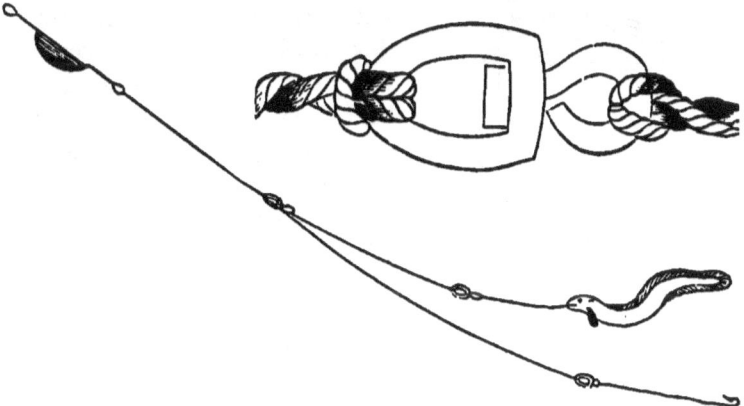

FIG. 46.—Boat-shaped Lead and Trace with Copper Swivels, and Swivel actual size.

shown in fig. 46, and strong hooks (No. 3, fig. 62, p. 210) for shore and the larger for offing fishing lashed on with brass wire to a piece of stout line, which wire is wrapped round to a distance of 8 or 9 inches above the hook, to protect the snooding from the teeth of the fish; or traced green hemp (as in fig. 47, p. 178) is used for the same purpose. To prepare the snood with wire (see p. 103); with hemp, as follows: Cut off a piece of strong line, medium-sized, or stout trawl twine 30 inches long, middle it, and secure it with tightly-drawn hitches over the flattened top of the hook. Hitch the hook over a nail, the latch of a door, or other firm point of attachment, and having

N

made fast another bit of strong line round the waist, secure the end of the snood thereto. Now take some green hemp (obtainable at a rope-maker's), middle it, and leaning back to make the snood as rigid as possible, or, as seamen say, 'as taut as a bar,' plait on the hemp tightly over the snood, particularly at the beginning of the binding and between the flattened top of the hook and the snooding. To fasten off, when you have nearly plaited on the whole, interlace the hemp between the double snood once or twice, and tie an ordinary knot (fig. 47). I much prefer this plan to wire for trots, and in Guernsey it is always used for hand-lines. A very general way of mounting Conger hooks, as well as Cod hooks, is with ten or a dozen thicknesses of stout twine; you hitch the hook over a nail or other point of attachment, fasten one end of the twine to the hook in the usual manner with a clove-hitch, then pass it round your hand and take another clove-hitch, and continue the process until you have the required number of thicknesses of twine. You complete the job by a succession of marline hitches until you reach within two inches of the bow you have formed by passing the twine round your hand. The strop thus formed should be 7 inches long, and to the eye you can conveniently bend on the snood.

FIG. 47.—Conger Hook, with snood traced over with green hemp.

The length of snood below the lead should not exceed two fathoms.

The best bait is undoubtedly Squid or Cuttle-fish, and if the latter, it should be first beaten with a stick, to render it somewhat soft, as it is naturally rather hard, and Congers are disposed to reject anything hard or bony. Cut a piece of this six inches in length and two in width, and having entered the hook at one end, turn it over, and pass the hook through the second time. (See 'Squid,' p. 194, for method of capture.)

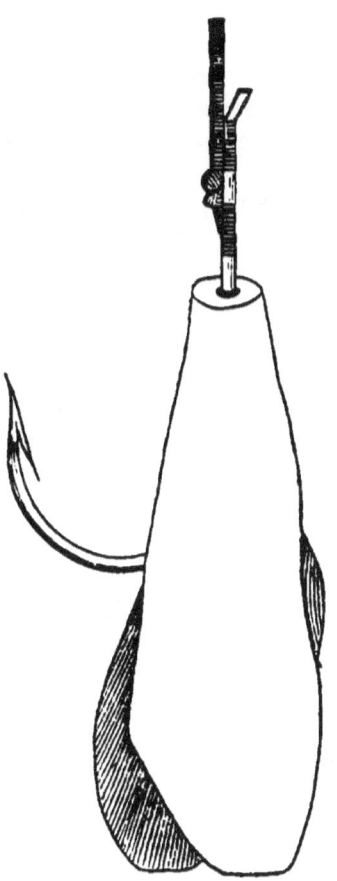

FIG. 48.—Hook baited with the tail half of a small Whiting or other fish.

Conger will also take a piece of almost any fresh fish, which should be first scaled and the backbone removed; for instance, a Chad about six inches in length makes a good bait, which being first scaled, enter the knife at the tail, and cut up to within an inch of the head; do so on the other side, and rēmove the backbone with the back fin, and cut off also the pectoral fins; give the head a blow or two with the lead to soften it somewhat, and entering the point of the hook down the throat of the fish, bring it out at the gills, now turn it over and pass it through the fish crossways. If your bait fish be eight inches or more in length, cut it in two diagonally, and make two baits of it.

Fig. 48 is an illustration of a hook baited with the tail half of a small Whiting, Mackerel, or other fish. The tail fin being cut off and the bait split halfway up, the backbone is to be dissected out thus far, the point of the hook is then to be entered at the tail-end, threaded down by the backbone, and brought out as shown in the woodcut.

If you are just clear of the rocks your leads should rest on the ground, but if not you must haul up sufficient to keep the hooks clear. Large quantities of Congers are also taken with bulters or long lines, fitted with two or three hundred hooks. One-third of an inch is a good thickness for the back line of a Conger-trot, and a very stout brass or copper swivel should be spliced into it at every twenty fathoms, which will be found very convenient for taking out kinks. It should be well wetted, stretched, and dried before use. (See 'Trot' and 'Trot-Basket and Hook-Holder,' p. 143.) The snoods should be three feet long and nine feet apart: a swivel on each is a great improvement.

Whilst fishing for Conger or other large fish you should always be provided with a strong gaff-hook, as large as a butcher's meat-hook, firmly secured to a handle two and a half feet in length, wherewith to lift on board large fish which might else endanger the tackle. At the top of the handle a knob should be left, and the wood be made smooth, that you may be enabled to allow the gaff to turn in your hand, for a large Conger can screw himself round with such violence that, if held rigidly, the dislocation of elbow or shoulder is by no means beyond the range of possibility.

The Conger, in common with the Freshwater Eel, is nocturnal in its habits, and rarely feeds freely in water of less than twenty fathoms' depth during the day; at night, however, they roam far and wide, approaching the shore quite closely, especially on the flowing tide in summer and autumn, and may be caught with a throw-out or leger-line (fig. 43, p. 140.)

A nutritious soup is made from Conger, thickened with oatmeal and flavoured according to taste, but the flesh is not in much repute, except amongst the poorer classes. A ten-inch cut, however, out of the thickest part of a Conger of 10lbs. and upwards, filled with veal stuffing, sewn up, and baked with a bit of fat bacon on the top, will be found a really good dish.

In unhooking Congers push the disgorger down on to the bend of the hook, take a turn of the snood round the stick, and twist out the hook.

The Skate.
(Raia batis.)

This member of the Ray family is very numerous on sandy and oozy ground both in the British and other seas, and is taken in large numbers in the trawl nets and also on trots, as well as occasionally by hook and line when ground-fishing for Whiting, Cod, or Conger. A Ray-fish trot or bulter should consist of a Cod bank-line sixty fathoms in length, with hooks Nos. 6, 7, or 8, fig. 62, p. 210, on double Salmon twine, laid up with the nossil-cock or fisherman's spinner (fig. 66, p. 216), in the following manner :—Cut six feet of the twine, attach the ends to two of the hooked spindles, hang the lead to the middle, and spin up the snood as shown and directed at p. 216. The little spinning jack is also very useful for this work (fig. 23, p. 70).

Bend on the snoods at intervals of 9 feet, by opening the strand with a marline-spike or pricker, introducing the end through the opening, and making a knot on it. Then, making a clove-hitch on either side of it, tighten them, and it will never draw. Bait with half a large Sand-Eel or a whole one of moderate size, pieces of Mackerel or Long-Nose an inch wide, half a Sand-Smelt, or pieces of Pilchard. Attach a twenty-pound stone to each end, and other stones of a pound weight at intervals, to prevent rolling. Shoot it on smooth ground, and you will take Rays and other fish. On a flat sandy shore, where there is a considerable rise and fall, you can use this without a boat, especially at spring tides.

Amongst the greater number of species and varieties of the Ray genus enumerated by Couch, Yarrell, Gosse, and others, there are, besides the Skate, two which are taken in large quantities in trawls, on long lines, and occasionally when hand-line fishing : these are the Homelyn and the Thornback. The Homelyn much resembles the Skate, and like it has smooth sides ; the Thornback has several large white buttons under the skin, which carry strong recurved spines, very similar to the claws of a cat. The flesh of the two former is excellent, and furnishes the crimped Skate of the London market ; but that of the Thornback is woolly and tasteless. As an illustration of the extraordinary change in the value of fish, I may mention that until within the last few years Skate was not worth sending to the London market; then, as other fish became scarcer, it was sold crimped at 4*d.* per pound, but the price has been often much higher.

The Fireflaire has a frightful double spine on its tail, with which it inflicts fearful wounds, but its liver is reputed to be excellent for scalds and sores. The best part of a Skate is the jowl. The flesh is much used as bait for Crab and Lobster-pots.

DOG-FISH AND SHARKS.
(*Squalidæ*).

Dog-fish are often very abundant round the British coasts, and Sharks are by no means uncommon, although the latter are neither so numerous nor voracious as those of warm climates. They are both looked on as the plague of fishermen, driving away other fish, devouring them on the hooks or in the meshes of the net, of which, as well as of lines and snoods, they make dreadful havoc. The Sharks sometimes swallow Congers already hooked, and are thus taken occasionally in the British Seas; there have been several instances in Guernsey.

THE HERRING (*Clupea harengus*), PILCHARD (*C. pilchardus*), SPRAT (*C. sprattus*), AND WHITE BAIT (*C. latulus*).

In 'Remarks on Nets' I have adverted to these fish. Pennant mentions that the Herring will take a fly, and they are often caught therewith in the Scotch and Irish lochs at the present day. I have caught a few at Plymouth whilst Atherine fishing. They are taken for bait by jigging on the Irish coasts; that is, by lowering naked hooks cast into a piece of lead and jerking up. It is of course unsportsmanlike, and only defensible as a means of procuring bait. The well-known Sprat is caught in moored nets in a tide-way, in seines, and in drift-nets of fine twine. White Bait are caught in bag and dip-nets in tidal rivers. Dr. Günther has undoubtedly proved them to be the fry of the herring. There are many other fish curious in their conformation and habits, but which are taken by chance in some of the methods described in this work. In the pages of Yarrell and Couch the student of Ichthyology will find them illustrated and described at length.

THE FRESHWATER EEL.
(Murœna anguilla.)

As the Freshwater Eel is very numerous in most harbours and tidal rivers, the following directions may not perhaps be considered out of place in the present work :—

They may be taken both by bobbing and by hook and line.

Bobbing is usually practised from a boat, and in the following manner : Procure forty or fifty large worms, and string them on worsted or coarse thread by passing a large needle through them from head to tail, then make a coil of them by wrapping them round your hand, and tying them across with a piece of strong twine or tape, which will not be so likely to cut them. Fasten the bunch securely to a piece of strong fishing-line or whip-cord twelve feet in length (fig. 49).

If unable to procure a needle sufficiently large, tie the worsted with fine waxed silk to a piece of iron or brass wire, seven inches in length, and of the thickness of a stocking needle.

You must provide a bell-shaped piece of lead of about three ounces in weight, cast with a hole through the centre, which slide down over the line as shown in the cut.

FIG. 49.—Lead and Clot of Worms for bobbing from a boat.

The best rod for this fishing is a clean cut off a fir plank eight feet in length, planed round, an inch and a half in thickness at the bottom and three-quarters at the top, where a small hole should be bored, and afterwards burned out with a red-hot wire, through which the line is to be passed : as the line will run easily through this hole, you can adjust it to the depth of water. The boat should be moored with two anchors in the following manner : first drop one anchor and push the boat about twenty yards down the stream ; now drop the other, and gather on the first, until the boat is midway between the two, with the bow up the stream ; haul in the rope of the second anchor taut, and your boat will now ride fair with the tide, and will not sheer about, which is of the first importance.

Regulate your line that it may be three or four inches less than the depth of the water, and sound the bottom with your bob every two or three seconds, lifting it off the bottom, and sounding again until you have a bite, which is easily perceived, as the Eel tugs very strongly : raise your line quickly but steadily, and your fish in dropping off will fall into the boat ; great numbers are taken by this method. Always bob in shallow water from two to six feet in depth, for if it is much deeper a great many will fall off.

N.B. If you use worsted, which is preferable to thread, as it gets entangled more readily in the teeth of the Eels, draw it three or four times over a piece of soap, and you will thread the Worms more easily.

I have usually had greater sport on the flowing than during the ebbing tide.

Clotting for Eels.—Clotting for Eels is very similar to Bobbing, and is adapted for rivers and brooks, where a boat is

FIG. 50.—Baited Clotting Pole.

not required. Your pole should be ten or twelve feet in length, to the top of which a piece of iron wire of the thickness of a quarter of an inch is to be fastened, eighteen inches in length, having an eye turned at the end, to which eye you attach your clot of Worms (fig. 50). The best time is immediately after heavy rain, when the water is coloured, or at night, for they will not usually lay hold until after sunset when the water is clear. In clotting you feel the bottom with the end of the wire every two or three seconds, raising it an inch or two from time to time, and when you feel a bite throw the Eel on the bank; in some parts a square box is used on the end of a pole, to drop the Eels into, but though useful in rivers, it is cumbersome to carry, and in small brooks is entirely unnecessary.

Hook and Line.—Eels may also be taken with hook and line, and very fine ones in ponds into which the salt water flows, often the case in land which has been reclaimed from the sea.

Half a dozen lines may be used at once, with two hooks on each line, which should be tied to twisted double gut, fine gimp, or fine snooding; tie on the hooks a foot apart from each other, the lower one a foot from the bottom, where a piece of lead, two ounces in weight, should be attached: the best-sized hook will be found to be No. 7 (fig. 62, p. 211) Limerick, and the bait Garden Worms, pieces of bait Crabs, or large Rag-Worms. In baiting with Worms, enter the point of the hook at the head, and thread on the Worm nearly to the tail, fastening a winebottle cork on each line, about four yards from the shore, which will show you when you have a bite; give a minute or two when you perceive it, and the Eels will hook themselves. A friend has taken large numbers of Eels in tidal harbours in the following manner:—He uses an ordinary perch rod, fine silk line, large quill, or small cork float, gut hook No. 6 or 7 river size, baits with a small piece of boiled prawn or shrimp peeled, and strikes directly the float is taken under water. He has a small mesh landing-net, into which he drops the Eels, and then takes hold of them outside; the net enabling a firm grip to be taken. Eels caught in this manner are nearly all hooked in the mouth, which saves much trouble in unhooking them. Many good Eels are taken with night-lines fitted like a trot for sea-fishing, and baited with a small fish, half a Lamprey, or even a Freshwater Eel, for they are perfect cannibals; for this night-line, fish hooks are especially made with a bow in the top which will admit four or five thicknesses of fine twine, which the Eel will find greater difficulty in gnawing through than if it were all in one.

An acquaintance who has paid much attention to Eel-fishing strongly recommends loosely twisted hair snoodings.

Each of your leger lines should be wound on a sharpened stick, which may be forced firmly into the ground whilst fishing. The best lines are of Whiting snooding, size No. 4 (fig. 11, p. 48), and if soaked in coal-tar and turpentine, and dried, they last a long time, and rarely become entangled; but for night-lines a stout line of the size of a small sash-line should be used, and the thicknesses of twine as described above, or twenty-four hairs. See No. 1 (fig. 11, p. 48.)

Large Worms may also be used, but they are more apt to be nibbled by small Eels, &c., than the other baits. A piece of Lamprey two inches long is good in tidal rivers.

To keep these trots from becoming entangled, it is usual to coil them in a box or Basket (see 'Trot Basket and Hook-holder,' p. 143), and when baited the hooks may be dropped into an old fig-drum, sewed on to the bottom of the basket in the middle. The lines best fitted for catching Eels in harbours from a boat are the dab and flounder lines, p. 116. The best of all baits is the Soft Crab, but pieces of Sand-Eel, Mackerel, Pilchard, Lugs, or Herring will also catch them. Eels are often numerous in harbours which have no brook or fresh water running into them.

Spearing, &c.—Fine Eels are constantly taken by striking an Eel-spear into banks of mud from the shore or a boat. These spears are made of thin blades of steel to open and clip the Eels between their edges, which are jagged to prevent their slipping out. Tongs made especially for catching Eels are used in some countries, and an old pair of scissors or garden shears with edges notched like a rat-gin, will answer the same purpose when turning over large stones in tidal rivers or ditches. A three-pronged kitchen dinner-fork is also useful.

If Lug-worms are used instead of Lob or Earth-worms, many Flounders are caught with the Eels in harbours and tidal rivers, whilst bobbing.

GENERAL BAITS FOR SEA-FISH

The Mussel.
(*Mytilus edulis.*)

This shell-fish is more in use for bait than any other kind, and abounds in many of the tidal rivers of Great Britain and Ireland, both above and below low-water mark; they accumulate in countless thousands on rocks, gravel, and mud-banks and a considerable trade is carried on in them with the metropolis and large provincial towns. In Scotland these banks are known as Mussel Scalps. They are to be procured from the fishwomen generally at about sixpence per peck, and may be kept alive almost any length of time by hanging them overboard in a basket or net. (See also p. 50.) I recommend every yachtsman interested in sea-fishing to procure a peck or two at the earliest opportunity, as at the anchorages in every bay on the coast Dabs may be caught with them, and if becalmed at sea, or he thinks right to lay to for an hour's fishing, he will find himself well provided with bait for either Whiting or Haddock, &c.

To open Mussels, you should take some lessons from the fishermen, and after a little practice you will accomplish it with ease.

This is the ordinary method. Take one in your left hand with the byssus or beard towards you, cut it off and introduce the knife (a round-topped one) as a lever to force the shells apart, which are to be so kept by the thumb of the left hand, whilst the round cartilage from near the open end is detached from the upper valve by a scraping movement of the knife, when the upper shell can be torn off, and the mussel be easily freed with the knife from the lower shell. It is a good bait for nearly all ground fish, and other kinds not unfrequently seize it. Mussels are found in most parts of the world.

THE LUG-WORM.
(*Arenicola piscatorum.*)

On many parts of the coast, in harbours, in coves, and between the rocks, you will frequently see small hillocks on the surface of the sand, and on close examination these will have the appearance of Worms, although composed of sand only; observe where the largest heaps are, and dig with a spade or three-pronged garden-fork, and you will find Worms four or five inches or more in length, and from the thickness of a swan-quill to that of the top of your little finger. (Fig. 51.)

They are good bait for all fish which feed at the bottom, but must never be cut, as they are full of blood, and of a soft substance like eggs or small mustard seed, which runs out im-

FIG. 51.—Lug-Worm.

mediately, leaving nothing but the empty skin, when the bait is of course spoiled. Be particular not to put in your box any pieces of this Worm, as they are not only useless as bait, but their blood will be sure to poison the rest. This bait should be used as soon after it is dug as possible, for it can rarely be kept after the second day. It shrinks up and becomes putrid. I have used it, whiffing for Pollack, baited as at p. 85, fig. 32.

THE RAG-WORM OR ROCK-WORM, ALSO CALLED THE MUD-WORM.
(*Nereis.*)

Of this Worm it appears there are two varieties: one inhabiting mud-banks, and rarely exceeding the length of three inches; another, attaining the length of six or seven inches, and found under stones overlaying sand, clay, and gravel, also in the cracks of rocks, and sometimes hiding under the tail of the Soldier or Hermit Crab, which has its abode in a Whelk-shell.

Many a ragged little urchin gets a considerable part of his living by procuring these Worms, digging them out of the

THE RAG-WORM OR ROCK-WORM. 189

greasy black mud at low water with his hands, and preserving them carefully in tray-like boxes 2 feet long by 1 broad, pitched at the seams to render them water-tight, in which boxes they are carefully tended with clean salt-water once or twice a day, and every particle of filth removed, together with such as may have been wounded in gathering, as their blood would kill the whole stock if allowed to remain.

When clean, these Worms are of flat form, as shown in fig. 21 (p. 68), and have a number of very short legs along their sides, giving them a serrated or saw-like appearance; they are almost always in motion, and are of a pale pink or salmon colour, some inclining to brown; all sea-fish, as well as Fresh-water Eels, greedily devour them. They are sold at Plymouth in small measures about the size of an egg-cup, at 1d. a measure, and three or four pennyworth are generally sufficient for a day's fishing for Whiting-Pollack, and should be kept in a box of wood 2½ inches deep, 1 foot in length, and 8 inches in breadth; having a cover, for if kept in a small box, heaped on each other, they soon die. Take care to place them in the shade, for if the sun shines on the cover of the box, they become sickly; also when you return from fishing, put them into a large box or tray, and never mix a fresh and stale lot together. These Worms are worth taking care of, as they are a choice bait.

The larger Rag-Worms are found by digging in stony ground, overlaying clay, sand, or gravel, as before mentioned, and are to be kept in sand nearly dry, or in the leaves of the sea-lettuce, which is found plentifully in harbours and sheltered coves in the summer and autumn. Put them in a box with this weed, as between the leaves of a book, and they will live several days. If you have any broken pieces, place them in a box by themselves, and use them first, or, as before observed, they will poison the others. In Yorkshire it is known as the 'Thirsk.'

THE WHITE SAND-WORM.
(Nereis versicolor).

This Worm is found in oozy sand in bays and harbours, in soil to all appearance similar to that inhabited by the Lug, yet but few Lugs are found with them, as they mostly live apart from each other. They are particularly useful for Atherine or Sand-Smelt fishing, but are taken also by other fish. Half an inch of this Worm is better than a larger piece for Atherine. A white Worm is used for Whiting-catching in Wales, identical, I apprehend, with this. It grows from 3 to 7 inches in length. It will live a week or ten days in salt water changed daily. A prejudice is held against this Worm by some fishermen, but Pollack, Bream, Atherine, Wrasse (and probably other fish), will all feed on it. At Dawlish, in Devon, no other bait is used when whiffing for Pollack, and it is baited as shown at page 85.

THE VARM OR SEA TAPE-WORM

is a very large flat Worm, and an excellent bait for Whiting, Pout, Bream, &c. When broken, always put the heads and tails in dry sand and in different boxes. (See also p. 100.)

THE EARTH, LOB, OR DEW-WORM.
(*Vermis terrestris*).

The use of this Worm for Pollack and Mackerel has been described at p. 85, fig. 32. To procure a quantity, search on grass-plots or flower-beds with a lantern on a wet evening or during a heavy dew, also in a paved court after sunset. Place them in a tub with some earth and plenty of moss on the top; they will live a long time if you examine them occasionally, and pick out those which may be dead or sickly. They are very useful where Rag-Worms are not obtainable, which is the case on large sections of the coast.

THE SAND-EEL AND LAUNCE.
(*Ammodytes tobianus* and *Ammodytes lancea.*)

These silvery little fish, of eel-like form, are very numerous on most sandy coasts, where they bury themselves during the receding tide, and whence they are frequently dug out in great numbers. The method of using them alive has been so fully entered into under 'Pollack' and 'Mackerel,' &c., that it is unnecessary to repeat it here. These little fish have a wide range, being found all round our coasts, those of France, Holland, Heligoland, Norway, North America, and possibly elsewhere in congenial situations. Their capture with the seine is fully described at pp. 74 and 228. Sometimes an iron rake is used to take them in loose sand, at others a small hook like a sickle or reap-hook, with a very blunt but jagged edge, that it may hold them without cutting them in two, which it will most assuredly do if the edge be at all thin or sharp. During moonlight nights many 'Launcing parties,' as they are called, are made for a visit to the sands at low spring tides, in summer and autumn, and sometimes quite a '*Saturnalia*' is held. In the island of Jersey, from the abundance of the Sand-Eel, one of the beaches has received the name of 'Grève au Lançon,' that is, 'Sand-Eel Beach.'

Almost all sea-fish devour them greedily.

Dipping Sand-Eels on the Surface.—At times the Sand-Eels collect in large shoals, and if discovered by the porpoises become so bewildered as apparently to lose all power of escape, either from the porpoises below or the gulls above, the former diving through them and munching them by mouthfuls, the latter dipping down and picking up the little silvery creatures with amazing rapidity. In order to avail themselves of such opportunities, fishermen provide a small meshed landing-net like a pool Shrimp-net, and, sculling up to the shoal, dip the net full to the ring, thus getting a large supply with little trouble. Such opportunities are rare. The small silvery fish known as the Mackerel-Brit may be taken at times in the same manner.

FRESHWATER EELS.

(Anguillæ.)

Eels are, I believe, universal in temperate climates, in almost every brook, drain, or tidal pond, on the largest continents or the smallest islands, and as they are frequently used as baits for Whiting-Pollack, and are also taken by Mackerel and Bass, they must not be passed by unnoticed. Those from $4\frac{1}{2}$ to 6 inches in length are the best, and the brighter in colour the more attractive. An excellent method of procuring them is to throw a bundle of osiers or withes into a muddy pit in a brook, drain, or tidal pond, when, after it has remained a few days, you will generally find in it a number of small Eels, fit for your purpose. If you require any at a short notice, take a fine Shrimp-net, and look in a harbour, tidal river, or small brook, for flat stones of a moderate size, and take up the stone in the net, when you will frequently have one or two Eels with it. You may often also procure small Eels by turning up the stones in a small brook, and catching them with a three-pronged kitchen dinner-fork, or nipping them with a pair of old notched scissors.

THE LAMPERN OR LESSER LAMPREY.
(*Petromyzon fluviatilis.*)

The Lesser or River Lamprey is usually from 7 to 10 inches long, and is so called to distinguish it from the Lamprey Eel, which attains the size of three or four pounds. They are very numerous in many English rivers and small brooks, where, during March and April, I have found them twenty or more together sticking to a stone, like leeches, from which circumstance they derive their name. They are good bait for Whiting-Pollack, better than Eels, as they are very much brighter under the belly. (See p. 82, fig. 26.)

They are useful cut in two for night-lines for large Eels, as is also the Pride or Blind Lamprey, and large numbers are used for Turbot and Cod trots. The seven little holes like shot-holes are very remarkable. A Shrimp-net or fine landing-net is the best adapted for taking them, when they should be kept in a bait kettle with a large stone or two for them to suck, and sunk under water.

Lampreys are not nearly as tenacious of life as ordinary Eels. A regular fishery for Lamperns is followed at Teddington on Thames.

THE LIMPET.
(*Patella vulgata.*)

Limpets are so well known as scarcely to need description, and may be used as bait when nothing better can be had. The soft part should be cut off and put in the sun for an hour before fishing, if possible, and will become somewhat firmer than if used at once. Sea Bream will take it well, also Whiting-Pout, and if the fish are well on the feed they will also take the hard part, but this is not ordinarily the case. Garden Snails are sometimes used with success.

THE WHELK.
(*Buccinum undatum.*)

The Whelk is much used as bait for Cod, and is procured by varied modes of capture. There is a very considerable demand for it in the London market, and great quantities are disposed of, ready cooked, at the fish-stalls in the poorer neighbourhoods. At Margate &c. boats are specially fitted out for dredging Whelks, and they are also taken on trots or long lines without hooks, the bait a number of small Crabs strung by aid of a needle on a twine snooding 2 feet long, made fast to the main line at about fathom intervals. Another method is to set dip-nets as for Prawns, with fresh fish instead of stale for bait. They enter Crab and Lobster pots in great numbers when baited with pieces of fresh Skate. It is necessary to break the shell with a hammer to extract the Whelk. Horseflesh is much used as a bait for Whelks.

THE CUTTLE FISH.
(*Sepia.*)

The Cuttle Fish is often taken amongst other fish in the seine or trawl-net, and is a good bait for Bass, Cod, Conger, &c.; the flesh seems something in consistence between jelly and leather, very tough and of a beautiful pearly whiteness, and it is this toughness which makes it so useful a bait for Bass-fishing off a beach, when the lead must be cast with all one's force to get it as far seaward as possible, clear of the breakers. it has in its back a bone of a shield-like shape, often found cast up on the beach, which was formerly much used as an absorbent, and as tooth-powder when pounded. The head of this creature is divided at the extremity into eight projections or horns, from inside which hang two, six or eight times longer, and the whole of them have a number of circular tubercles, by help of which it clings to and sucks into its throat any unfortunate fish it may succeed in capturing, and proceeds to devour it by help of a horny, parrot-like beak placed at the entrance thereof. In its inside is a small bag filled with an ink-like liquid, which is its means of defence when attacked; this it vomits forth in a dark cloud, and blackens the water for some feet around it. This liquid was used for writing by the ancients, and it is believed to form the chief ingredient in the Indian ink used by artists, as a very large kind is found in the Eastern seas. Clean these fish by pulling off the head, and splitting them sideways, remove the skin, backbone, and ink-bag, and wash them in salt water. They should be cleaned as soon as dead, and if put in a cool place will keep a day or two; sometimes they are salted, but are certainly not as good as fresh. Sausages were made from them by the Greeks and Romans, and they are eaten at the present day in some parts of the kingdom and on the Continent.

THE SQUID.
(*Loligo vulgaris.*)

This kind is much more numerous than the first named, and they are found in large shoals. The body is of a somewhat cylindrical shape, semi-transparent, and of a greenish hue whilst alive, changing to speckled brown, and the bone long, thin, and more transparent than thinly-scraped horn, but equably flexible. From the resemblance of this bone to a quill-pen, the Squid has been called the pen-and-ink fish, the ink being contained in a bag in the interior of the body.

It is better bait than the large Cuttle for Conger, Cod, or Bass, as it is not so hard and quite as tough. The Squid are often taken for bait in the following manner: Take half a Gar-fish or Long-Nose, or for want of it any small fish, and lower it to within a few feet of the bottom by a fishing line: if there are any about, they will at once seize upon it, when you must draw them steadily to the surface, and being before provided with a stick 6 feet in length with four hooks No. 7 (fig. 64c, p. 213) lashed on the end back to back, hook the fish near the tail if possible, and with the same stroke drag it under water, by which means you will escape the shower of ink which they almost always vomit forth at such times.

Nearly all the barb of the hooks should be filed off, or you will find it difficult to unhook them. Hooks without barbs are specially made for catching Squid for the Newfoundland Cod Fishery. Both in Spain and Newfoundland, Squid are taken in large quantities by a jigger, made of pewter, having a dozen or more hooks cast into it grapnel fashion at one end. The piece of pewter is about $3\frac{1}{2}$ inches long, with a hole at the upper end to attach the line. In the dusk of the evening, the jigger is lowered over the side of the boat, and jigged up and down. It is scraped bright to attract the Squid, which embrace it with their arms, and are then caught by the hooks. A Spanish fisherman, some years since, took a quantity of Squid in this manner at Plymouth, and it is a method which might be widely introduced to procure Squid for bait. Several of these jiggers were in the Exhibition of 1883. There is a smaller kind also not so frequently seen, with a short rounded body, known as the Sepiola or little Squid, and another the Flying Squid (*Ommastrephes*), so called from the fact of rising out of the sea and sometimes falling on the deck of a vessel. A piece of Squid $2\frac{1}{2}$ inches long, cut tapering, is a good whiffing bait for Pollack and Bass. The Squid attains a monstrous size at Newfoundland.

THE SUCKER OR POULPE.
(*Octopus vulgaris.*)

The Sucker is the most hideous of its kind, consisting of nothing but a head with eight arms and large staring eyes; they are often found under rocks and stones at low water, whence they are drawn out by iron hooks, to be used as bait for Conger.

This species is much more abundant on the French side of the Channel; in Guernsey it is known under the appellation of 'Pieuvre,' in Normandy as 'Minaur,' and has obtained a world-wide notoriety through the work of M. Victor Hugo, 'The Toilers of the Sea.' It is widely spread in the seas of the world, and is found in Australia, New Zealand, the Mediterranean, British Columbia, &c., and varies in diameter when its legs are spread out from 8 inches to 10 feet. Its flesh is harder than the preceding kinds. Crabs and Lobsters are its constant prey, and many shells are found at the entrance of its haunt under a stone or hole in a rock. In Guernsey to disable it they turn its cap inside out, and carry it on a stick after capture. Large numbers are used by the French fishermen for Conger trots, and about 5 inches of a horn of this creature skinned will answer as a whiffing bait for Whiting-Pollack and Bass. (See cut of Eel bait, fig. 25, p. 82.) It is excellent for Cod-fishing, and as a bait for Snappers on the Australian coasts. In British Columbia and Jersey they are roasted and eaten, and are cooked in various ways on the Continent.

SHRIMPS AND PRAWNS.

Shrimps and Prawns are often used as bait when alive for Whiting-Pollack, and dead for Mullet and Smelts. Boiled Shrimps peeled will take Eels, Dabs, and Flounders, in most pier harbours on the coast, when used as bait for fine tackle with the rod. Large quantities are taken by trawl-nets in Boston Deeps, on the Lincolnshire coast, and in the mouth of the Thames, &c., for the London market, and the method of catching them with hand-nets in sand and rock pools is known to every sea-side visitor.

For hand-nets, see fig. 74, p. 243, and following pages.

Small traps or pots are much used in Dorsetshire and Hampshire, which are very similar to Crab-pots; and in Devonshire they are taken in hoop-nets baited with stale fish, at night, or during the day, after the water has been discoloured by a storm. (See p. 246.)

THE COMMON GREEN OR SHORE CRAB.
(Carcinus Mœnas.)

This small Crab is found in great numbers in all harbours as well as on the open coast, and is an excellent bait for Flounders, Freshwater Eels, and various sea-fish, as Bass

Bream, &c., when about to cast its shell, in which state some are to be found the whole year.

To obtain these soft Crabs, see p. 120.

In their ordinary state they are much used to bait Lobster-pots, the back shell having been first removed, and many Wrasse or Rock-fish, otherwise called Conners or Curners, are frequently taken in the pots at such times.

Green Crabs may be caught in any quantity in hoop-nets baited with a bit of meat or any garbage. When two crabs are found under stones in company, the lower one is fit for bait.

THE HERMIT OR SOLDIER CRAB.
(Pagurus Bernhardus.)

This curious animal, having no shell to protect the tail part of its body, takes up its abode in that of a Whelk.

This tail part is a good bait for Whiting-Pout, Cod, Haddock, &c., and must always be put on whole, or it is spoiled.

Soldier Crabs are frequently found in Crab-pots, and where they abound may be taken in hoop-nets baited with a piece of any fish. (See also fig. 76, p. 246.) Numbers are caught while trawling and dredging.

A large Rag-Worm is sometimes found living in company with this Crab in the tail part of the shell.

THE SOLEN OR RAZOR FISH.

The empty shells of this fish are constantly met with by seaside visitors on every sandy beach. They are both eaten and used as bait for ground-fishing, and are to be procured by the aid of the spear (fig. 52, p. 198), and described by H. K. in No. 1053 of the 'Field.' This consists of a piece of iron wire about one-sixth of an inch in thickness and $2\frac{1}{2}$ ft. in length; one end of this is heated in the fire and then beaten out flat to the thickness of about $\frac{1}{20}$ of an inch for 2 inches of its length, and then with a file a triangular head is cut on the flattened part, projecting equally on both sides, and about half an inch in width (see figure); the other end is then firmly fixed into a small cylindrical piece of some hard wood, such as ash, about 5 inches long and one in diameter,

THE SOLEN OR RAZOR FISH.

to serve as a handle. To use this 'spear,' the point should be inserted in the hole (left by the Razor as it descends in the sand), which should be very gently probed in all directions, the flat head of the spear being held with one edge uppermost till the direction of the hole is ascertained, when it should be allowed to run down to the end; then, by a turn of the wrist, the flat of the spear should at once be brought at right angles to its former position, and the spear immediately be withdrawn steadily from the hole, when, if the operation has been properly performed, the Razor will be found on the end of it, and can easily be removed. The best time is at low water of spring tides, and the farther from the shore the more abundant the Razors generally are. By this means I have frequently obtained more than 300 in less than an hour to be used as bait. Some little practice is necessary to acquire the knack of inserting the point of the spear at the proper angle, which always descends in a slanting direction in the case of the common species. In another species it descends vertically. A steady firm pull (not too quick) is necessary to draw them out of the sand, to which they firmly hold by their foot. If not drawn up at once, they obtain such a hold that it then requires very great force to dislodge them; so much so, that frequently in this case the edge of the spear-head will cut through the shell, and the spear-head will be drawn out without the Razor. When a strong wind has been blowing, I have frequently seen numbers of Razors with about one-half or two-thirds of their shells protruded above the surface of the sand, and at such times many may be caught by the hands alone; but it requires some force to draw them out of the sand, to which they firmly hold by their foot, which they expand in the hole. Another correspondent (A Ballybrack Boy) advises running backwards along the edge of the ebbing tide (rather awkward), and when a Razor fish is pressed and

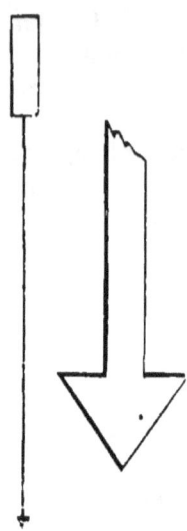

FIG. 52.

the foot removed, it spouts up water, showing you where to thrust down your spear. Another method of taking these fish consists in throwing a large pinch of salt into the hole made by the fish, which immediately comes up partially above the surface of the sand, and must be quickly secured. The holes left by the Razor Fish in the sand are very like a key hole. Take about a pound of salt with you, and dropping a good pinch into the holes of five or six at a time, pour a little water with your hand on top of it. This washes the salt down on to the Razor Fish, which at once rises an inch or more above the sand, so as to enable you easily to grasp it and draw it up. Dozens may be thus taken for food or bait.

Fish Baits.

No class of creatures prey more upon each other than fish. Pieces of Mackerel, Pilchards, Herrings, Gar-fish, &c., are therefore extensively used for Mackerel, Whiting, Hake, Bream, and Conger, &c., as already described under these several heads.

Artificial Baits.

The Spoon-Bait is widely used. In addition to these, I may mention the India-rubber band and Captain Tom's Spinning Sand-eels and Lug-worms, as well as the Silver-spinners, by several makers. See p. 83, figs. 27, 28, and 29.

Use none but brass swivels for sea-work, as those made of steel corrode very quickly.

Knots, Splices, and Bends.

The Overhand or **Common Knot** (fig. 53).—This is useful on the end of a line to prevent unravelling, and I recommend that one should be made on the end of the sidstrap of ground-lines over which the snood is to be looped.

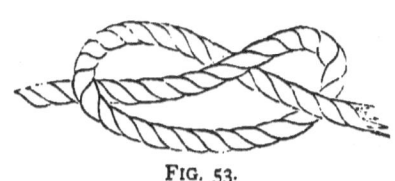

FIG. 53.

Regarding these matters, if in a difficulty consult some old seaman.

A Bend (fig. 54).—The method of connecting two ropes or lines, and additional security is obtained by passing the end round the bow and underneath its own part a second time.

FIG. 54.

For facilitating the joining and casting-off of seine-net ropes, &c., an eye or loop is often spliced in one of the ends.

The Bowline Knot (fig. 55).—One of the most important knots to fishermen and sailors, and particularly useful to amateurs whenever it is requisite to make fast a boat to a ring likely to be submerged on a rising tide, as the lower part of the knot in the engraving may be made sufficiently long to reach the level of high water if desired, when the knot being cast off the end of the rope may be hauled through the ring, and the boat set at liberty.

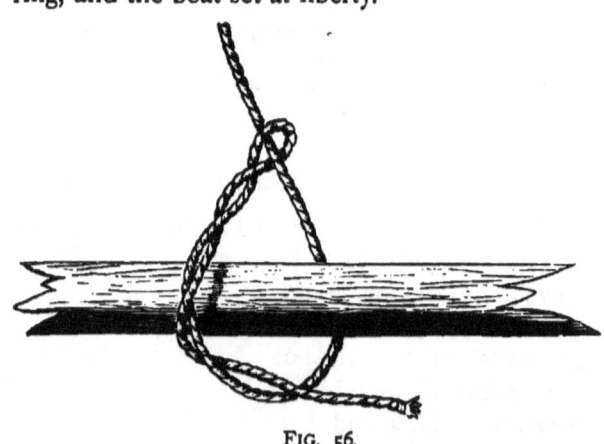

FIG. 55.   FIG. 56.

The Timber Hitch (fig. 56).—A speedy method of securing a piece of timber, spar, &c., whilst afloat, and very useful also in attaching a stone to a bulter, or for the purpose of mooring a boat in place of an anchor on very foul rocky ground, with an additional hitch at its side.

The Killick, or **Sling-stone** (fig. 57), does the duty of an anchor on rocky or rough ground, the loss of which it often obviates.

The Yoke Anchor.—This is also known by the appellation of the 'killick' in many places, and consists of a triangular frame of wood, enclosing a heavy stone. The piece of wood at the bottom, forming the base of the triangle, projects beyond the legs about 5 inches, and the ends act much in the same way as the flukes of an anchor. The legs pass through two holes bored through this piece of wood, which forms the base of the triangle, and a pin through each prevents their slipping out again.

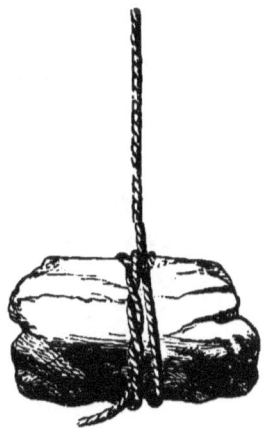

FIG. 57.—Stone Killick.

This holds better than a stone on a mixed bottom, although it is not equal to an anchor on smooth ground. It is generally used on a rough bottom, and a short piece of half-worn rope should be spliced into the fork of the triangle, and the cable be bent on to this. If this yoke anchor gets fast in the bottom, a heavy strain will break the weak piece of rope, and thus the cable will be saved. The cost of making one of these yoke anchors is very trifling, and the risking of a real anchor is avoided. The stone here shown should also always be slung with a weak piece of rope, and the cable be bent on to it, to avoid risk. Varieties of the wooden anchor combined with a stone appear to have been in use by maritime nations from remote periods. In Norway a wooden cross is used, from the arms of which four legs rise, and are secured together, at a height of about 2 feet 6 inches, enclosing a stone of a sugar-loaf shape to give sufficient weight. This will hold well upon soft ground, and acts like a mushroom anchor, which are used for mooring the lightships. These contrivances may be looked on as the parents of the iron anchor, and doubtless originated at a very remote period, far beyond the reach of history.

An Eye-Splice.—A loop or eye on the end of a rope is made by untwisting three or four inches of the line or rope, and having opened the twist or lay with a marline-spike, pricker, or stiletto (best adapted to fishing-line), inserting the middle strand first and then the others successively in the openings to be made between the strands. This being repeated, the ends are to be drawn tight, the splice to be beaten or rolled on the floor, and the ends cut off not quite close until the splice has taken some strain in use, after which it will never draw.

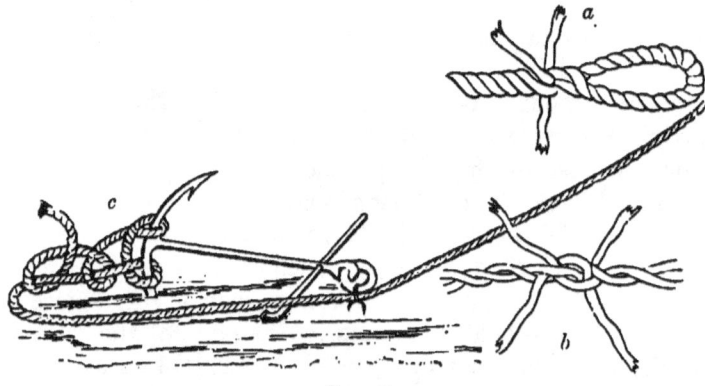

FIG. 58.

a, Eye-Splice commenced; *b*, Short Two-strand Splice; *c*, Anchor scowed.

An eye-splice with a line or two strands is even more simple than the above, and is formed by merely crossing the ends through the openings in the strands.

The commencement of an eye-splice with the ends once through is shown (fig. 58 *a*).

A Short Splice with a Two-strand Line (fig. 58 *b*).—Very useful in joining the two-strand snooding Mackerel-lines, and formed by untwisting a couple of inches of each end and interlocking them with each other, after which the strands are to be opened, and the ends of each to be crossed three times in the apertures.

Crowning or Scowing the Anchor (fig. 58 *c*).—An excellent plan where the bottom is doubtful or known to consist of sand or gravel with rocks interspersed. It is performed by making fast the cable where the shank joins the arm, and stopping it down to the ring with a single rope yarn, or piece of twine, for if the anchor gets fast in a rock you will probably clear it, as the yarn or twine will break and the anchor be lifted by the whole strain coming suddenly on the crown.

Slipping the Cable.—As it sometimes happens that the particular fishing-ground you desire to visit is directly in the track of passing shipping, it is requisite on the approach of any vessel to be ready to remove out of the track more quickly than raising the anchor will admit of. I therefore recommend some kind of a buoy, such as a stump end of a mast, or broken oar, or a small bag of cork, always to be kept bent on to the inner end of the cable, which being thrown overboard, the boat is at once set at liberty, and all risk of collision may be avoided by backing astern by a few strokes of the oars. The vessel having passed, the buoy may be picked up, and the position at once resumed. In connection with this, I may as well draw the attention of my readers to the

Belaying Thwart.—The Belaying Thwart is a short piece of rough inch-and-one-eighth plank, fixed a foot from the stem of the boat, and provided with two strong belaying pins and a mast-clamp. One of these pins is useful for belaying the cable, which can be instantly cast off and slipped if required on any sudden emergency; the other serves for the halliards when the mast is stepped forward for sailing short distances. A ·ring for the bowsprit is also fixed thereto. Many accidents,

FIG. 59.

some fatal, have happened for want of this simple arrangement.

The Anchor Bend.—The method of joining anchor and cable by passing the cable end twice through the ring and then securing it, as shown in fig 59, which appears sufficiently to explain itself.

If the anchor is kept continually bent, the end is usually attached to the cable by a seizing.

Hooks.

No part of the fisherman's outfit is of more importance, and no part has so little attention paid to it, as the Hook. I am not going into a minute description of the method of making them; I will merely state that the best are made of cast steel drawn into wire, which is cut into lengths, the barbs raised with a lever knife, and the points filed to shape; they are then bent into proper form round a mould, the top of the shank flattened if for sea-fishing, or filed taper for fly-fishing, hardened, tempered, and finished by tinning, black varnishing, or blueing. Good hooks are somewhat difficult to obtain; the makers will readily supply them, but dealers want a cheap article, and consequently every device is resorted to to save a penny or twopence in the hundred. Many hooks are only made of iron case-hardened, which are useless with heavy fish. There are hooks now sold which to appearance are excellent, of Limerick shape and good finish, but the metal is worthless, and they break with a very small strain.

While on the subject, I may mention that hooks for special purposes are sometimes left soft.

The fishermen at Dover, Folkestone, &c., lay long trots for Silver-Whiting: the ground is very foul, and the hooks used by them are made of soft iron wire tinned; if they catch the rocky ground, or are taken by rough fish such as Rays, they bend, but are afterwards easily put in shape again; at the same time they are quite strong enough to catch the Whiting, which seldom run there over a pound or pound and a half. These hooks are of French manufacture.

Whipping and Bending on Hooks.—There are several methods of fastening hooks on to the lengths of the different materials used for snoods. For freshwater fishing, the usual plan is to whip hook and gut together with fine silk well waxed; but for sea-fishing, snoods of hemp, flax, or silk are frequently employed, and these are tied or bent on the hooks as described below.

Within the last few years, however, the use of silkworm gut, either single or twisted, has been extended to sea-fishing, and it is now generally used in Mackerel and other fishing, where the bait is near the surface, when much depends on the tackle being fine. In deep water the snoods may be used stronger with advantage.

The snoods generally used by fishermen are of two sorts: one sold in small knots of about 10 yards each, which are cut into lengths for use: the other sort is made by doubling one or more strands of fine twine and twisting them together. The first are fastened on by taking two half hitches round the top of the hook, forming the common clove-hitch; these are pulled as tight as possible, and then a third half hitch is passed over the lower part of the hook, so that the line points upwards, as shown in fig. 60 A. The knot lies closer and neater if about an inch of the end of the snooding be unravelled and wetted before tying it to the hook.

The method adopted with the double snood is to pass the snood through the double end, thereby forming a loop; this is passed over the top of the hook, drawn tight, and a single half hitch made below it; this is an excellent fastening, being neat and strong (see fig. 60 B, p. 206). The best snoods for Pout, Whiting, and fish up to four or five pounds weight, are made of Shrewsbury thread (No. 25) doubled as above. No. 18 may be used by those who like tackle a little stronger: this in careful hands is equal to fish of eight or ten pounds weight. To make a 16-inch snood, take a piece of thread 42 inches long, tie the ends to two of the hooks of the twisting-machine (fig. 23, described at p. 70), hang the lead on to the middle of the thread, and turn the handle a few times round, with or against the sun, as may be necessary, taking care that the twist in the thread be increased and not diminished; then place a small piece of wood or wire between the two strands close to the lead and gradually raise it as the snood twists, turning the handle gently at the same time. Two or three trials will show the right amount of twist; too much should not be used, or the material will be so punished as to break with a very small strain. If Shrewsbury thread cannot be obtained, shoe-thread, or the sort called whitey-brown, will answer, but not so well as the first named, as they do not possess that amount of stiffness which is so desirable in all snoodings to prevent entanglement,

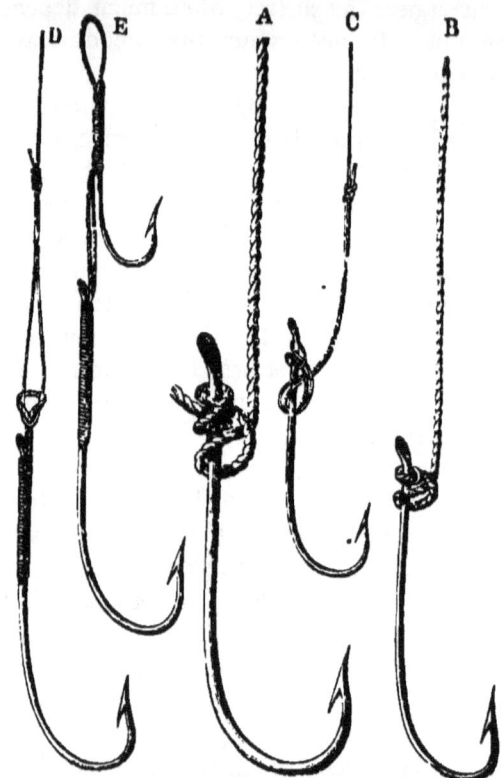

FIG. 60.—Whipping and Bending on Hooks.

to which limp snoodings when wet are particularly liable. Snooding can be made in six-feet lengths also, either with the twisting-machine or by the nossil-cock, or fisherman's spinner (fig. 66, described at p. 216), still very much in use at fishing villages along the coast.

In making these longer snoodings—or, as they are provincially named, *nossils*—two persons are necessary, as the machine, whether the twisting engine or the nossil-cock, must be fixed about 7 feet above the ground, and one should stand on a stool or chair (in using the jack), whilst the other attends to and regulates the twisting of the strands, as before described.

The shorter snoodings are used with the Kentish and Southampton fishing gear (figs. 2 a and 3, pp. 35, 38), the longer with the boat-shaped, Newfoundland, Dartmouth (fig. 4, p. 39), and Grapnel or Creeper Sinker Rigs (fig. 5, p. 40), these being the two classes of long and short snooded tackle or gear. The latter term 'gear' is that in universal use amongst sailors and sea fishermen, and to which I recommend all my readers to accustom themselves in conversation with men connected with the sea and its belongings, if they wish to be understood.

Hooks upon single gut are not whipped on as in freshwater fishing, as from the rough treatment they meet with the gut would soon chafe and break. One plan is to make a loop at the end of the gut; it is first soaked in warm water, then turned back about three inches and a common overhand knot made, but the loop is passed a second time through the ring of the knot before it is drawn tight. This is called the gut knot, and when properly made it looks something like the figure 8, and cripples the gut less than any other tie. It is then fastened by putting the end through the loop, drawing it tight, and making a half hitch below, exactly as with double snooding (see fig. 60 C); or it is looped on, as in fig. 60 D. In this a small loop of silk or flax snooding is whipped to the top of the hook, as described below, the gut loop is passed through the loop on the hook and over the lower part, forming a sort of hinge-joint. This is called the Nottingham bend, and has great advantages; one hook fitted in this manner will outlast ten tied on freshwater fashion. Fig. 60 E shows two hooks whipped to a double piece of fine snooding or strong silk; the gut is looped on Nottingham fashion. These hooks are used when baiting with small Eels, Lampreys, Mud, and Earthworms: they are sometimes whipped on to single gut, but are not nearly so durable, the mouth of most salt-water fish being lined with sharp teeth, most trying to fine tackle.

To whip a hook on to triple gut, wet the gut and make a

knot very near the end, take some fine sewing silk well waxed, place the end close to the knot, and take a few spiral turns up the gut, then whip downwards with close turns until about a quarter of an inch above the knot, place hook and gut together, and continue the whipping to the end of the gut. To finish off, lay the silk against the shank, leaving a loop which must be passed round the bend three or four times, so that the silk is between these turns and the hook, then pinch the hook between the finger and thumb, and draw the silk tight, when the end may be cut off. Fig. 61 shows this better than a long description.

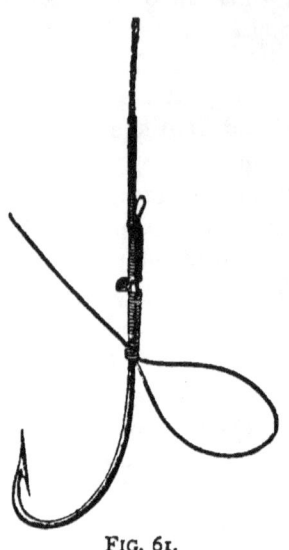

FIG. 61.

Gimp is fastened to hooks in the same manner, except that there is no knot made in it.

For general purposes those of round bend (pp. 210, 211) are commonly preferred, but for certain special kinds of fishing I make use of the Limerick. From 1 to 4 both exclusive for Conger and Cod ; 1 and 2 for hand lines ; 3 and 4 for trots and bulters ; 4 and 5 for ground fishing for Bass ; 6 for Ray fish ; 7 and 8 for Turbot trots ; 9, 10, and 11 for Mackerel, Whiting, or Pollack ; 12 for small Whiting, large Pout, and Plaice ; 13 and 14 for Dabs, Fluke, or Flounders, and Freshwater Eels, so often taken with the latter fish ; No. 7 Limerick for Mullet on the French side of the Channel, and also for Smelts (Atherine) ; but on the English coast No. 8 Limerick should be preferred for the Atherine or Sand Smelt, as they are generally much smaller than on the French coasts.

3/0 Limerick I use for fishing at anchor for Mackerel. The hook of a round bend or Kirby I consider excellent for flyfishing for Bass, finding them very strong and well pointed ; they have no flattened top, and are therefore well calculated for fly-making. I refer now to the second hook, marked 3/0.

The large outsized hook (partly enclosing those just described) is an excellent hook for Hake or Bonita-fishing, but for Albicore the hook should be certainly twice as large. Of late years Hake hooks are made eight inches long in the shank with an eye instead of a flattened top. The snooding thus escapes the sharp teeth of the Hake. Some of my readers may possibly think 4 and 5 unnecessarily large for ground-fishing for Bass, but I can assure them they are by no means too strong for this kind of work in a heavy surf or in the mouth of a bar-harbour, where you are almost sure to meet with them, as well as occasionally large Conger, if baiting with Squid or Cuttle-fish. A very great strain is thrown on the hook when a Bass of from ten to fourteen pounds weight is fast, the tide ebbing at the rate of perhaps six or seven miles an hour. In such situations the water is often coloured, which causes the Conger to feed; usually in shallow water they do not move much before sundown.

Redditch, in Worcestershire, is, with its surrounding district, the seat of the hook manufacture, with which is commonly associated that of needles.

The cuts of hooks in this work have been drawn and numbered from those commonly used by the fishermen in the West of England, and called round bend; but as there are several other patterns in great favour in different localities, the numbers of which vary very much, I have thought it best to construct a table showing the equivalent sizes of all the principal sorts.

As there is no recognised universal system observed in the numbering of hooks, much confusion often occurs in ordering them; the best method is to mention the number of 10ths of inches they measure across the bend. The Limerick and Kirby hooks of the two smallest sizes are of better quality wire than that used in the Exeter bend, and are therefore preferable.

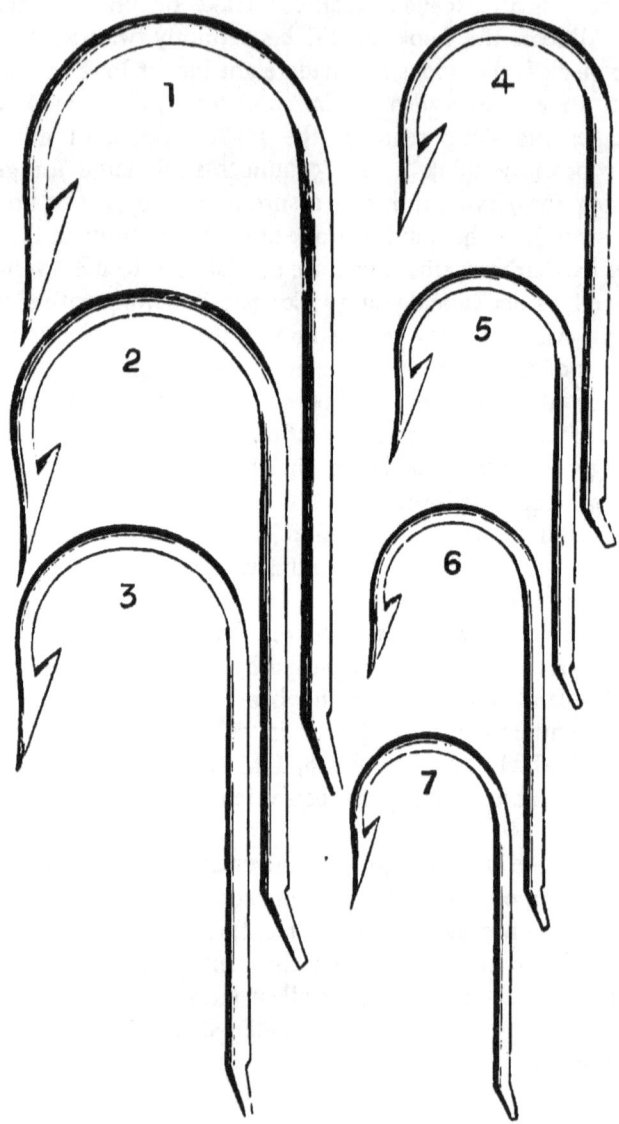

FIG. 62.—Exeter Round-bend Hooks.

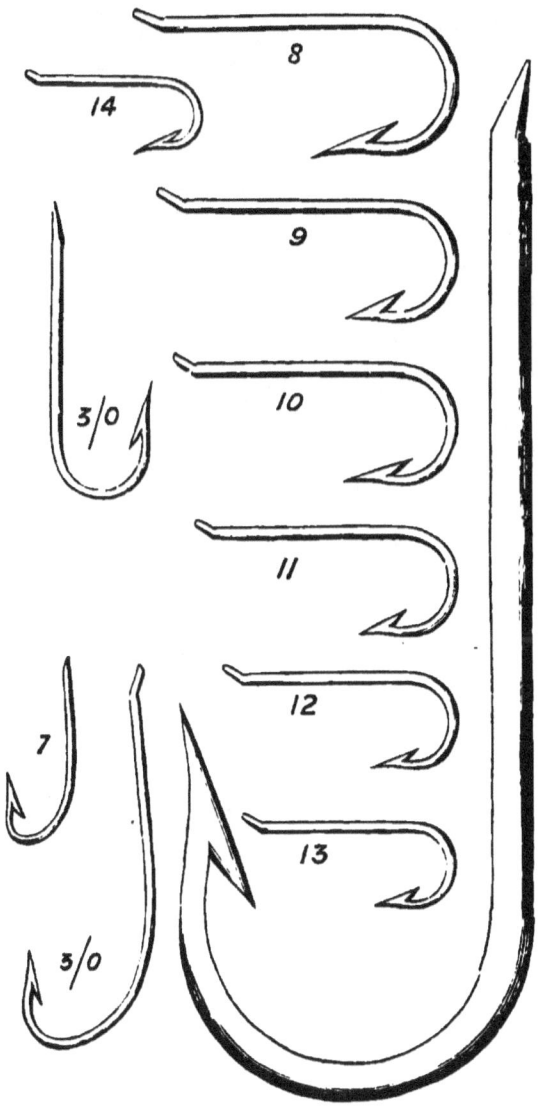

Fig. 63.—Exeter Round-bend Hooks.

TABLE OF HOOKS.

Cuts of hooks in 'Sea-Fisherman'	Walker's round-bend hooks	Kirby bend sea-hooks	Harwich sea-hooks	Norway sea-hooks	Kirby bend Whiting hooks	Kirby and Limerick river hooks	Size across bend in tenths of inches
No.	No.	No.	No.	No.	No.	No.	
—	—	—	—	13	—	—	17½
—	—	—	1	—	—	—	17
—	—	—	2	—	—	—	16
1	1	—	—	12	—	—	15
—	—	1	3	—	—	—	14
2	2	—	—	—	—	—	13½
—	—	2	—	11	—	—	13
—	3	—	4	—	—	—	12½
3	—	3	—	—	—	—	11¼
—	4	—	5	10	—	—	11
4	—	—	6	9	—	—	10
—	—	4	—	8	—	—	9½
5	5	—	—	—	—	—	9
6	—	5	—	—	—	—	8½
7	6	6	7	7	—	—	8
—	7	7	8	6	—	—	7
8	8	8	9	—	—	5/0 r	6½
—	9	—	—	5	1	4/0	6
9·10	10	9	10	—	2	3/0	5½
11	11	10	—	4	3	2/0	5
12	12	11	11	3	4	1/0	4½
13	13	12	12	2	5	·1	4
14	14	13	13	1	6	·2	3½
—	15	15	14	—	7	4	3
—	17	17	—	—	—	7	2½
—	19·20	19·20	—	—	—	9	2

REELS, GAFFS, AND BAIT TRAY.

The best materials for reels (fig. 64 *d*) are either teak, oak, or mahogany, with half a bottle-cork attached, to receive the hooks, as in the accompanying cut.

The most convenient sizes for portability will be for the horse-hair Pollack lines and the Bridport snooding Mackerel lines, six inches by five, and for the hemp lines for Whiting-fishing, eight inches by six; but if in the habit of using a large boat for the offing-fishing, of from say five to ten tons, your reels had better be eighteen inches by fourteen, as you will use a great length of line, and your lines will dry more quickly than if made up in a smaller compass.

Hair-lines are best kept in a net bag, the mesh sufficiently small to prevent the corners of the reels protruding. As the hemp lines are not commonly taken on shore from a large boat, good-sized reels will not be found productive of inconvenience; but in boats of moderate size, or for amateurs travelling, eight inches by six is quite large enough for packing comfortably. The large reels will be strong enough if made of deal.

Never hang your lines to dry on an iron nail, as they will be sure to rot at the rusty spot.

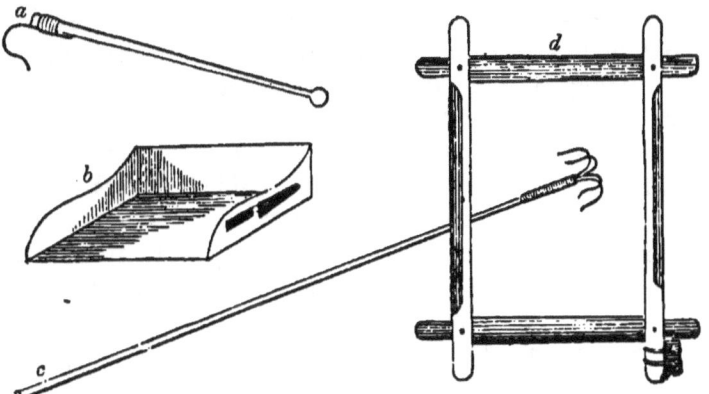

FIG. 64.—Reel, Gaffs, and Bait Tray.

Gaffs.—The long-handled gaff (fig. 64 c) is used for taking Squid, as described in the article on that fish, p. 194.

The Short-handled Gaff (fig. 64 a).—This is used to take on board any fish large enough to endanger the tackle or snooding, and for a seaside resident should be made by a blacksmith, $2\frac{1}{2}$ inches wide from the point to the shank, and securely clenched or riveted through a handle $1\frac{1}{2}$ inch in diameter at the hook, and tapering to 1 inch at the top, where a knob or head should be left to prevent it slipping through the hand in lifting in a large Conger or other monster of the deep.

Ash, black-thorn, or beech make a good handle, but a clean piece of red pine without knots may serve if the others are unprocurable; length from 2 to $2\frac{1}{2}$ feet.

For travellers, a couple of Hake hooks (the largest shown in the engraving of hooks, fig. 63,) will be found the most convenient, but be mindful of filing off the barbs before leaving home. One of these may with some waxed twine be speedily lashed on to any odd stick.

Landing-Net.—A landing-net is generally much more convenient than a gaff, particularly with fish of moderate size, yet too large to lift out by the line only. A good form for rough boat-work may be made with a piece of galvanised iron wire, the thickness of a pencil and 5 feet long. Five inches at each end should be turned at right angles, and the middle bent into a circle which will be about 16 inches in diameter. Bind each end separately with twine well waxed, then take a few turns round the two ends together, and then lash them securely with stout fishing-line to an ash stick 3 feet long. If tied on without this precaution, the ends of the iron will be always slipping and coming loose. The net itself should be made of salmon twine, on a mesh full two inches round, which will give knots one inch apart. Begin with 60 stitches, when about 20 inches deep reduce the number of stitches, by netting every 5th and 6th stitches in the row into one; this will bring the number to 50; net two plain rows, and in the next row net every 4th and 5th stitches together, and after two more plain rows every 3rd and 4th together; after one more plain row all the rest of the stitches may be drawn together and tied securely.

This net may be stiffened by soaking it before lashing to the ring in a mixture of one part coal-tar and two parts turpentine. A little of the same mixture should also be applied to the lashing on the handle.

The net is preferable to a gaff when two persons are in the boat, as one can assist the other, but if alone, the gaff is handier, as you can haul the line, and hold the gaff in the right hand at the same time as soon as the fish comes near the boat.

The Bait Tray (fig. 64 *b*).—This is a very useful adjunct to boat-fishing, and a great assistance in keeping a boat clean, especially when ground-fishing, as the bait, consisting of Mussels, Lug-Worms, or pieces of Mackerel, Pilchards, &c., can be conveniently kept thereon and cut up as required.

THE FISH BASKET.

A foot in length, and the same width as the seats round the stern of the boat, will be found the most convenient dimensions for a small boat of fifteen feet length and under, but for a larger craft a tray of eighteen inches in length will be more advantageous—in fact, if three or four are provided, so much the better for cleanliness.

On one side (the right) a sheath for a knife should be nailed, and on the other a second for a sharpening stone.

The tray should be of deal and painted, to prevent the slime penetrating the pores of the wood.

The Fish Basket (fig. 65).—These baskets are of an oblong form, and framed on a stout stick $1\frac{1}{2}$ inch in diameter, of unpeeled osiers, and protected on the bottom outside by four or five stout rods or bars projecting beyond the wickerwork; they are known in Guernsey as 'Paniers à coup,' and are provided with a piece of rope untwisted at its ends, and woven into the sides of the basket, by which it is carried over the left shoulder.

FIG. 65.

They are the most convenient baskets I have met with for boatfishing, as they stow well in the side of a boat, are not easily upset, and are withal comfortable to carry, even up a rugged cliff path. I am told that a similar kind of basket or creel is in use in the northern part of the kingdom. For an amateur, they might be made of peeled osiers, and painted straw colour if desired.

A useful size I find to be eighteen inches in length at the bottom and a foot in width, and on the top fifteen inches in length and nine inches in width ; depth of the basket, including the thickness of the back stick, eleven inches in the inside; length of the back stick, eighteen inches, through which holes are bored to receive the D-shaped rod which forms the edge of the basket, at right angles to which two other holes are bored to receive another rod of similar shape, forming the back of the basket. These baskets are not provided with covers, which are unnecessary for sea-fishing. Price, from 2s. 6d. to 3s.

The Nossil-Cock, or Fisherman's Spinner (fig. 66).—This simple little machine has been in use amongst fishermen from time immemorial, for the purpose of laying up their twine or shoe-thread snoodings for Whiting-catching, or those of silk for Mackerel, before gut had come into general adoption. It consists, as here shown, of an oblong frame of ash or other strong wood with four hooked spindles within it, which are made to revolve rapidly by the aid of an endless cord passing through each end of the frame, with a single turn round each of the spindles. Two nossils or snoods can, as shown in the cut, be spun at one time, but it cannot be used single-handed, as one man is required to pull on the endless cord, whilst the other,

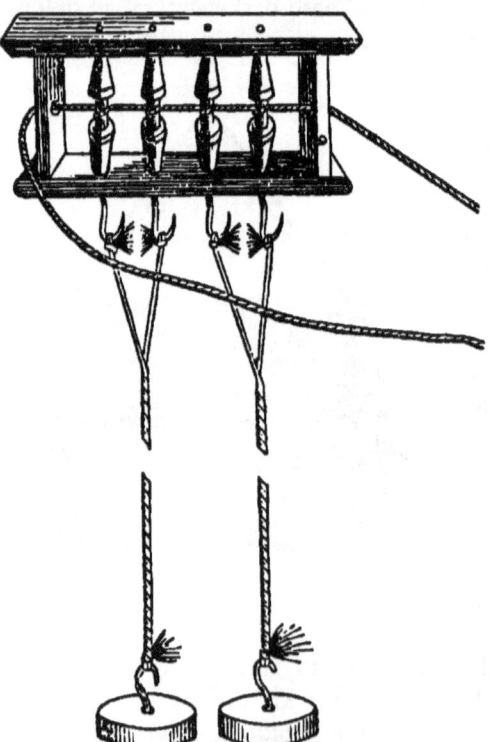

FIG. 66.
The Nossil-Cock, or Fisherman's Spinning Machine.

either with a piece of stick or his finger, prevents the too rapid twisting of the strands.

The man pulling on the cord ceases immediately he perceives the strands have twisted up as far as the spread of the hooks will allow, but the leads continue to revolve until the momentum is spent, when each man should seize one of the leads and rub the nossil down with a piece of upper leather provided for the purpose. The weight of these leads should be two pounds each. Should the man pulling on the cord not cease doing so when the strands have twisted as far as the spread of the hooks permits, the nossil will at once cut itself in two and its twist become irregular, and therefore useless. Previous to being used, these pieces of line are termed nossils, but after hooks are bent to them they are commonly termed sids or snoods, in the vernacular of Dorset, South Devon, and Cornwall.

These nossil-cocks are usually a foot long and about six inches in width, and the wood-work about one inch in thickness by two in width.

When in use, they are commonly fastened to a beam by two three-inch screws through the holes in each end of the frame. For amateurs' use, the little brass jacks, to be bought at most tackle shops, will be found more convenient, as they can be worked single-handed for spinning short snoods or horse-hair lines. I have, however, thought it best not to omit this primitive affair, as it may probably be useful to some emigrant, if this work should perchance fall into such hands.

In the chief seaports of the kingdom good snooding is to be purchased at so reasonable a price that it is simply absurd to spin it, but where it cannot be obtained the few hints I have given may,.I hope, be useful, as by following them it may be made out of either shoe-thread or ordinary thread at a short notice. The Shrewsbury thread spun double makes good snooding for Whiting and Pout-fishing on rough ground (Nos. 25 and 18—25 is the finer). To any intending emigrant taking an interest in fishing I strongly recommend the purchase of a small jack or twisting machine, the cost of which is only a few shillings.

The finer cotton lines make excellent snoods.

The Lester-Cock Trot.—The Lester-Cock is a piece of board about three feet long, a foot wide, and an inch in thickness, fitted with a mast and sail, which must be stepped one-third from the fore end; this being launched from the beach when the wind is off shore, drags after it the trot or long line with baited hooks at intervals, which will probably be taken by either Bass or Pollack. Best bait, Squid or Cuttle in long strips, or Mud-Worms; also living or dead Sand-Eels. This mode of fishing is much followed by the Greeks.

The Sunken Lester-Cock.—This is useful for ground fish, and is formed of two zinc cylinders eighteen inches in length and two and a half in diameter, connected by two slips of flat wood, about four inches from each end, to the middle of one of which a mast and sail are attached. Each of these cylinders has a small hole on the upper and under side, consequently the whole affair sinks after the cylinders become filled with water.

These sunken Lester-Cocks are useful for towing out a trot for ground fish, and will sink more rapidly by attaching a bag of sand (if necessary) to overcome the buoyancy of the slips of wood and mast.

The Otter.—The otter is a piece of light wood, of say two feet length, one foot depth, and an inch and a half thickness, provided with sufficient weight in the shape of a lead or iron keel to cause it to float perpendicularly, and is used to convey a trot-line seawards, but has this advantage over the two preceding methods, that it can be used without wind, provided you have some little extent of coast or beach available to walk along as you tow the otter.

To use a nautical term, it is slung on the 'sheer' by a double bridle, to which the end of the line is attached.

For an illustration, the reader is referred to the Otter Trawl (fig. 73, p. 240), which is provided with two of these so-called otters.

A great improvement on the above would be to fasten the line to the end of the otter itself, and afterwards connect it with the bridle with a piece of fine twine; when a fish is hooked you will by the act of striking the fish break the twine, and the strain coming on the end of the otter you will easily haul it on shore.

The Lester-Cocks and Otter are useful aids to a person unprovided with a boat; and it oftens happens there is excellent fishing where the shore is too wild or iron-bound to keep one.

The Sunken Lester-Cock is probably the most killing of the three, and once submerged may be left to itself for some time.

It has been used with much success in the Island of Alderney.

Neither of these methods is available if there be much floating weed, and recourse must then be had to leger-fishing, for which see under 'Bass,' p. 139.

Crabs, Lobsters, and Cray-fish.—These are taken in the beehive-shaped baskets or pots, which are some of the first objects usually meeting the eye on the shore of almost every watering-place or fishing village.

These pots are weighted with stones, and then being baited with pieces of Ray, or other fish suspended in the interior, are lowered by buoy-lines on or close to rocky ground in various depths and distances from the coast, sometimes even as far as ten miles from the land. On some parts where the bottom is not too rough a kind of drum net trap or pot is used, extended on a hooped framework. Crabs prefer fresh bait, the other crustacea, such as Lobsters, &c., stale.

Clothing.—All clothing should be of wool, or of material chiefly consisting thereof, as it absorbs perspiration and prevents catching cold.

A Crimean flannel shirt, coat, trowsers, and waistcoat of blue serge, angola or woollen stockings or socks, and early or late in the season, or whenever additional warmth is needed, a short Guernsey or Jersey frock is recommended.

All the above can be freed from salt-water, &c., by washing, and kept clean and sweet.

If caught in bad or wet weather, a suit of waterproof, with a sou'wester hat, will be found very comfortable, which should be large enough, especially the trowsers, to jump into them, so to speak.

Those who object to waterproof trowsers should wear high leggings or overalls, and a fisherman's waterproof petticoat, sufficiently long to shoot the wet clear of their knees when seated.

Stout half Wellington boots without nails will be found preferable to heavy sea-boots, as they leave your movements unembarrassed.

In warm weather, whilst fishing, a pair of shoes may be worn without stockings or waterproof clothing; but as soon as you have ceased 'shift your rig' for your dry clothing immediately.

Cork Seats.—A few cushions of painted canvas, stuffed with cork shavings, fifteen inches by eleven, or the same width as the boat's seats, and four inches thick, will be useful to sit on at times, and if edged with some stout Cod-line in loops will answer as life-preservers.

Ocean-fishing.

The following particulars on this branch of sea-fishing, communicated by a friend who has paid much attention to the subject during voyages extending over a long period, will probably not be unwelcome to individuals who for a time may have occasion to make their home upon the bosom of the 'vasty deep.'

Ocean-fishing has two great divisions, hooking and harpooning, the first being nothing more than whiffing or trailing on a large scale, the second consisting in striking the fish with the instrument known as the 'Grains,' or with the harpoon. The fish most frequently taken are Bonita, Dolphin, and Albicore, and the gear or tackle as follows: the hook for Bonita of the largest size shown in the cut of hooks, fig. 63, p. 211, attached to six feet of copper bell-wire, and a very stout Cod or log-line, the bait a piece of raw pork-skin cut of the shape shown in fig. 67, nearly twice the length of the hook, through which the hook is to be passed twice, and the top of the bait secured with a bit of twine round the wire at the head of the latter, so that it may not slip down on the bend, which would spoil the appearance of the bait in the water. The bait being split in two, and a narrow gore or slice cut out of the middle, the two tails will hang down beyond the hook and play in the water, which action of the bait may be further increased by jerking the line, and making it leap out of the water, when the fish will spring to catch it, and sometimes even take it in the air.

When the fish are playing across the bows of a vessel, a line is often dropped from the jib-boom with success, especially if a life-like motion be given to the bait by dipping it up and down, which indeed the action of the vessel is often sufficient to effect.

Albicore are taken with the same kind of gear and bait, but the hook must be two sizes larger; it is usual to strike very large fish of this kind with the grains or a very stout gaff, for many are lost by breaking the hook in the attempt to weigh them out of the water with the line alone.

Artificial baits are also sometimes used, namely, the spoon bait, made by cutting off the handle of an albata spoon and boring a hole at each end, and attaching a wire snood at one end and a very strong double hook at the other.

It would be quite worth while for any emigrant or other individual making a foreign voyage to get a few flies made Salmon-fashion on the largest-sized hooks figured in the engraving, for I have had such frequent success with flies for Pollack and Mackerel that I cannot but think that some gaudy construction of the kind would answer for ocean-fishing, particularly for the Bonita, as it is a large fish of the Mackerel family.

FIG. 67.—Pork-skin Bait for Ocean-fishing.

'**The Grains.**'—The instrument known as the Grains consists of five harpoons in one, which may be either used in a line with each other or may be unscrewed and arranged four-square, which is frequently preferred; it is attached to a stiff light ashen staff, with a ball of lead at the top, which gives force to the blow and turns the fish up when struck. The fish is hauled on board by a small but strong line bent on to the grains, one or two hands being ready, watching the actions of the striker.

The Bonita and Albicore are the chief fish taken by this method, although others are occasionally met with.

If a log of timber is found floating at sea and covered with barnacles, it is often surrounded with fish attracted by the various small Crabs &c. which also make it their habitation; if the weather be calm, a quantity of fish may be 'grained' by the aid of a boat.

The Triangle Net.—On such occasions quantities of fish may be taken by a bag net on a frame of three iron bars, 12 to 15 ft. long, lashed together in the form of a triangle, supported at two of the angles by small casks. This net is worked by a rope attached to a triple bridle, a part bent on to each angle of the frame, which, in a perpendicular position, is to be hauled under any floating piece of timber. A boat-load has often been caught at one haul.

Porpoises.

These are occasionally harpooned under the bows of sailing vessels.

Remarks on Nets.

There are several varieties of nets used in sea-fishing, of which the chief are Trammels, Seine, Trawl, and Drift-nets, the construction and working of which are so very different that a particular description of each must be given in order to afford anything like a true conception of the subject.

The Trammel.—The appellation of this net is doubtless of French origin, for '*trammel*' is evidently '*trois mailles*,' or three meshes, which exactly describes the net. (*Vide* ' Life in Normandy,' vol. i. p. 163.) It consists of a loose net of small meshes or sheeting, between two tighter nets of larger meshes called the walling.

It is thus made and of the following materials :—

The twine known as the Shrewsbury thread is found to answer better than any other that has been tried, as from its fineness and pliability it meshes the fish remarkably well, which a hard twine fails to do.

The size No. 18 is preferred in the island of Guernsey, and is manufactured by Messrs. Marshall & Co., Shrewsbury.

A flax-twine used in Kent for drift Sprat-nets has also been tried and approved of.

The size of the mesh should be three inches and one-eighth, and the depth forty-two meshes or eighty-four rows.

Supposing the net to be eighty fathoms in length, when mounted it must only occupy half this distance, namely forty fathoms, and on every length of six inches of the roping four meshes must be taken in the needle at once and secured.

On each side of this fine netting is fastened a net with meshes of eighteen inches in length when stretched tight, including three knots, and of Salmon twine : these large meshes are secured to the rope or rawling at distances of twelve inches on the head and foot ropes, and being only half the depth of the middle net leave a large amount of slack which allows the fish to pocket themselves, as shown in the second engraving.

Fig. 68, p. 224 represents a side view of the trammel when set in the water, or as it appears when hung up to dry upon a wall, the large meshes plainly visible crossing the smaller, which being slacker fall down to the foot-line. As there is a walling on both sides of the small mesh, the fish pocket themselves when swimming from either side. Plain sheet-nets are used in some localities, and are erroneously called 'trammels.' At others they are used with one walling only ; to be strictly a 'trammel,' there must be three nets side by side. At Poole the middle only is of three nets, the ends single sheets. Square wallings originated, I believe, in Spain, but are now frequently used in England.

The cork required for the head-line should be cut two inches square and three-quarters of an inch in thickness, from

THE TRAMMEL.

which cut off the corners, that they may not entangle in the meshes. These corks must be placed on the head-line at intervals of eighteen inches, and the pipe-leads abreast of them on the foot-line : the dimensions of these perforated leads are one and three-quarter inches in length and half an inch in diameter. These leads will be of the required weight to sink the corks, which will raise the net like a wall from the bottom

FIG. 68.—Side view of Trammel-net.

to the height the large meshes will allow, namely five feet three inches, which is found sufficient for the capture of Red Mullet, Soles, Plaice, Craw-fish, Crabs, Lobsters, &c. (See the article on 'Red Mullet,' in addition to the present remarks, p. 174).

The best time to set a trammel is just before sunset, as the fish mesh themselves more frequently on the approach of darkness, although some may be taken even in the daytime.

The most likely places to shoot a trammel are in the eddy

of a large rock, if there be sandy ground at the foot, also in sheltered bays and coves of the shore and deep-water harbours, and in fact anywhere, if the locality promises sufficient shelter to warrant leaving it moored without undue risk. On some parts of the coast where the rocks rise like a wall out of the water, although it may be impossible, or at least very difficult, to keep a boat, a trammel may be worked to advantage by taking the opportunity of a fine day to drop a heavy anchor, say of fifty-six pounds, at the distance of fifty fathoms from the rocks, attached to a piece of chain long enough to reach the perpendicular height of spring tides; to the end of this chain secure a six-inch block, and provide a rope of an inch and a half in circumference, long enough when doubled to reach a few feet above high-water mark on the rocks. By help of this rope you can haul out your net whenever you think fit, and haul it in also to examine it and unmesh the fish, and may thus supply yourself with choice fish, unprocurable by other means, supposing your locality is at a distance from a fish market.

To set a trammel from a boat the following arrangements are required: Two buoy-lines with corks at intervals and stones at the ends about twenty-five pounds weight must be provided, to one of which, close to the stone, make fast the head-line of the trammel, and at the breadth of the net, above the stone, make fast the cork-line, being careful not to stretch the trammel up too tight, lest the strain be taken by the net-work instead of the buoy-line, which, being stronger than the netting, ought to take the whole of the strain (fig. 69, p. 226).

Place the buoy-line carefully in a coil with the stone by itself in the middle of the boat, and proceed to drop the cork-line in the stern-sheets as near the stern as possible, but the lead-line should be in advance of it about three feet, when the slack net will naturally take its place between the two. When you have thus arranged the whole net, place the second buoy-line conveniently, together with the stone, on one side of the net, having first secured the head and foot-lines, as mentioned above, to the buoy-line, which you are now to throw overboard, and then proceed to lower the stone with care and deliberation to the bottom. It is always better that two should be in the boat on these occasions, as one can pull slowly whilst the other pays out the net; for one person to pay out the net conveniently, a moderate run of tide or light breeze to drift the boat is required.

A trammel should always be shot with, and not across, the

FIG. 69.—End view of Trammel-Net.

tide, for if the latter mode were adopted, the force of the current would tend to depress the net towards the ground, and thereby injure its efficiency. Before arranging the trammel in the boat, make a rigid examination of the gunwale, lest any nail, splinter, or other obstacle prevent the escape of the net, and thereby break the twine of the smaller meshes. I have sometimes found even the leather on the rowlocks objectionable on this account, to obviate which, fasten on with tacks about five feet of old cloth or canvas over the gunwale, which will effectually cover all inequalities.

A trammel of forty fathoms' length will be found quite large enough for general use, and if two of these nets be required I think they should not exceed thirty fathoms each, as they are then very convenient for river-fishing, and for sea-work a long net is at once made by joining the two together.

The price of a good trammel of the Shrewsbury thread, ready to put into the water, is about 1*l*. 5*s*. or 1*l*. 10*s*. per ten fathoms, but in France they are procurable at a lower price, yet of such inferior twine that the Guernsey fishermen who use trammels to a considerable extent, have nearly all discarded those of French manufacture.

The Shrewsbury twine is a patent material, and has been in use for the best quality of flue-nets (the fresh-water appellation of the trammel) for a long period, but its adoption as a material for sea-work is comparatively recent, having been introduced to the notice of the Guernsey fishermen many years since, by a gentleman visiting the island for the sake of sea-fishing.

Trammels and other nets should be spread on a clean shingle beach or grass field, or hoisted up to dry after using, and all weed picked carefully out; they should likewise be barked, in common with other nets, at least once a season.

All broken meshes should be at once repaired, as 'a stitch in time saves nine.'

Many yachts on coming to anchor of an evening in a roadstead set this net; it should, if shot at six or seven, be hauled at about half-past nine P.M.; it may then be shot again, and hauled at daylight. If left the whole time without examination, the fish will probably be devoured by Crabs, Squid, &c.; to which the Red Mullet generally are the first to fall victims.

The Seine.—The word 'seine' has been adopted from the French, and signifies any draught-net which forms a bag; they are made of various length, depth, and mesh, according to the purpose required, and are extensively used for Salmon, Mackerel, Pilchards, Gar-fish (otherwise Long-Noses), Smelts, Atherine, Mullet, Flat-fish, Herrings, Sprats, &c.

The 'seine' may be considered the most ancient description of net known, and the method of enclosing a space by shooting it in a semicircle, and then drawing it towards shore, seems naturally to suggest itself first to the mind of the fisherman.

In the miraculous draught of fishes on the lake of Gennesaret the words are, 'they *enclosed* a great multitude of fishes,' clearly evidencing the manner of using to be precisely similar to our own practice at the present day, and that the net in question was a 'seine' or draught-net. (Luke v. 1–11.)

Other instances from the Old Testament mention drag or draught-nets—or, in other words, 'seines'—from which, and various relics of antiquity, we may conclude it to be the source from which all other nets have sprung.

The seine consists of three divisions: the bunt, or centre, and the arms or two sides.

The bunt is much deeper than the arms, in order that it may form a considerable bag to receive the fish; and to render it still more capacious, the net, or calico, or canvas, or whatever material may be used to form the bunt or bag (for the terms are synonymous), is gathered in setting it on to the rope; and at each end of the net a pole or spar is attached, weighted with sheet lead at the larger end, which keeps these pole-staves (as they are termed) in a perpendicular position, thereby causing them to spread the net to the best advantage.

The Sand-Eel Seine (fig. 70).—The Sand-Eel or Launce, as elsewhere observed, is very numerous in the sandy bays and harbours of the French and British coasts, and being not only delicate eating, but without doubt also the best bait for sea-fish generally when used alive, a description of the seine for its capture will probably not be unwelcome to my readers. It should not be less than 20 fathoms or 40 yards long, the part marked in Italic capitals, *B, U, N, T,* forming the bag or bunt, consisting of unbleached calico 20 feet in length and 12 in depth, and the 20 feet length, when attached to the rope, must be gathered, that it may not occupy more than 16 feet of space, which will cause it to bag well.

Three gores of very fine netting, as fine as can be made, 18 inches in width, are to be inserted at equal distances from the edges of the calico and from each other, which gores are to reach to within 18 inches of the top and bottom of the calico.

The bunt of calico is preferred to one of netting, because it offers no apertures for the fish to get their heads fast, and the gores are inserted that the water may escape more freely, for it is found that the calico without gores is of exceedingly heavy draught, as might be expected.

As a difficulty sometimes occurs in procuring these gores of fine netting, a very coarse kind of canvas, of a light make, and very open between the threads, is substituted, and answers the purpose well at a less cost.

On each side of the calico three fathoms of one-inch meshed net is

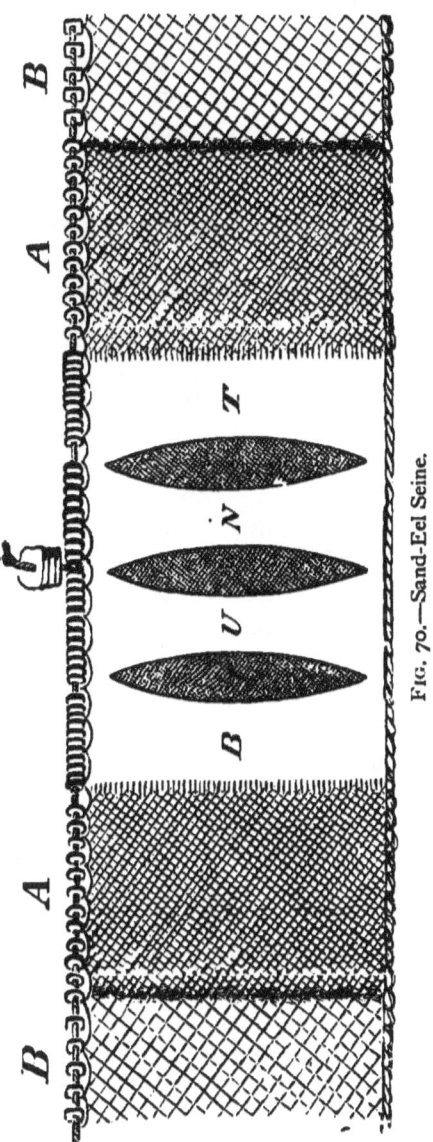

FIG. 70.—Sand-Eel Seine.

to be placed (marked A in the cut), which is to be gathered sufficiently to bag equally with the calico, and the remainder or arms of the net may consist of three-inch mesh (marked B), which may gradually diminish in depth from its commencement at the one-inch mesh, where it is eighteen feet in depth, until it reaches the pole-staff, where eight feet will be sufficient; but if the shore be very flat and shallow, the net need not exceed five feet at the end. A net of this description will cost about 12*l.*, and in the island of Guernsey is owned by several fishermen; or if by one only, other fishermen having the benefit of the net pay tenpence a week to the owner, each man attending to work the net when a supply of Sand-Eels is requisite, which is commonly every fourth or fifth day.

The importance of the Sand-Eel as a living bait cannot be too widely known, and I trust that ere long it will be as commonly used as on the coasts of the Channel Islands. The method has been introduced into England with complete success, and has reached the Scilly Islands.

Two boats are required in obtaining the Sand-Eels: one casts its anchor on the top of a rock out of the water and veers out a few fathoms of rope, whilst the other, having one end of the seine attached to the first boat by a rope, proceeds to shoot the net in a semicircle in the eddy of the rock, and casting anchor hauls in about half the rope, when the men get up their anchor, and going on board the first boat haul the seine alongside, keeping the boat broadside on to the net, whilst one of them continually thrusts down an oar to frighten the fish back towards the bunt, that they may not escape under the boat's bottom, the only means now left open to them.

The fish being hauled alongside in the calico bunt are now dipped up in buckets or hand-nets, and poured into the 'courges' or baskets, to be kept until wanted (for which see figs. 19 and 20, pp. 66, 67).

The method used in England of hauling the seine on the beach cannot but be prejudicial to the Sand-Eels, as they are washed to and fro in the surf, and are more or less injured thereby, so that they are found not to live on the average more than half the time as compared with those taken in Guernsey. I refer more particularly to the south coast of Devon, where they are taken more as an article of food than bait, although they are used as a dead bait in that locality, and have been tried by some few amateurs as a living bait with much success. In procuring any from fishermen, endeavour to have them placed in your courge before the seine is drawn out of the water, and attend or send a man in a boat to fetch and tow the courge to a place of security—generally a boat's moorings.

The Night or **Small Seine.**—This, as implied by its appellation, is commonly used after sunset, and, as regards size and weight, may be carried in a boat under fifteen feet in length.

It is shot out in a semicircle and hauled on to the beach, and may be managed by four men.

Most of the seines made for the use of amateurs or for use by yachts' crews, have hitherto consisted of hemp twine, and have been much too stout and heavy. I suggest for the future they should be entirely of cotton, as being much lighter and easy to work and stow away. A generally useful net is 60 fathoms long, 15 feet deep in the bunt, and 8 feet at the ends of the arms, the mesh of course increasing in size gradually as the distance increases from the bunt or middle.

In or near the mouth of a tidal river, if the bottom be gravel or sand unencumbered with large stones, it is often very successful, and the take commonly consists of Plaice, Flounders, Salmon Peel, an occasional Salmon, Sea Trout, and Red and Grey Mullet, hundreds of which latter often escape by leaping over the cork-line.

In hauling this net much care should be taken to keep the bunt-cork in the middle, as this shows the net is being equally hauled; the foot-lines must also be kept close to the ground, especially as the bunt comes near the shore, for then the Mullet &c. will be almost certain to dash towards the beach, seeking some available outlet.

If the water be shallow, it will be well to hold up the cork-line if you have additional help sufficient to spare for that purpose. From low-water to half-flood is commonly the most successful time, particularly when the tide suits after sunset. In many creeks a trammel may be used as an auxiliary net to a seine for Mullet as a stop-net.

The Mackerel Seine.—This differs in nothing from the preceding, except in size of mesh and length and depth of the net, and a boat rowing not less than four, and sometimes five or six oars, is used to work it.

In favourable localities the net is hauled on to the beach, but where no such convenience exists it is drawn up into the boat until the fish are brought alongside, which are dipped up in maunds or baskets and emptied on board, which latter method is termed 'tucking,' as distinguished from 'seining,' when the net is hauled on the beach.

This term 'tucking' appears to be taken from the 'tuck-net,' a small net used to empty a Pilchard seine, which is an immense net used only to enclose the shoal or school of fish, as hereafter described.

The Mackerel approach the shores during the summer months, and a constant look-out is kept for them by some experienced hand noted for his keenness of sight, who is also generally provided with a telescope.

No fish is more fitful or uncertain in its movements than the Mackerel, as they suddenly appear on the surface where least expected, in pursuit of the 'Brit' or Mackerel-bait, which are either the young of the Herring or some other member of the family of *Clupeidæ*; they also pursue the Sand-Eels or Launce with like avidity, and unless their movements be closely watched a very poor season will be the result.

The capricious nature of this fish is well conveyed in a West country proverb, 'If you'd Mackerel catch, you must Mackerel watch'—a maxim evidently well understood and acted on by the fishermen, both as regards their capture with the seine and with hook and line.

The seine Mackerel fishery is by no means confined to professional fishermen on the SW. coast of England, many tradesmen and mechanics being part owners both of boats and nets, and assisting to work them when the Mackerel 'play' or show themselves on the surface. This leads me to digress,

THE MACKEREL SEINE.

whilst I endeavour to describe one of these occasions, to many of which I have been witness.

It is not always the custom to keep the Mackerel boats afloat for many consecutive hours, even during the season, unless the fish are unusually abundant, as it would interfere needlessly with the ordinary avocations of the owners ; they are, however, kept perfectly ready for launching, the net carefully disposed in the stern compartment, partitioned off for the purpose, and the oars crossed in their respective rowlocks, the boat's bow pointing seaward, ready to launch instantly on notice from the look-out from the neighbouring cliff, whose messenger we will suppose to have arrived at the top of his speed, shouting as he runs, with the lungs of a Stentor, 'The Mackul plays, the Mackul plays ! Hooraw ! hooraw ! the Mackul plays !'

The effect is perfectly electrical : the mason and his labourer cast aside trowel and hod, the tailor his shears and thimble, the shoemaker his last, and each and all join in shouting, as they rush to the beach, 'The Mackul plays ! Hooraw, the Mackul plays !'—in happy ignorance alike of Carpenter and Lindley Murray, or the evident concern depicted on the countenance of sea-side visitors, who, hearing the cry for the first time, anxiously inquire the cause of the commotion ; but on being informed also rush to the beach to see the fun, and not unfrequently to lend a hand in hauling the net on shore.

If the fish are still playing on the surface, and just in front of the boat, which naturally happens, sometimes, in the chapter of accidents, no delay takes place ; the boat flies down over the greased ways, placed in its track on the shingle, the crew partly tumbling in as she goes, partly scrambling on board as she dashes through the surf, and seizing the oars 'give way with a will,' endeavouring to head the fish, so that they may not be able from want of time materially to change their position before encircled by the net, which with lead and cork-line is flung overboard with a rapidity truly astonishing by the steersman and his mate, whose movements are directed by the look-out on the cliffs, who 'hoizes' or waves his hat or a flag, shouting at the same time to shoot the net east or west, or as may be required.

The whole of the seine having been shot, the men by no means relax their exertions, but row ashore with all speed, carrying with them the end of the rope up the beach, and set to work to haul the net to land, another set of men being in the same manner occupied hauling on a rope attached to the end first thrown into the water. It often happens that a distance of two hundred yards or more intervenes between the ends of the seine and the land, and in order to deter the fish from attempting to escape by passing the ends of the net every effort is made by those in the seine boat and another in attendance to frighten the fish back into the body of the net, and guard the openings by throwing stones, splashing with and thrusting down the oars.

Meantime, the pole-staves at the ends of the seine approach the shore, and being landed, the fish are comparatively safe, provided the weather is fine and the sea without much swell; and the cork and lead-line of each arm of the net being now available, four lines of men unite in hauling the seine, which, gradually approaching the shore, gives unmistakable signs that the fish have not been missed, exhibiting sundry unlucky Mackerel strangled in the meshes or struggling to escape.

The space becoming every minute more contracted, the body of fish, like a dark blue cloud, is seen rushing wildly to and fro; the bunt or bag-cork in the middle of the net is only fifteen yards from shore; all hands redouble their former exertions; the net is almost in the landwash or breaking of the wave; the 'Brit' or small fish of the Herring tribe, forced to accompany the Mackerel by the draught of the net, friz out of the water in a pearl-like shower; shouts of 'Keep down the foot-line' are heard; the crowd is commanded in language more energetic than elegant not to scramble (trample) over the net, and the bunt or bag grounds on the shingle, being immediately surrounded by the fishermen, who hold up the cork-line, that the fish may not be washed out of the net.

The sight is such that, once seen, it will never be forgotten; a quivering, heaving, struggling mass of silver and blue combined is before you; every bystander is covered with scales, which fly like rain from the captured fish in their ineffectual struggles, as they are dipped up in large baskets and emptied into the first cavity on the beach above the reach of the water.

The seine being of large dimensions, the capture is by no means confined to Mackerel, but generally includes representatives of every variety found on our coasts, whether flat, round, or shell-fish, as Plaice, Soles, Dabs, Dories, &c., and occasionally a Salmon, or Salmon-Peel, Mullet, Bass, Gar-fish, Squid or Cuttle-fish, and Crabs; and the spectacle usually attracts a large crowd of both residents and visitors, should the net be hauled in front or close to any of the many watering-places on the SW. coast of England.

The Pilchard Seine.—This is the largest and most expensive description of seine used on the shores of the kingdom, and with its three boats—the seine-boat, the volyer, and the lurker—and all apparatus complete, involves the outlay of a considerable sum of money, from eight to nine hundred pounds; consequently they are owned by companies.

A 'huer' or look-out man is required to signal the fisherman from the cliffs when the fish show themselves, and to direct the master-seiner's attention to the course and movements of the fish. The manner of enclosing the fish &c. is very similar to that practised in the Mackerel fishery; but the fish being enclosed, the ends of the net are connected, and being moored by large grapnels, locally termed 'greeps,' the Pilchards are taken in a smaller net called a tuck-seine, used from the volyer, which proceeds over the cork-line within the enclosure for that purpose.

The seine-boat and volyer are rigged with large main and small fore-lug sails, and are capable of carrying about fifteen tons dead weight; the lurker is a spritsail boat of moderate size, that it may carry out the orders of the commander-in-chief, the master-seiner, who commands in person.

The same excitement is evinced in the Pilchard as in the Mackerel fishery, but the interests involved are infinitely greater and the feeling proportionate; for if the 'Pilchard harvest' fails, a trying winter is the almost certain lot of the Cornish fishermen: should the season turn out well, he will be

THE PILCHARD SEINE.

easy in his mind and comfortable in his circumstances, and may probably even be enabled to discharge any small debt he may have incurred through disappointment in other branches of fishing.

There is a large foreign as well as home consumption of this fish, both in a fresh and salted condition, and for the former they are salted and laid in piles, regularly built up on a slightly sloping floor, in buildings denominated from their use 'Pilchard cellars,' during which a large amount of oil exudes from the fish; after which they are washed and packed in casks, and subjected to pressure, which extracts a still further amount of oil, all of which obtains a ready sale.

The Pilchards are in this condition known as 'fair maids,' probably a corruption of 'Fumados;' from which, it would appear, they were formerly smoked before exportation to the Mediterranean, for the shores of which they are commonly destined.

The Pilchard is identical with the French Sardine, the latter being the Pilchard before it arrives at maturity.

The Pilchard swarms on the coasts of Ireland and Guernsey: on the former, I am told, they are entirely disregarded; on the latter they are sometimes captured with Garfish nets, and as they sell well, it is probable a special fishery may be eventually established.

Pilchards when salted for home use are usually cleaned by pulling off the head and removing the inside, splitting the fish with the finger and thumb, and leaving the scales, if possible, undisturbed.

They are now salted in a round cask or butter jar, the bellies being first filled with salt.

I consider them much improved by the addition of one and a half or two pounds of coarse moist sugar mixed with the salt, supposing you cure about four hundred fish. Factories are now established at Mevagissey and Fowey to cure the Pilchards in oil after the French fashion.

They are useful as a bait for Whiting when fresh ones are not procurable, and, whether for bait or eating, are commonly soaked in fresh water twelve hours previously.

THE TRAWL.

The Trawl (fig. 71).—The Trawl is a large bag-net dragged along the bottom by a boat or vessel where the ground is free from rocks, and captures a great variety of the best quality of fish found in our seas, namely, Turbot, Soles, Plaice, Dories, Red Mullet, and Whiting; also Hake, Cod, Dabs, &c., and not a few large Oysters, Crabs, Scallops, &c.; to which may be added numerous specimens of great interest to the marine botanist and zoologist.

I have given three engravings of trawls: the first, the net itself as it appears spread out to dry on a beach or grass-field, when it has the appearance of a large net-bag, its wide mouth partly encircled by a stout rope termed the ground rope,

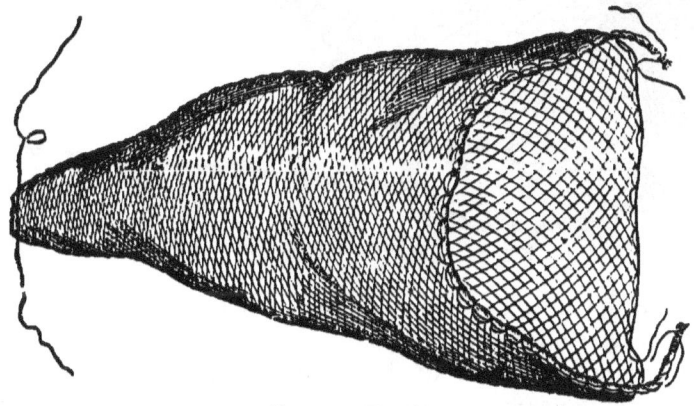

FIG. 71.—Trawl

inasmuch as it scrapes along the ground when in use, and from this acquires its name.

In the sides of the trawl the reader will observe four darkly shaded spaces, which are termed the pockets, and are formed by sewing the back and belly of the net together from the sides as far as the abrupt termination of the shading, thereby making four inverted bags or pockets to entrap the Soles in their attempts to escape, which are frequently rendered futile by this arrangement, as the numbers found therein sufficiently testify. Other fish are sometimes taken in these pockets, but Soles predominate. Some trawls have five or six pockets, but the ordinary number is two only, one on each side.

The small end of the trawl is termed the bunt or cod, and has a draw-string through the lower meshes, by which the bunt is carefully closed before the net is shot or thrown overboard.

Fig. 72 shows the beam and trawl irons, or heads; the beam is used to keep the mouth of the trawl open, and is supported by the irons, which are provided with square sockets to

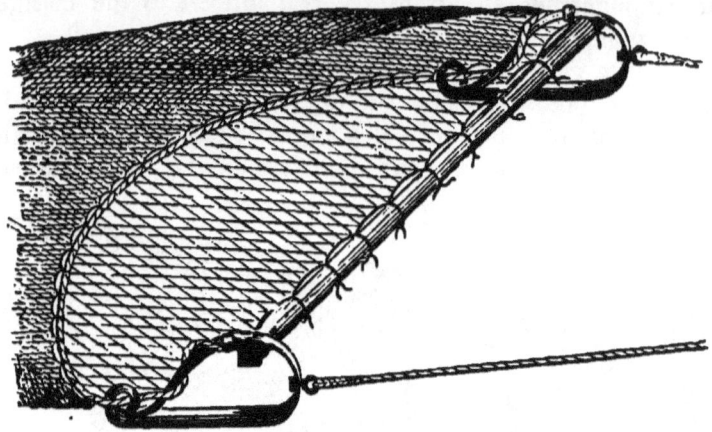

FIG. 72.—Trawl, with Beam and Irons.

receive the ends, and are made very stout on the lower part, both to give additional weight to sink the whole and to keep the ground-rope close to the bottom, that the fish may not escape underneath, which flat fish generally endeavour to do by flapping their side fins and covering themselves with sand or ooze when in danger or difficulty.

Where the ground is very soft the rope scrapes them out of it frequently, yet many escape by darting away in front, particularly when the wind is light.

In the illustration of 'the trawl, with beam and irons,' the beam sockets are represented inside the iron, but they are just as often placed on the top, there being no rule for these minor arrangements.

Hake are not commonly taken unless during a strong breeze, that is to say, sufficient wind to tow the trawl at a good rate.

The ends of the ground-rope pass through the eyes of the

irons, and a turn or two being taken, they are lashed to the iron near the beam sockets.

Two swivel eyes are attached to the front of the irons, into which the spans or bridle are spliced, which being connected with a strong rope, known as the rode or warp, the whole apparatus is then complete.

A reference to the accompanying cuts will render the above description easily intelligible.

Trawls are towed by craft of various sizes, from boats of twenty feet in length to vessels of sixty tons, the gear being of proportionate weight and dimensions.

The length of trawl-beam for a twenty-foot boat will be twelve to fourteen feet, that for a vessel of fifty or sixty tons, thirty-six to thirty-eight feet.

The larger craft are very powerful vessels, easily handled and excellent sea-boats, and remain out for weeks at a time, sending their fish to market by the finest and fastest cutters which skill and experience can combine to construct for the purpose.

These are termed 'Carriers,' and on their outward trip take provisions and water for the 'trawling fleet,' with many tons of ice in which to pack the fish on their return. Steamers are now also used as 'Carriers.' A good deal of trawling is now done in steam vessels specially built for the purpose, and many tugs are registered as fishing vessels also, and carry a large trawl, which they use to fill up their time whilst waiting for employment in towing sailing-vessels.

Cutters are more effective as trawlers than any other rig, the mainsail giving them great power of towage; all trawling-craft fishing near the shore deliver their fish daily, if possible.

In the North of England, instead of the ugly, unhandy beam, they have for small trawls, say up to the size of twenty-five feet beams, substituted $1\frac{1}{2}$-inch galvanised wrought-iron pipes, which are always clean, and take up very little room on deck.

Size of Mesh.—The size of the mesh of trawl-nets must be not less than $1\frac{3}{4}$ inch from knot to knot, and any other net added to a trawl must have meshes of 2 inches from knot to knot. From knot to knot means along the side of the square.

The Otter-Trawl (fig. 73).—The otter-trawl has received its name from the two boards which take the place of the irons in keeping the mouth of the trawl extended.

This kind of trawl is often preferred by amateurs to the beam-trawl, as it is much more portable, for the otters being detached, the whole may easily be stowed away on board.

The beam is here (see. the cut) superseded by a number of corks, which support the upper edge of the trawl and keep the mouth open.

The otters of a trawl which I had the opportunity of in-

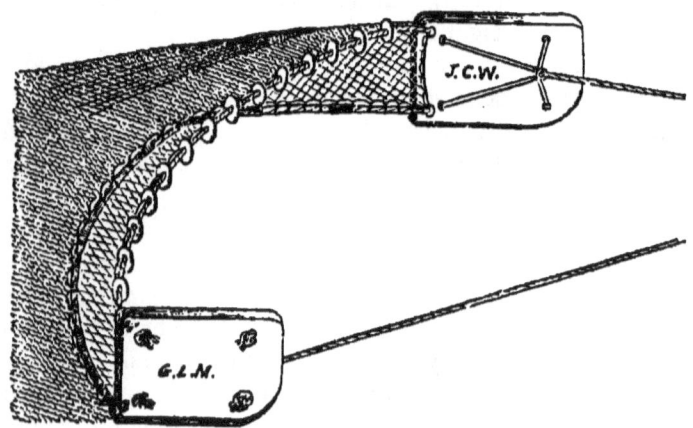

FIG. 73.—Otter Trawl.

specting were of the following dimensions : length two feet two inches, width one foot six inches, and two inches in thickness, of elm, weighted with iron keels just sufficient to sink them, the iron being curved like the fore-foot of a boat, that it may pass easily over the ground.

This trawl was highly spoken of by the crew as very successful, and was used in a vessel of about twenty tons.

I have heard it alleged that it is not equal to the beam trawl, as it is said to collapse if towed at all across the tide ; but I do not make this observation from my own experience. The effectiveness of an otter-trawl is much increased by lashing a piece of galvanised chain along the ground-rope. These trawls have come into favour amongst yachtsmen on account of a beam not being required, which is a very ugly thing on board a yacht. A correspondent from the eastern counties speaks very highly of the otter-trawl, and avers that he takes much more fish than when he used the beam. The regular beam-trawl, however, is too firmly established ever to be superseded amongst fishermen.

My correspondent's words are as follows:—'The use of the otter-trawl is very simple; there is only one thing to be cautious about, which is that you must weight the foot-rope exactly right, neither too heavy nor too light; if it is too heavy you get such a quantity of mud, weed, and stones that you require powerful tackle besides all hands to get the net on board, but you at the same time catch plenty of fish. If you do not weight it sufficiently your net does not drag the bottom nor open properly, so you catch no fish, as the cork head-rope and the leaded foot-rope come together. I found my net, after altering the leading twice, go quite right, and catch double the quantity of fish that the professional trawlers here catch with the beam-trawl. My head-rope is forty-two feet long, foot-rope the same; at every half fathom of foot-rope I wrap a piece of sheet-lead round once and a half; when all is on, I serve the whole foot-rope over with one strand of an old Manilla hawser, which makes it very thick and prevents it cutting into the mud too much; I think if I did not it would pick up a fourpenny bit.'

The Dredge is a very important implement, the amount of money earned by its use being very great. It may be briefly described as a heavy scraper with a net at its back, that part rubbing on the ground being formed of iron wire-rings connected together, and of such a size that gravel and small shells will escape, whilst Oysters and larger objects are retained. The dredges vary in number according to the craft or the nature of the ground; about 30 inches is the average width. For grounds which have been little dredged, or new and unbroken grounds, a very heavy kind are used with teeth; these are called breaking-up dredges, as they will break through a bank of Oysters and loosen the ground preparatory to using the smaller dredges, which would pass over a bank of this sort with little or no impression. The cutter is the favourite rig of the Oyster-dredger, the swing of the main boom giving great power and steadiness of towage. One of these vessels will tow three of these dredges at once, which on being overhauled contain Oysters, Scallops, empty shells, Hermit Crabs, corals, and many curious specimens of marine zoology. The dredges are got on board by aid of a winch, and a buoy is kept ready for each dredge, which can be slipped if needful in case of hooking the bottom. A large number of these vessels belong to Colchester, Whitstable, Burnham River, Portsmouth, Jersey, &c., and on any fresh bank or bed being discovered flock to it from their respective ports. Large numbers of Oysters have of late years been dredged in mid-channel or between Beachy Head and Fécamp, and to the westward, and there can be little doubt that other banks are yet to be found by persevering search.

There is too little restriction placed on Oyster-dredging, and such a number of boats flock to any newly-discovered bed that it very soon becomes exhausted; whereas, if a fair number of boats only were permitted to work, the ground would be kept clear for the fish to deposit its spat, and a regular supply would be afforded. For many years, however, we have been 'killing the geese which laid the golden eggs,' and it is no wonder Oysters are becoming yearly more scarce and dear. Ostreaculture, as far as hatching fresh brood is concerned, we have been far from fortunate in. The most promising result recorded is in Hampshire, where, at Hayling Island, things look remarkably well, large numbers of young Oysters having been hatched. Very praiseworthy efforts have been made in Guernsey, but they are a dead failure. It is averred by a native seaman that the water is too clear round this island, and very different from that on the French coast, where ostreaculture is a success—for instance, at St. Brieux a considerable deposit is always in progress; but here little or nothing, on account of the strength of the tide. This may be worthy of some attention.

Shrimp and Prawn Nets.—The Pool Net (fig. 74) is the smallest of the three, and is stitched on to an iron, grooved to

THE STRAND-NET.

protect the stitches, through holes punched at intervals of two inches.

The iron should be of a spoon-like shape, eighteen inches long, by fourteen wide, and the form of the material itself should

FIG. 74.
The Pool Shrimp Net.

be elliptical, as this will admit of deeper grooves on each side than any other, and will offer very little resistance in passing through the water.

A piece of round bar-iron, five-eighths of an inch thick, beaten a little to render it somewhat elliptical, will be stout enough for the purpose, which must be drawn down at the larger end into a six-inch spill, and securely fastened into an ashen staff, nine feet long, and one inch and three-quarters in diameter at the bottom, tapering to an inch and a quarter at the top.

If a staff slightly curved can be procured, it will turn much more easily in the water when sweeping around the pool.

An iron ferrule an inch and a half in width must be provided for the end of the staff, which, as well as the ring, is better galvanised to avoid rust.

The best twine is the Shrewsbury No. 18, and the top row should be double, having a leather thong or stout piece of bell-wire run through; this is to be drawn tightly into the groove below by a lacing of finer wire or strong string passed first down and then up each hole in succession, until the circuit of the iron is made.

The net is useful in rock pools for Prawns, and also in the small sand pools, commonly formed at the foot of rocks, sometimes scattered over sandy beaches.

All nooks and corners must be carefully searched, and the same spot tried over more than once.

The Strand-Net (fig. 75, folding for portability).—This is a very useful net on a flat strand or in large shallow pools on sandy beaches, where it is pushed before the shrimper wading.

On the flats and in pools free from weeds the ordinary Grey Sand-Shrimp will be taken, but where these flats are covered with the long tape-like sea-grass (*Zostera marina*), so frequent in many harbours and roadsteads of the coast, they are replaced by Prawns, for which this net is equally well adapted.

There are different methods of fitting these flat strand-nets, but I consider that which I have represented in the cut to be the lightest and most effective, and it can be easily made to fold up and pack in line with the staff. I find it more convenient than any other.

For a man it should not be less than three and a half feet in width, but for a boy two and a half only, as the larger will be found of too heavy draught. The width of the board four inches, thickness three-quarters of an inch, the front thinned down to an edge of a quarter of an inch, that it may skim the sand closely, and even be capable of entering just beneath the surface where the bottom is sufficiently soft.

A stout chock of elm two and a half inches thick at the back and somewhat of a heart shape, having a mortise to receive a staff, must be securely screwed to the centre of the board at a slight angle. This staff should be seven feet long, two and a quarter inches in diameter at the bottom, and one and three-quarter inch at the top, having a small cross-head or handle.

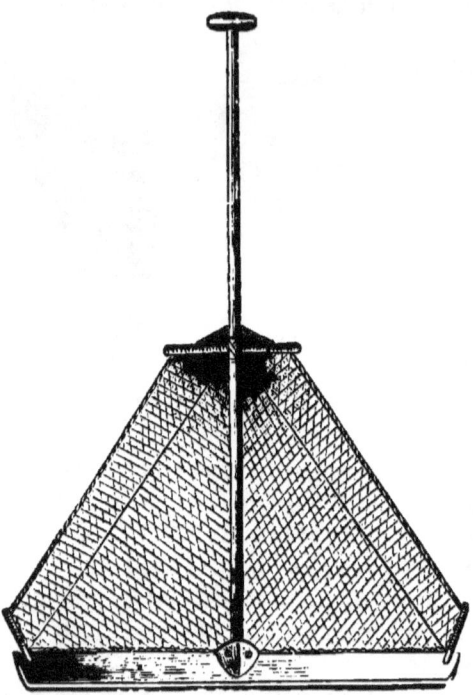

FIG. 75.—The Strand-Net.

A piece of clean deal without knots is strong enough, and of course lighter than ash.

Two strong pieces of stick being fastened in the ends of the board, inclining slightly outwards, the net is supported on a stout cord tightly stretched between these and a piece of stick lashed crosswise on the staff.

The net is secured to the thick back edge of the board by loops of thin leather nailed to this edge with copper tacks, at intervals of about four inches.

The Shrewsbury thread is quite strong enough for this net also, but should be edged with double or stouter twine, through which a cord should be run for convenience in fixing.

If desirous of packing the net snug for travelling, you have merely to cast off the cross piece and unship the staff, when you can easily withdraw the sticks from the ends of the board, and the whole affair will not occupy more space than a gig umbrella.

Where the sea-grass abounds you will of course push the net in the direction of the lay of the weed, or you will double your labour with little result.

The best receptacle for Shrimps or Prawns is a common fishing basket, but the better kind of basket is not so well adapted for this purpose, being usually too shallow and having too wide an opening; in fact, the front and back should be of equal height, and the mouth somewhat narrower than the bottom.

Shrimps and Prawns should be sought for during spring tides; that is to say, from three days before new and full moon until three days after, when the greater reflux of the water renders their haunts more accessible than at neaps.

Never wade after Shrimps or Prawns with bare feet, lest you tread on the back-fin of a Weever (*Trachinus draco*), a little fish which has the habit of lying hid in the sand with the said prickly back-fin just beneath the surface, wounds from which are most painful. The author of 'Fish and its Cookery' says: 'The most effectual cure for a wound of this kind is to make a strong brine, and then plunging in the wounded part to keep the brine as hot as the patient can bear it.' Mr. Couch says that smart friction with oil soon restores the wounded part to health. An old pair of shoes will entirely obviate the risk. The common

Weever is from three to five inches long, but the greater Weever is of more elongated form, being from ten to twelve inches in length. These fish are exceptionally taken with hook and line, being caught in seines and trawls shot for other kinds. Pennant observes, 'The first dorsal fin consists of five very strong spines, which, as well as the intervening membranes, are tinged with black; this fin, when quiescent, is lodged in a small hollow. The covers of the gills are armed with two very strong spines.' These may inflict painful wounds.

The Baited Prawn-Net (fig. 76).—This description of net is much used on the south Devon coast, being baited with stale Rock-fish (locally called Curners), Gurnards, Horse-Mackerel, &c., a dozen or twenty being shot at short distances apart, and raised occasionally by aid of a buoy-line attached to each.

They should not exceed two feet diameter, depth sixteen inches.

An iron hoop is superior to those of wood, round bar three-quarters of an inch, if used on an open exposed coast, but of half an inch in more sheltered localities; if wood be used, they must be weighted with lead or stones, a clumsy method.

Iron hoops should be galvanised, being much cheaper in the long run, although a little more expensive at first.

Hermit Crabs for bait or Green Crabs may be taken in these nets; also Whelks, particularly if baited with a piece of fresh Ray or Skate, to be placed between the double string shown in the cut crossing the net, and then secured by a leather slider on each side, or a piece of string tied once or twice round.

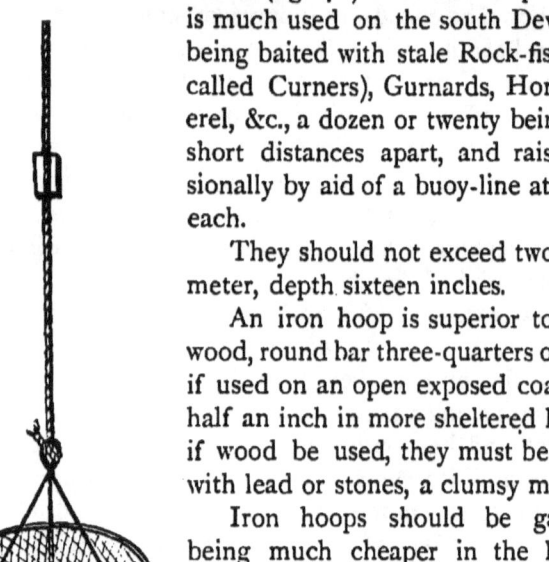

FIG. 76.
The Baited Prawn Net.

On such parts of the coast as afford facilities for so doing by a moderate rise and fall of tide, and where a succession of pools exists among the rocks, or the rocks are flat, and long canal-like channels run up towards the land, these nets can be used without a boat, with a short piece of line, and a single cork at the end, of the size of a large bung-cork; these nets, being sunk alongside the rocks, may be raised by aid of a forked pole to catch the cork. Try on a rising tide.

In boats, this fishing is quite near the shore, and is carried on in small boats, each provided with a dozen or more nets, slung by a buoy-line, which are placed a few fathoms apart, and baited in turn. A few Crabs and Lobsters are sometimes captured at the same time, and numbers of Velvet-Crabs, which, although never eaten in England as far as I am acquainted, are, together with the Spider-Crabs, regularly sold in the markets in the Channel Islands and France. The Velvet-Crabs are commonly called 'Fiddlers' in many parts of the kingdom.

Freshwater Eels may be also taken by baiting with Herring, Pilchard, or Mackerel offal, &c., in harbours.

The quantity of Cray-fish and Lobsters to be taken in this manner in South Africa and on the North American coasts is very great.

Small pots similar in form to Crab-pots are much used, made of fine osiers very closely placed; they are commonly termed Shrimp-pots. It is somewhat remarkable that small Prawns almost universally receive the appellation of Shrimps.

Drift-Nets.—Drift-Nets are extensively used for Mackerel, Pilchards, and Herrings. They are attached to each other, and are shot in a long continuous line to the length of one thousand or fifteen hundred fathoms, more or less, in proportion to the size of the boat.

The nets are provided with corks on the head-line and weights on the foot-line (commonly stones), and are very frequently lowered to the depth of two or three fathoms, being sustained at this depth by large buoys of cork or inflated skins placed at intervals, with a small keg at the junction of the nets.

This is done as a precautionary measure, that vessels passing over them may not hook them and carry them off on their keels, and it also enables the fishermen to adjust the depth of the nets as may be desired. At many parts of the coast no weight is used on the foot-lines.

These nets are shot at sunset and commonly remain until daybreak, and the boat, being attached by a rope to the last net, keeps the whole on a stretch, and the fish, either unable to see the net or else rendered heedless by the obscurity, strike into the meshes, which are of a size just sufficient to receive the head of the fish.

Neither Mackerel, Herrings, nor Pilchards mesh well on moonlight nights, or when the water 'fires' or becomes phosphorescent.

The nets should be all in the water and the boat riding at the end of the warp as soon as the sun is set or very little after, for the first meshing of the fish often takes place before dark, after which, should the 'brime' or 'fire' show itself, the fish will not be likely to strike the nets again till just before dawn. Hake, Conger, Cod, large Pollack, and Coal-fish may all be taken, as the boats drive along with the tide, by hook and line.

Of late years our drift fishermen—or drivers, as they are commonly called—make much more lengthy voyages than formerly in quest of the fish, particularly in the Mackerel and Herring fishery.

Soon after Christmas, the boats or vessels (for they are entirely decked when the hatches are on) proceed to Plymouth, or even farther west, and meet the Mackerel coming up Channel, and forward their takes to London by rail, returning to their own part of the coast when the fish make their appearance there.

The western Mackerel drift-fishery yielded 30,000*l.* in 1867.

Whilst the Mount's Bay men from Cornwall also seek the Yorkshire shore, to partake of the Herring fishery during the summer, and the boats from various localities migrate to the shores of Sutherland and Caithness for a similar purpose, French vessels of sixty tons and upwards bend their course to the westward of Scilly and Cape Clear, salting their captures on board and returning to port when loaded, a practice not followed by British boats, salted Mackerel being of comparatively small value in the English market. Kinsale in Ireland has become an important station for this fishery.

The chief portion of our drift-boats are fitted with their masts to lower, in common with those of our French and Dutch neighbours, which eases much of the strain on both boat and net when riding in a gale of wind.

Moored Herring-Nets.—An ordinary drift-net for Herrings is, in sheltered bays or deep lochs of the coast, moored by an anchor or heavy stone at each end, and left for the night to itself. This is a common practice in the Scotch lochs, and is likewise followed in Babbacome Bay, Torbay, and Plymouth Sound, as well as Cawsand Bay, on the south coast of Devon, and elsewhere. Babbacombe Bay is the chief rendezvous of the Herrings on the south coast of Devon (although they are not confined to this neighbourhood), and consequently boats assemble there from October to January from the different ports and villages to the east and west, as the southerly trending of the land affords considerable shelter from the prevailing south-west winds. Pollack are also taken in moored nets, placed in deep water.

Peter-Nets are straight nets attached to the shore by one end, the other end extended seaward by an anchor. They are of the same depth as the water, and have corks on the head-line, and stone or lead sinkers on the foot. Pollack, Salmon-Peel, and Red Mullet are caught in them.

Drum-Net.—This net is very useful in preserving fish alive, particularly in hot weather. It is a net bag made of fine twine with a small mesh, distended by three cane hoops about twenty inches in diameter and eight inches apart. The bottom is drawn up flat, and a piece of sheet lead, of half a pound weight, stitched on to it. The top is much smaller, and three seven-inch hoops form a neck through which the fish are passed into the net as caught. A loop of strong line is fixed across the upper small hoop to suspend it by. When in use it is hung in the water with the neck above the surface.

Tanning Nets.—The following method is both simple and convenient:—

For a net of twenty-four pounds' weight previously untanned, take four pounds of pulverised catechu and boil it until thoroughly dissolved in eighteen gallons of water, adding thereto, if procurable, about two hatsful of young oak-bark pounded small, and either put the net into the boiler, leaving it to steep two days and nights, or into a barrel and pour the hot liquid over it.

If the net be put in the boiler, it should be quite covered with the liquid, or the hot metal may scorch it. The price of catechu is about fivepence a pound, and it is frequently used without any admixture of oak-bark.

Catechu contains from forty-five to fifty-two per cent. of tannin, more than double the amount contained in oak-bark.

Catechu can be obtained from ship-chandlers in the chief ports of the kingdom.

The boiler should be one of copper, or if of iron galvanised, to avoid rust.

For a net previously tanned, use a pound less catechu. Some of the Cornish fishermen pour out their tanning liquor into large vats with coal-tar, and this ingredient is found to preserve the nets much longer, when added to the liquor.

Landing or **Hand-Net.**—This is described at pp. 91, 214.

BOATS AND BOATING.

Several works of merit have of late years been published on these subjects, but as they have no particular reference to boats as connected with sea-fishing, I think it necessary to enter into the subject somewhat in detail.

Irrespective of means of propulsion by rowing or sailing, there are two kinds of boating, that is to say, Beach or Surf Boating, and Harbour Boating.

On all open beaches there is generally some amount of run or swell, caused by the ceaseless undulation of the ocean, which expends itself on the shore in such localities, and is known under the appellation of 'the surf.' This is much more dangerous and difficult to deal with on a flat sandy shore than on a steep pebbly one, as a boat, either in landing or launching, will on the former be exposed to a succession of breakers, whilst on the latter it is only the last wave which breaks. From the want of shelter upon all open shores, it is never safe to keep a boat at anchor at night, for the sea gets up so rapidly, should the wind set in from seaward, as frequently to prevent putting off to fetch it from its moorings; it therefore follows that your boat should be of a moderate size, that it may be launched and hauled up with facility. There should be but little keel and a flattish bottom amidships, for with this form she will float off more quickly in launching, and in landing will not strain as a boat with a sharp bottom, which in anything like a swell on the shore would be almost certain to force out the plank next the keel, called by shipwrights the 'garboard strake.' (See figs. 77 and 78).

FIG. 77.
Midship Section of Beach Boat.

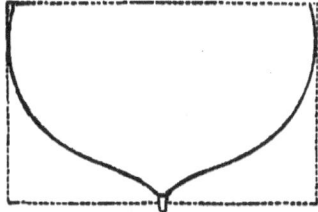

FIG. 78.
Midship Section of Harbour Boat.

Fig. 77 is the section of a beach boat intended to represent the form best adapted to a flat sandy beach; but where it consists of shingle, or of sand and shingle mixed, the beach will be found to be much more inclined, which will admit of the bottom of the boat being somewhat sharper, and of a shape between the sections of the beach and harbour boats in the illustrations.

The most useful sized boat for an amateur fisherman for beach boating and fishing is one of the length of thirteen feet six inches, and four feet ten or five feet in breadth, clench or clinker built (the planks overlapping each other), with ribs or timbers of oak or American elm, and the plank young English elm, of course fastened with copper-nails, every one of which should be riveted to ensure security.

The breadth of beam is certainly very considerable for so small a boat, but is necessary to ensure stability under sail, which in a boat of light draught of water cannot be obtained without it; in addition to this, the boat will be stiff with a moderate amount of ballast, which is a matter of consequence, as the ballast has to be carried down and up on every occasion you go to sea for a sail. For rowing, all beach boats are quite stiff enough without ballast.

I am now of course speaking of what I may term a single-handed boat for general purposes, to row or sail, according to circumstances; but for offing fishing, when never less than two should be on board, eighteen feet will be a convenient size, beyond which they will be found cumbrous to amateurs. A capstan should be provided for getting up a boat of the last-named size, for a man's strength can be more effectually exerted at a capstan bar than in any other manner which has been adopted for heaving a strain.

Various kinds of ballast are in use; namely, lead, iron, stone, and water ballast, of which lead is the most effective, *but very likely to take unto itself wings*, when cast into weights of as little as fifty-six pounds, which are quite heavy enough to be thrown out on the beach, when running on shore in a surf. Next comes iron, which is more in use than any other, and sufficiently heavy; but in this metal I do not recommend the pig ballast with a hole in each end (for this necessitates a rope handle or becket where quick handling is required), but the square half-hundredweights, such as grocers and coal-merchants keep, having a square sloping hole in the centre crossed by a handle. This, as it does not project above the surface of the iron, is not at all in the way of the feet when moving about in the boat.

Barrels or breakers of an elliptic shape take up too much room in small boats, but if four tank boxes be made two feet square, of three-quarter inch plank when planed, and rounded in the bottom to fit the boat, and of a depth of six or seven inches in the middle part, they will stow, when filled and soaked, a considerable weight of water. This, with two half-hundredweights, and a stone of about forty pounds weight, used to moor on rocky ground, will amount to near three hundredweight, and should be sufficient, with an anchor, ropes, oars, &c., to ballast a boat of this size.

Water ballast is undoubtedly the safest ballast when properly managed in this manner, as should the boat through an accident be filled she immediately gets rid of her ballast, and her crew may hold on to her until assistance arrives. So important has this been deemed, that in our navy and coast-guard service no other than water ballast is on any pretence allowed, and it would of course be much better to use no other kind in all boats; but the truth is, there is rarely sufficient space to stow the necessary weight, and a little stone or iron must be had recourse to, to make up what is wanting. The tanks for water ballast will be found well adapted to a beach-boat, as they can be filled after you are afloat, and emptied on coming ashore. They should rest firmly against each other on the timbers of the boat, without any bottom boards intervening, and be secured by a stout batten, firmly nailed to the timbers at the outer edge of each, which will prevent their shifting when the boat heels over to the breeze. When there are no battens, it is usual to have a board nailed on its edge, on the central bottom board, which prevents the ballast shifting in a squall or puff. The number of fatal accidents which have resulted from neglect of the above precautions exceeds belief.

Beach-boats should never have an iron band on the keel, as it causes them to drag heavily in launching and hauling up, but instead of this a false keel or shoeing, as it is called, of holly or African oak, with a crooked piece for the fore-foot and heel. This shoeing is to be fastened on with oak pegs (by shipwrights termed tree-nails, *vulgo* trunnels), as they will wear away equally with the shoeing, and offer no impediment in passing over the ways, which for small boats consist of pieces of flat oak, holly, or beech, five feet in length, by four inches in breadth, and with a little grease lighten the work of launching and hauling up very considerably.

For small boats of this class, only two rigs are admissible, the spritsail and lugsail, by which I mean the working lug, the tack of which being fastened down close to the mast, the sail does not require lowering when the boat is put about.

Both the sprit and working lug are handy sails, as they can be set or lowered with great despatch, from the small amount of rigging connected with them in boats of this size; but in large boats the sprit is fast becoming obsolete, as the spar, from taking nearly the whole diagonal of the sail, is necessarily very large and cumbersome, and has been known to part the heel strap (by fishermen called the snorter) and drive a hole through the bottom in its descent, which circumstances have led to the more frequent adoption of the gaff and lug sails of late years. The difference between sprit and lug sails will be easily seen from the accompanying cuts.

All spritsails should be fitted with a tackle purchase, which will enable you to set the sail very flat, and also prevent the heel of the sprit slipping down—a constant source of annoyance without it. Spritsails set rather more closely by the wind than either gaffs or lugs.

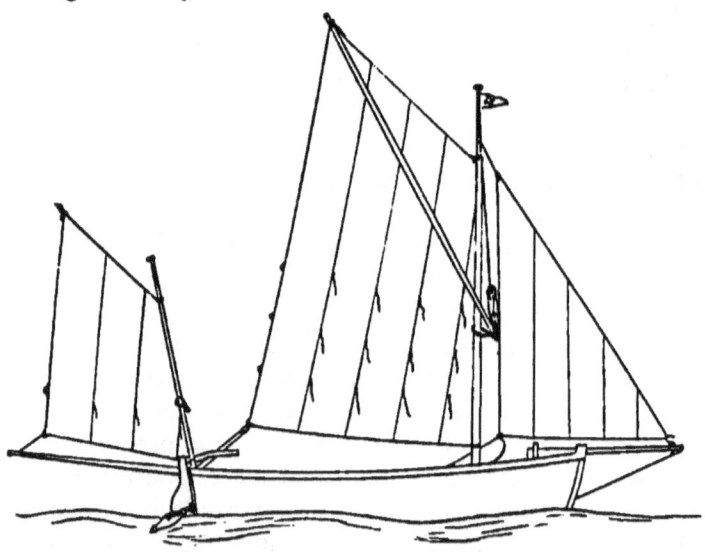

FIG. 79.—Spritsail-Boat with Mizen.

Spritsail-Boat with Mizen (fig. 79).—In fishing for Whiting-Pollack, it is necessary when under weigh to go through the water at a very moderate rate, for which the mizen and jib will be found quite sufficient in a strong breeze with wind against tide; under these circumstances, in boats unprovided with a mizen, it is difficult to retard the boat sufficiently by reefing the mainsail closely, which has still too much power for your purpose. In squally weather it is sometimes requisite to take in the mainsail altogether, when your boat is still under command if rigged with a mizen, whereas without it you would drive considerably from your course, or perhaps be compelled to anchor. I therefore think that the small amount of additional spars and gear involved in its use, is more than compensated by the accompanying advantages.

These observations on mizens are equally applicable to both the sprit and other rigs. (See the cuts.)

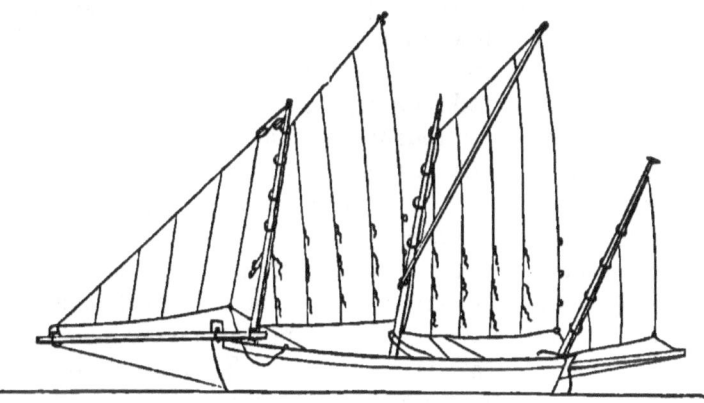

FIG. 80.—Guernsey Spritsail-Boat.

Guernsey Spritsail-Boat (fig. 80).—This is a very useful rig up to a certain size, beyond which the sprits become heavy and troublesome. The dimension of a useful boat of this class are 12 ft. keel, 6 ft. 5 in. beam, 17 ft. 3 in. over all, draught of water 2 ft. 10 in. The foremast being stepped well forward, one triangular sail alone is required. She is represented under full sail, but is easily handled also when the canvas is reduced as follows : first, without the mizen, in this case a 'Mudian cut or shoulder of mutton sail ; secondly, with a smaller jib and one-reefed mainsail ; thirdly, with the same jib, two-reefed mainsail, and one-reefed foresail ; fourthly, with three-reefed mainsail and two-reefed foresail ; fifthly, with two-reefed foresail, no mainsail, and the mizen and small jib (very handy for Mackerel-fishing or railing in a fresh breeze, when, if considered desirable the mainmast may be struck); sixthly, under a reefed mainsail and moderate-sized jib; seventhly, under a whole or reefed foresail only; eighthly, under the two spritsails, the mainsail to be reduced by one reef, so as not to have too much after-canvas; the small jib, however, is a great recommendation in steadying the boat and preventing her flying up into the wind, which they always are liable to do without some little head-sail when the mainsail is set. The sprits are set up with a tackle consisting of two blocks. These boats draw a good deal of water aft and not much forward, are stiff under canvas with a moderate amount of ballast, go better to windward than any of their size I have met with on the English coast, tack as quickly as a 'Mudian, and are very strongly built. As they have often to make very short tacks among the rocks, and also to work the eddy tides in the lee of the rocks, to make their passages to and from their fishing grounds, it is absolutely necessary they should answer the helm quickly, and they certainly do 'stay like a top,' owing to their beam, deep draught aft combined with light draught forward, and considerable rudder power, hung as the rudder is on a raking stern-post. In the older boats, this rake was often carried to a ridiculous extent, but in those more lately built, although the rake is still considerable, it is much diminished, and with no perceptible disadvantage, for the beam, joined to the difference of fore and aft draught, renders them quick enough in stays for all practical purposes, and the additional length of keel makes them run more steadily off the wind. They are all carvel-built, but chiefly framed with elm, and not often with oak, which is far from abundant in the island. Copper fastenings are in most cases used to the top streak, which, however, must be nailed with iron, as copper is not found to hold well in the gunwale. Some few are copper fastened only as far as the water line, and of late years galvanised iron nails have been introduced, but as a rule copper is preferred for fastenings. It has been previously mentioned that the sprits are hoisted by the aid of two blocks : the upper block should be a double one, with a tail, usually formed, in the boat here represented, by a piece of double rope, which being spliced together at the ends forms two eyes; one of these goes over the mast-head, and rests on a shoulder cut for it, the other receives the double block. In nautical language this particular tackle (*vulgo tayckle*) is termed a tail-jigger, on account of the rope or tail by which the double block is attached to the mast-head. The lower block has either a hook, or is connected by a strap with a ring or traveller, having a loop of rope termed a 'snotter' or 'snorter,' which receives the end or heel of the sprit after the point has been introduced into the eye formed in the bolt-rope at the peak of the sail. By aid of these tackles the spritsails may be flattened out more than almost any other kind of sail, and the sprits are secured in their places by belaying the tackle-fall to a pin in each thwart. Great care must be taken that the snotters are of good sound rope, for if carried away, the sudden descent of the sprit might knock a hole through the boat's bottom. With ordinary care, however, there is no cause for apprehension, and a new snotter occasionally will obviate all risks. In large boats the downward pressure of the sprit is very great. This, and the superior handiness of the gaff, has caused the latter to be adopted as the boats increase in size.

I had once a good-sized boat rigged with a large sprit-sail as the mainsail, like the illustration p. 254 ; but not in those my early days of aquatics being up to the mystery of a 'tail-jigger,' my sprit was always slipping down, no matter how much I wetted the mast to keep it in its place. The fact is, any sprit-sail larger than a moderate-sized mizen should be fitted as a matter of course with a tackle, having one, two, or a double and single block, according to its size.

Lugsail Boat (fig. 81).—This is also a very handy rig, and extensively used on the coast. It is the working lugsail previously referred to, and does not require to be lowered in going about.

A boat of thirteen feet six inches in length over all is quite large enough for one man to handle in average weather when there is no harbour, and in calms can be pulled at a tolerable rate ; it affords room for two or three people to fish at the same time without incommoding each other, and with anything like careful management will go through a heavy sea.

s

It is frequently a matter of importance to place your boat exactly in a certain spot, which is extremely difficult to effect in a large boat, and often impossible in a tideway: therefore the great advantages which a boat of moderate size has over a large are, I think, sufficiently evident.

A boat for general fishing where a harbour exists may be fifteen feet in length, but if it exceeds this, it will be found too heavy to work in calm weather single-handed.

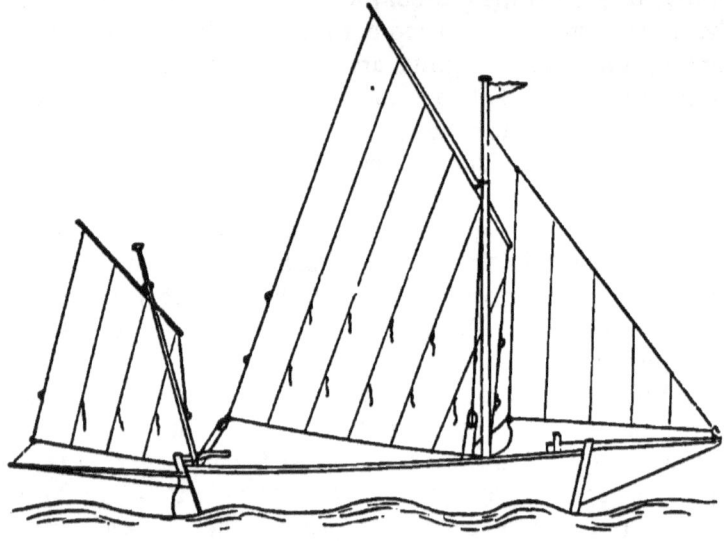

FIG. 81.—Lugsail Boat with Jib and Mizen.

Most of the fishing described in this work can be obtained from a quarter of a mile to two miles' distance from the shore: but as Whiting-fishing is often carried on at greater distances, say from six to twelve miles, those who wish to follow it will require a craft of not less than from sixteen to twenty feet keel, twenty to twenty-five in length over all, and from three to five tons, rigged as a dandy or yawl, and if decked to the mast with a narrow water-way round her, so much the better. Many of these boats are yachts in miniature, being built and fitted with great taste.

The Yawl or **Dandy** (fig. 82).—The yawl is frequently preferred to the cutter for fishing purposes (trawling or dredging

THE YAWL OR DANDY. 259

excepted) both by amateurs and professional fishermen, as the mainsail is generally used without a boom, or if a boom is used it is not a fixture, as in the cutter, but fitted to ship and unship when required.

This rig has of late years found increasing favour with yachting men, even for vessels of large size, as the diminished weight of the mainsail renders them more manageable. In large vessels the mainsail is permanently bent on to the boom, as in cutters.

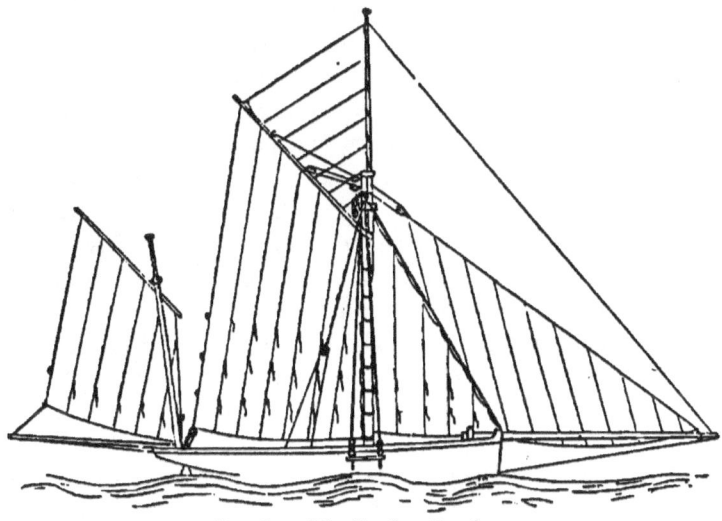

FIG. 82.—The Yawl or Dandy.

Yachts of this rig for fishing purposes often run up to the size of fifteen tons builders' measurement, in which case they are not decked over entirely, but have a large open well, in the edges or coamings of which a wink or winch is fixed wherewith to get up the anchor, for which a hawser-laid rope is used, as being more manageable than a chain for great depths. These vessels should not have a water-way or side-deck more than twenty inches wide, or it will be inconvenient to stand in the well to haul the lines. If they exceed twenty inches wide, they should not be less than three feet, as you can then fish from the deck; and that you may conveniently do so, an additional rail should be fixed by iron stanchions through the top-rail and secured at the heel in a staple firmly driven into the bulwark stanchions, this rail to reach breast high. Fishing over the low bulwarks of a small yacht whilst standing on the deck is a most fatiguing affair. The most comfortable boat to fish in is one of such a depth, as you stand on the platform, that the deck will reach the waist of a man of ordinary stature, and if deeper, a false platform should be raised above the other. This is about the depth of the majority of fishing-boats which work with hook and line. A boat of this depth affords sufficient shelter without embarrassing a man's movements, which is a great consideration. If it were not for the appearance, a fishing-boat would always be better without a counter, because the lines are more readily accessible, and can be more easily cleared in case of a fish sheering amongst the other lines and causing a foul, to which a great liability exists when Mackerel-fishing under sail, from the sudden darting of the fish from side to side, and which can be only prevented by the absence of all obstacle to freedom of action on the part of the fisherman. If a boat is built, therefore, specially for fishing, let her have no counter.

The Itchen River Rig.—The Itchen River or Southampton Rig is much used by the fishing-boats of that town, and is frequently adopted in pleasure boats of twenty feet length and upwards.

It consists of a yawl mainsail without a boom, travelling on an iron horse across the stern, with a jib or forestay sail on a short bowsprit or iron bunkin, which sail does duty for both jib and foresail.

These boats in strong winds, provided there be not too much sea, will work to windward under the head-sail only, a great advantage in squally weather.

With their two sails set they are full powered, but in light weather they sometimes set a sharp-headed topsail on a spar, which does duty as both yard and topmast.

I have seen some small yachts of this rig with a jib and mizen in addition to the above-mentioned sails, which appeared to answer remarkably well; but for match sailing they often bend a cutter's mainsail.

The Itchen river boat requires less rigging than almost

any other, and is, therefore, well adapted for the combined purposes of pleasure and fishing.

The cost of such a boat will be about 75*l.*, length twenty-one feet, breadth of beam eight feet, draught of water three feet seven inches. Excellent boats of this class are built by Payne, of Southampton.

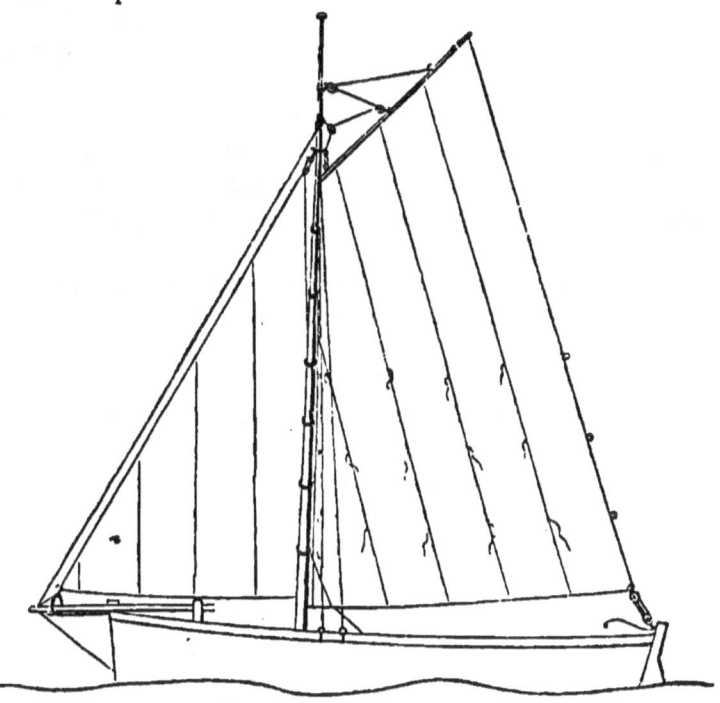

FIG. 83.—Itchen River Rig.

Prices of Boats.—The charges for building boats of moderate size are usually ten or twelve shillings a foot, measuring the extreme length, and the cost of a boat thirteen feet six inches in length complete will be nearly as follows:—

	£	s.	d.
Thirteen feet six inches, at twelve shillings	8	2	0
Sails, Oars, Anchor, and Rope	5	0	0
Fishing Gear	2	0	0
Sundries	1	0	0
	£16	2	0

Second-hand boats of this size from 6*l.* to 10*l.*, according to condition, &c.

Open pleasure-boats of from sixteen to eighteen feet length will cost from 20*l.* to 30*l.* Second-hand, from 12*l.* to 20*l.* Anything from eighteen to twenty-two feet keel is more expensive in proportion to size, and if fitted as a yacht will cost from 70*l.* to 140*l.* new; and, second or third hand, may be bought from as little as 15*l.* up to as much as 80*l.*, according to condition and existing demand.

A yacht of fifteen tons is the maximum size manageable by a man and a boy, and can be so contrived as to be made very comfortable; about 10*l.* a ton is the price of a craft second-hand, as a rule, but opportunities often occur of purchasing at a lower figure.

For a fishing-boat where a harbour exists, the size may range from five to fifteen tons for an amateur.

CENTRE BOARD BOATS.

Many boats, from the ten feet dinghy upwards, are now fitted with centre boards or plates, enclosed in a water-tight case or trunk, and working on a pivot at the fore end. This arrangement gives considerable sailing power to a boat of light weight, as it to a great extent prevents the lee way or side-drift. For a boat of small size kept on a beach, it is desirable that the plate should lift out of the case to diminish weight in hauling up. A good-sized sailing craft thus fitted can be kept for use from a very shallow harbour, and the finest sailing boats of this class I have met with are at Brighton and Shoreham. For this and all other varieties of boats and yachts of moderate size, I refer my readers to 'Yacht and Boat Sailing,' by Mr. Dixon Kemp.

REMARKS ON BEACH OR SURF BOATING.

Launching from a Steep Shingle Beach.—In calm weather with a smooth sea you may launch stern-foremost; but when there is much swell, which is frequently the case even in fine weather, prefer the bow, as your boat will not ship nearly so much water going off head against the sea.

The beach-boats of Yarmouth and Deal have always been celebrated for their sea-going capabilities, and their crews, known as beachmen, are a fine hardy race of fellows, accustomed to launch in all kinds of weather to assist vessels in distress. In the west of England there are also very good beach boats, particularly at Chisel, Portland, Beer, Sidmouth, and Budleigh Salterton. At the last named, I saw in 1866 some of the best boats of a medium size, eighteen to nineteen feet long, and from seven to eight beam, I have met with anywhere. These boats are built for the offing Crab-fishermen, by Exmouth builders. Their chief fishing-ground is from six to ten miles at sea.

If the beach be of shingle and steep, heave your boat down until the water nearly reaches her, having previously placed in her track three or four greased ways, upon one of which she is supposed to be resting at the water's edge, and if the swell runs along the beach from either one side or the other, point the head of the boat a little to meet it, and watching for a smooth, as the water runs up, away with her, and tumble in over the stern, shoving off with an oar to get clear of the breaking wave with all despatch. If you have a companion, let him get on board before you heave afloat, and stand prepared to shove off and pull immediately the boat moves. Your ballast will of course have been placed on board at the water's edge ; but if you have tanks for ballast you will fill them more readily after you are afloat, for which you should provide a funnel tub, with a couple of inches of gutta-percha tubing, one and a half inch in diameter.

Launching from a Flat Sandy Shore.—On beaches consisting almost entirely of sand, there is frequently a very long flat, except just at the top of high water, when the above directions are also applicable.

As I found it very laborious getting my boat down and up a beach of this kind near which I resided for some seasons, I had a simple kind of carriage made (fig. 84), consisting of a frame about eight feet in length by five in breadth, supported on two rollers instead of wheels, fixed a foot from each end, and equal in length to the breadth of the framework, under which the axles of these rollers worked in sockets driven up from the under side, and secured on the top by nuts let into the upper surface. These rollers did not sink into the sand as wheels would have done, and much facilitated the work : but on dry sand above high-water mark I found ways of hard wood five

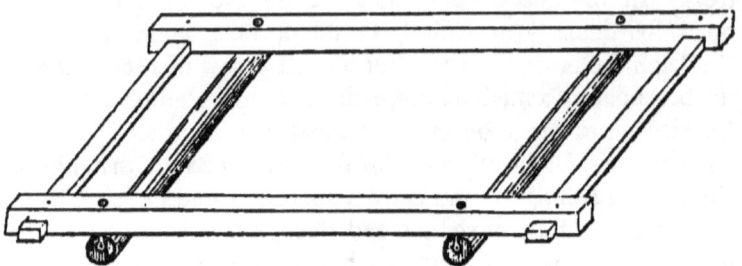

FIG. 84.—Boat Carriage for a Flat Sandy Shore.

feet in length and four inches in breadth preferable, as the rollers drove the sand before them and choked themselves. On this shore the tide ebbed about six hundred feet, but watching the tides I rarely had half the distance to go over.

It often happens that there is excellent fishing off a coast of this kind, consequently if undeterred by a little labour very good sport may be had. Boats for such coasts should be built of fir for lightness.

On the coast of Ireland boats called 'Curraghs,' constructed of a framework of wood covered with canvas, are in use ; and a smaller description for Salmon-fishing are found on the Severn and Wye &c., and termed ' Coracles.'

In launching off a flat sandy shore, get the boat nearly afloat, and having crossed the oars in their places, drop the handles inside; if alone, stand at the stern in the water holding the boat at right angles to the waves, and when you perceive a smooth, shove off and pull seaward as fast as possible. If caught by a breaker, hold the boat up with the oars but do not force against it, and she will split the wave with her bow without damage, unless an excessively bad one.

Beaching or **Landing.**—In landing under sail keep your boat straight before the sea, and if she steers wild take in the mizen if you carry one, or bring a little more weight aft ; let your companion be ready with the painter,[1] sitting on the middle thwart (not further forward), and the instant before the boat strikes put down the helm and sheer her up against the beach aground. Your friend should by this time be over the bows, holding the boat by the painter, to prevent her listing out towards the sea, and you will get rid of your iron or stone as fast as possible, casting it out of the way of the boat up the beach, when you will leap out, and having thrust a way underneath the stem, will be able to haul the boat out of the reach of the water. It is customary to have a hole bored underneath the stem, by passing the painter through which you will be able to lift and pull at the same time.

The fishermen generally use a single block, making the rope fast to a post driven into the shingle high up the beach, which is a very good plan ; but still better is a small capstan and chain having a hook at the end, which can be easily attached to a loop of iron bolted low down on the stem. When running ashore under sail do not lower the canvas until the boat is out of the water, as the sails will materially help in keeping the boat in against the beach, to ensure which the main sheet is often belayed on the weather side. In rowing ashore, watch for a smooth and pull sharply in, and having the painter ready act as above directed.

I have been induced to enter into beach boating somewhat at length, as there are many places on the coast without harbours, such as Brighton &c., where much boating is done during the summer months ; but at the same time I should not advise visiting these open shores specially for the sake of boating or fishing, which can generally be followed much more agreeably where there is a good harbour.

Harbour Boating.—Harbour boating is far preferable to beach boating, for the following reasons : it is not necessary to haul up on coming ashore, consequently the ballasting and unballasting is got rid of, your boat is generally, if not always, afloat, which admits of the bottom being sharper, or, as shipwrights term it, of greater rise of floor, and this gives her great

---

[1] Painter, three fathoms of two and a half inch rope spliced into the ringbolt inside the stem.

advantage in plying to windward; in addition to this, the ballast can be neatly floored over, which renders the boat much more comfortable. Various kinds of bait are usually procurable in most of our larger harbours, which have to be fetched frequently from a distance for the use of the beachmen, and consequently are not always to be had. In many of our harbours a running mooring can be fixed, by help of which the boat can be hauled in or out at pleasure, a great advantage for which no one would object to pay a moderate yearly fee. For harbour boating, the sprit or lugsail rigs are quite as well adapted as for beach work, but as a carvel or smoothly-built boat can here be used, it is preferable to clench-work, as it can be kept clean with much greater facility. A boat of thirteen feet six inches to fifteen feet in length will allow of more variety of fishing than a larger for the reasons before mentioned ; but as many prefer sailing to fishing, they will find no difficulty in suiting themselves in most of our larger ports with any size they may fancy. In case of keeping a large boat for offing-fishing, a smaller one will also be requisite for shore work.

General Remarks.—Whilst sailing under high cliffs when the wind is off shore caution is necessary, as in such localities it is very uncertain, being a constant succession of flaws and calms, and on passing any opening in these cliffs, as the entrance of a valley or ravine, the flaws are felt with increased violence, which is frequently perceived as far as a mile from the shore. In a small boat the main sheet should always be held in the hand, ready to let fly and take the pressure off the boat if necessary ; but if the flaws should not be of extraordinary violence, it will frequently suffice to put the helm down a little and luff up into the wind, as it is called, by which time the flaw will probably have passed. A moderate amount of sail should always be carried in uncertain weather, and far more accidents happen to amateurs in smooth water sailing with offshore winds than at any other time, generally from heedlessness and want of common precaution, from which even many of those who get their livelihood on salt water are not altogether exempt. Large deep-bodied boats are longer in feeling the effect of a flaw of wind, as their form gives them great hold of the water, and there is generally time to shorten sail by tricing up the main tack and lowering the foresail, and, if a yawl, the mainsail also, as the mizen is sufficient to keep the boat close enough to the wind in smooth water.

STAYING is the evolution of coming about against the wind, and is performed by putting the helm down gradually and slackening the jib-sheet, at the same time shifting the mainsheet to the other quarter of the boat ; the jib-sheet must now be hauled in a little on the weather side, until the head of the boat has fallen off sufficiently to fill the sails, when the lee jib-sheet is to be hauled aft and belayed. It is a great error to put down the helm suddenly, as the way of the boat is thereby much deadened and the power of the rudder nearly lost ; in fact, a boat often misses stays (particularly in a heavy swell) from this cause.

Many boats will come round without easing the jib, which is an advantage. In yawls, cutters, &c., as the foresail pays off the head of the boat, the jib-sheet is hauled aft in the act of staying.

WEARING is the evolution of going round on the other tack before the wind, and is performed by putting the helm up or a-weather, changing your position and setting your back against it to keep it so, which leaves your hands at liberty to cast off the main-sheet ; this is to be hauled taut amidships until the wind comes on the other quarter of the boat, when it is to be eased over and made fast as before. Wearing in a strong wind requires much caution in all craft, but particularly in small boats, for if the mainsail be not steadied by hand, but allowed to sally across from one side to the other, the boat may be upset, or if a spritsail, the head of it may get over the top of the mast, and cause risk and trouble to clear it.

BELAYING THE MAIN-SHEET.—The improper performance of this is a fruitful source of accident, and in small boats should be so managed that it may be let fly instantaneously if required. The best arrangement for small craft is to introduce two wooden pins through the stern or transom board of the boat, one on each side ; the ends of the pins should project through the inside long enough to allow of the main-sheet being passed under the inside end after it has been passed under the end outside. When the main-sheet has been passed under the inner end of the pin, a bow or bight is pushed under, which is held securely by the haulage of the sail: this can be immediately cast off by a sharp pull on the free end when necessary. (See fig. 85). In squally weather or under high lands in anything like a breeze, I merely pass the main-sheet under the outside end of the pin, holding it in my hand, ready to let fly on the instant. Much of my experience having been in sailing under the high cliffs of South Devon, the necessity of caution in no ordinary degree has been impressed on me, from the frequent casualties in these localities. Large boats have a tackle purchase fixed on each quarter, or a horse of bar iron, which should be galvanised; the horse is preferable for amateurs.

FIG. 85.—Belaying the Main-sheet.

BEATING TO WINDWARD in a heavy sea, watch the approaching waves, and should any appear likely to break aboard, yield the helm a little to the boat, which will, as sailors term it, 'ease her when she pitches;' she will thus pass lightly over it and ship little or no water, when you can again keep your course.

SCUDDING or RUNNING in a heavy sea, carry as much sail as your boat can bear comfortably, but no more, and diminish your after sail by lowering your mizen, or your sprit if your boat is of that rig, or dropping the peak of a gaffsail-boat a little; you will thus be enabled to keep before the sea, for should she broach-to—that is to say, under these circumstances present her side, instead of her stern, to the pursuing waves—she will very likely be filled.

The Coble.—This description of boat has been in use from time immemorial on the north-east coasts of England, and is well adapted for beaching from its light draught of water. The ordinary dimensions are twenty feet six inches long and five feet beam. The bow is very high and sharp, the stern low, and with no keel aft. The rudder is very deep, running underneath the bottom, so that in scudding before a heavy sea when the stern rises the rudder is still sufficiently submerged to keep the boat from running out of steerage and broaching to, a constant source of accident in other boats They have one mast with a considerable rake, on which they set a lugsail and sometimes also a jib. They are favourite boats of the fishermen and pilots, and when landing are backed in stern foremost because this part of the boat draws hardly any water, as from the stern forward half the bottom is flat. Astern of a vessel they are always towed stern foremost, and are often pulled in the same fashion. They are peculiar boats to handle, but are safe enough when skilfully managed, and some go so far as to say that no other boat, a life-boat excepted, could be used on the above coasts.

Safety Fishing-boats.—The accompanying woodcuts, pp. 273, 274, having attracted my attention in the 'Illustrated News' of March 30, 1867, I communicated with Richard Lewis, Esq., Secretary of the Royal National Life-boat Institution, who was kind enough to place the blocks at my disposal, and at the same time to forward the following article on the subject. I have given a place in this work to the matter, with the intention that should any of my readers take an active interest in the welfare of our hardy fishermen, so far as to forward the object of the Institution, or wish to obtain such a craft for their own use, they may at once be enabled to refer a boat-builder to the plans and sections.

As I find my work is circulating not only through the kingdom but also in the Colonies, I trust the information contained in this article may be useful in the development of our fisheries both abroad and at home. The following article refers *in extenso* to this subject.

IMPROVEMENT OF FISHING-BOATS.

In consequence of the frequent loss of life through the foundering of fishing-boats on the coast of the United Kingdom, it appeared to the Committee of the National Life-boat Institution that the safety of the larger class of open and half-decked fishing-boats on our coast might be greatly increased by enabling them to be made temporarily insubmergible, in the event of their being overtaken by gales of wind when at long distances from the land.

No doubt was entertained by practical persons on the coast, who were consulted on the subject, of the need of such improvement, and of the feasibility of the plan proposed to effect it; but the coast boatmen being an inert class, not readily departing from what they have been accustomed to, it was not thought likely that they would themselves initiate any such changes, however needed.

The Committee, therefore, decided to build a few pattern boats, and to place them at some of the principal fishing stations, in the hands of experienced and trustworthy boatmen, to whom they would be lent or let at a small percentage on their earnings, for a period of twelve months; at the end of which time they might be sold, and would remain in the several localities where placed as samples, from which the other local boats might be improved in a similar manner.

As these boats would be seen by large numbers of fishermen from different places at their chief ports of rendezvous during the fishing seasons, it was considered that it might not be necessary to build any large number in order to make them generally known, and that a short period would suffice for those to whom they were entrusted to form a correct estimate of their properties.

In the event of the experiment proving successful, it was believed that a great boon would thus be conferred on the fishermen and other boatmen of certain classes on the coast, as not only would numberless lives and boats be saved that in course of time would otherwise be lost, but that the boats would often be able to remain at sea and safely continue their fishing in threatening weather, instead of returning to the shore, at great pecuniary loss to their crews, as is now too frequently the case.

Five of such boats were accordingly ordered—three to be built in Scotland, one at Yarmouth, and one by the builders to the Institution in London. Two of the boats built in Scotland, one at Peterhead and the other at Anstruther, have been tested and are now at work, having already afforded the utmost satisfaction to their crews, as will be seen from the following extracts from letters received at the Institution. Captain A. Sim, Hon. Sec. of the Lossiemouth Branch of the Institution, writing from that place on March 18, states :—' The safety fishing-boat sailed from Granton Harbour on Wednesday, the 13th inst., at 6 A.M., and was here the following day at 5 P.M., after lying to for some time off Peterhead, thus making the voyage in thirty-six hours —no bad test of her sailing qualities. She has been very much admired here by all the fishermen ; in fact, the seafaring population are unanimous in their opinion that she is just the thing for this coast, and I trust she may be the beginning of a new era in decked boats.'

Wm. Boyd, Esq., Hon. Sec. of the Peterhead Branch, also writes on the same date :—'You will be glad to hear that the new safety fishing-boat gives very great satisfaction. John Geddes, the life-boat coxswain, lay alongside the Lossiemouth boat in the Frith of Forth, and declares that she is a fast sailer, having accomplished the run from here to Granton in thirteen hours. She works well and satisfactorily, but she had not experienced such bad weather as would thoroughly try her safety powers.'

The interior fittings of the boats have been so arranged as not to interfere with their everyday work, yet so as to enable them to be quickly made insubmergible.

This object has been effected, as is clearly shown in the eight diagrams and explanations prepared by the Society : 1st. By making the usual forecabin a water-tight compartment, the access to it being by a water-tight hatch in the deck, instead of an open door at the side. 2nd. By making the usual compartment at the stern also water-tight. 3rd. By running a side deck along either side, as in barges and in some of the smaller class of yachts, called well-boats. Thus leaving a large open main hatchway, of sufficient size for conveniently working the nets, yet which, by the aid of coamings and hatches, and a water-tight tarpaulin, stowed away in the hold or forecabin in fine weather, could in a few minutes, on the occurrence of bad weather, be securely covered over so that no water could get access to the hold on a heavy sea breaking over the boat.

The inspection of a common coasting-barge with her hatches on and covered over, will convey an exact idea of the simple manner in which the above arrangements are carried out.

The size of these boats, viz. length 40 feet, width 14 feet, depth amidships 7 feet, has been selected as the most convenient size for use both in line and net fishing. A sixth boat, however, 45 feet long by 15 feet wide, is about to be built for Anstruther, where the fishing-boats go as far as 100 miles from the land to fish, and have lines on board of the total length of 23,500 yards, or nearly $13\frac{1}{2}$ miles, which require a large space to stow them away all coiled in baskets, besides a cargo of fish.

The Committee of the National Life-boat Institution entertain sanguine hopes that this experiment will be ultimately productive of much benefit, both by saving life and property.

In conclusion, it may be stated that the National Life-boat Institution has been engaged during the last two or three years in perfecting this model of the safety fishing-boat. Captain J. R. Ward, R.N., its Inspector of Life-boats, has visited during that period some of the principal fishing stations on the coast of the United Kingdom, with the view of eliciting from the most experienced fishermen practical suggestions, to be incorporated in the construction of the boat; so that thus she may be correctly termed an *omnium gatherum* safety fishing-boat. It may also be mentioned that the drawings of the boat have been furnished by Mr. Joseph Prowse, of Her Majesty's dockyard, Woolwich, who, with the kind permission of the Admiralty, superintends the building of all the life-boats of the Institution.

THE SAFETY FISHING-BOAT OF THE ROYAL NATIONAL
LIFE-BOAT INSTITUTION.

The accompanying figures show the general form, the nature of the fittings, air-compartments, shifting-coamings, and hatches of one of the safety boats, 40 feet in length and 14 feet in breadth.

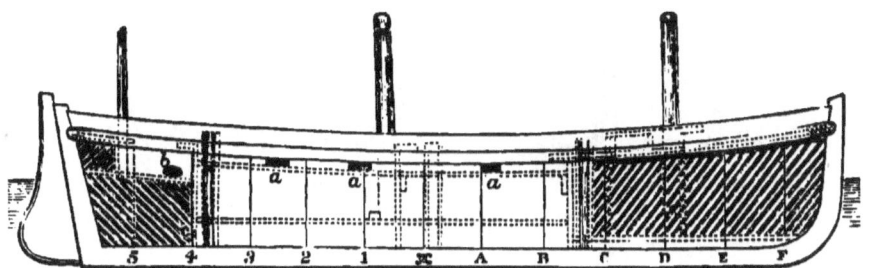

FIG. 86.—Sheer Plan.

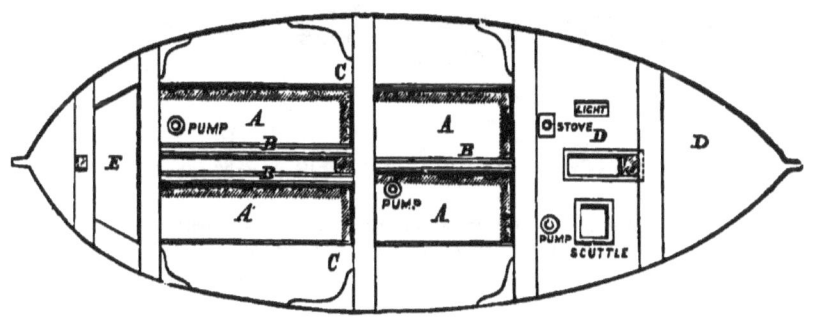

FIG. 87.—Deck Plan.

In figs. 86 and 87 the elevation and deck-plans, and the general exterior form of the boat are shown, with the sheer of gunwale, length of keel, and rake of stem and stern-post.

The dotted lines of fig. 86 show the position of the compartments, bulkheads masts, pumps, thwarts, and shifting flat or deck. *a*, Scuttle in boat's side above the side-decks. *b*, Scupper in the boat's side above the stern deck. *c*, Screw plug, to drain the stern compartment.

In fig. 87, A represents the open hatchways of the main-hold, to be covered with portable hatches and a water-tight tarpaulin cover in gales of wind ; B, shifting-coamings for the hatches ; C, the side-deck ; D, the forecastle-deck ; and E, the stern-deck.

T

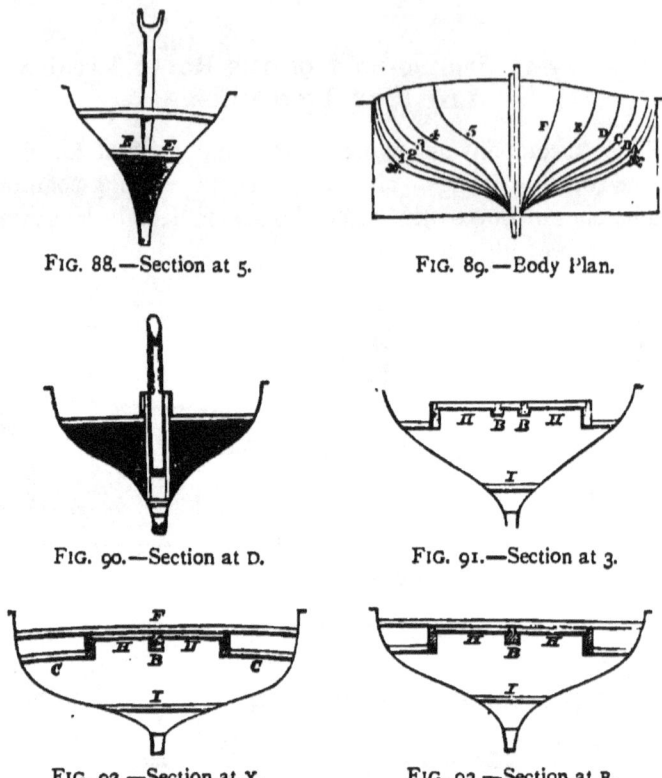

FIG. 88.—Section at 5.   FIG. 89.—Body Plan.

FIG. 90.—Section at D.   FIG. 91.—Section at 3.

FIG. 92.—Section at X.   FIG. 93.—Section at B.

Fig. 88 represents a section at the after air-compartment, showing the thwart, and crutch to receive the mast, and the stern-deck.

In fig. 89 the exterior form of transverse sections, at different distances from stem to stern, is shown.

Fig. 90 represents a section at the fore air-compartment, showing the thwart and mast.

In fig. 91 the shifting-coamings over the main-hold are shown, with the portable hatches (H) in place, and (I) the shifting deck or flat.

Fig. 92 represents a midship transverse section. F the thwart, C the side-deck, B the shifting-coaming over the fore-hold, (I) the shifting deck or flat, and H the hatches in place.

Fig. 93 represents a section abaft the foremost bulkhead, showing the shifting-coaming (B) and portable hatches (H) in place, and (I) the shifting deck or flat.

SAFETY FISHING-BOATS.

To the Editor of the 'Star.'

SIR,—It has often been observed, on the subject of the annual average take of herrings on the Scottish seas, that there have been fluctuations and variations—good seasons and bad—all, however, reducible to one uniform cause of success or failure. The one element which determines the result is weather. In rough and stormy seasons the boats cannot get to their work or get through it, and therefore the catch is diminished, but the harvest is always there if the reaper can reach it.

I have been led to crave the valuable aid of the 'Star' to make more generally known, and to call public attention to, the benevolent object of the Royal National Life-boat Institution, with which I have the honour to be connected, in devising and reducing to practice a plan of a safety fishing-boat, which has, under my own observation, proved most successful as a convenient boat for the prosecution of the white and herring fisheries on our coasts, and is calculated to enable these fisheries to be prosecuted with greater safety to those engaged in them than has yet been attained by any other means of which I am aware. The element of safety in these boats to which I have referred is attained mainly—1. By making the usual fore-cabin a water-tight compartment, the access to it being by a water-tight hatch in the deck instead of an open door at the side. 2. By making the usual compartment at the stern also water-tight. 3. By running a side-deck along each side. In this way a large open main hatchway is left, of sufficient size for working the nets, but which may be covered when necessary by portable hatches and a water-tight tarpaulin cover, to be used in gales of wind. These hatches and tarpaulin are easily stowed away in fine weather, and on the occurrence of bad weather the main hatchway can be securely covered in a few minutes, so that no water can possibly get access below should a heavy sea break on board. In fact, as will be seen, the boat is capable of being converted into a decked water-tight vessel within a few minutes after the occurrence of an emergency, and of being almost instantly restored, if desired, to the condition of an ordinary fishing-boat with a large open main hatchway.

Your columns attest the fact that scarcely a season passes without the occurrence of some terrible disaster, which in all human probability might have been averted by the adoption of an expedient by which open fishing-boats might be preserved from the risk of a sea falling on board of them. Living as I do in one of the most exposed positions of this exposed and stormy coast, I am unfortunately able to bear ample testimony to the benefits which would arise from the general adoption of such an expedient, and supported as I am by the opinions of practical persons, to the evident utility of the very simple plan which I feel that I have somewhat imperfectly described. I can recall to my recollection at least one occasion within the last ten years when an ordinary fishing-boat was, in the presence of numerous but helpless spectators, swamped within a short distance of the shore, and all hands perished. Again, I was not long ago told by a fisherman that, in running for the harbour from the fishing-ground on a dark and stormy night, he heard a cry of distress, as he was borne swiftly before the tempest past the wreck of a swamped boat, one end of which—probably the fore cabin—rudely and imperfectly representing a water-tight compartment—maintained its position above the water after the other portion of the boat had been submerged. Upon this precarious safety stood the sole survivor of the crew, who hailed with the energy of a drowning man the welcome advent of apparent rescue. But the wreck slowly disappeared before my informant was able to stay his boat, with the intention of rendering assistance. And who that witnessed them can forget the scenes enacted on these coasts in August 1849? The fleet of boats, being unprotected to any extent by decks, and therefore unable to live in the heavy sea which suddenly overtook them at the fishing-ground, crowded in disorganised masses towards the entrances of the nearest harbours, and amongst a fearful list of other casualties a fishing-boat, which was descried from this brought up at anchor, rode in safety for a time, but was swamped and her crew drowned, through her increasing proximity to the breakers on the approach of low water.

I venture to think that such calamities as these, with a number of others, some of which have been seen, but many of them occurring in dark and stormy nights have been witnessed by no human eye, would, under Providence, have been averted, had there been a general adoption of the simple principles which I have described as entering into the construction of the safety fishing-boats devised by the Royal National Life-boat Institution. Already five of these boats, with safety fittings such as I have described, have been built by the Institution and placed at selected stations—one of them at this port—in the hands of experienced and trustworthy fishermen, to whom they are let at a small percentage on their earnings, in order that they may serve as models from which the local boats may be improved in a similar manner.

I need scarcely say that the efforts of the Institution, which has expended a large sum of money in bringing the plan I have described under the notice of fishermen and others in a practical form, are entirely devoted to the preservation and safety of life, and that it is not to any extent actuated by the expectation of pecuniary gain. Yet it may not be out of place for me to say, as a proof of the utility of these boats in the advantageous prosecution of the fisheries, that I have just remitted to the Institution a sum of 26*l*. 17*s*. 5*d*., being one-twelfth part of the earnings of the safety fishing-boat here, with which I am connected, since the month of May last. I trust, therefore, that I may be allowed to express an earnest hope that the plan of the safety fishing-boat, of which I rejoice to be able to speak so favourably, may be adopted in the course of time wherever the lives of fishermen are now exposed in open boats to the perils of the sea. I feel assured that those who advocate its adoption not only advance the cause of humanity but promote the welfare of the fishing communities, which must always maintain an important position in the maritime interest of the world.—I remain, sir, your faithful servant,

WILLIAM BOYD, *Hon. Secretary*,
Royal National Life-boat Institution.

PETERHEAD, N.B., *Nov.* 8.

ON THE MANAGEMENT OF OPEN ROWING-BOATS IN A SURF, BEACHING THEM, &c.

The National Life-boat Institution some time since collected information from 128 different places on the coasts of the United Kingdom regarding the system of management of boats in a surf and broken water pursued by fishermen and other coast boatmen. It has appeared to the Committee of the Institution that the information obtained in this manner and in other ways may with advantage be published and circulated, for the guidance of those who may have insufficient experience in the management of boats under such circumstances. Rules for the management of boats in a surf and broken water naturally fall under two heads, viz.: 1st. Their management when proceeding from the shore to seaward against the direction of the surf. 2nd. Their management under the opposite circumstances of running for the shore before a broken sea. Before stating the course to be pursued under each head, we may remark that it is an axiom almost universally acknowledged that there is, as a general rule, *far more danger when running for the shore before a broken sea than when being propelled against it* on going from the land; the danger consisting in the liability of a boat to broach-to and upset, either by running her bow under water, or by her being thrown on her beam-ends, and overturned broadside on.

Rules of Management.—*In Rowing to Seaward.*—As a general rule, *speed must be given to a boat rowing against a heavy surf.* Indeed, under some circumstances, her safety will depend on the utmost possible speed being attained on meeting a sea. For if the sea be really heavy, and the wind blowing a hard on-shore gale, it can only be by the utmost exertions of the crew that any headway can be made. The great danger then is, that an approaching heavy sea may carry the boat away on its front, and turn it broadside on or up-end it, either effect being immediately fatal. A boat's only chance in such a case is to obtain such way as shall enable her to pass, end on, through the crest of the sea, and leave it as soon as possible behind her. If there be rather a heavy surf, but no wind, or the wind off shore, and opposed to the surf, as is often the case, a boat might be propelled so rapidly through it that her bow would fall more suddenly and heavily after topping the sea than if her way had been checked; and it may therefore only be when the sea is of such magnitude, and the boat of such a character, that there may be chance of the former carrying her back before it, that full speed should be given to her. It may also happen that, by careful management under such circumstances, a boat may be made to avoid the sea, so that each wave may break ahead of her, which may be the only chance of safety in a small boat; but if the shore be flat, and the broken water extend to a great distance from it, this will often be impossible. The following general rules for rowing to seaward may therefore be relied on:—

1. If sufficient command can be kept over a boat by the skill of those on board her, avoid or 'dodge' the sea, if possible, so as *not to meet it at the moment of its breaking or curling over*.
2. Against a head gale and heavy surf, *get all possible speed on a boat on the approach of every sea which cannot be avoided*.
3. If more speed can be given to a boat than is sufficient to prevent her being carried back by a surf, her way may be checked on its approach, which will give her an easier passage over it.

II. *On Running before a Broken Sea, or Surf, to the Shore.*—The one great danger, when running before a broken sea, is that of *broaching-to*. To that peculiar effect of the sea, so frequently destructive of human life, the utmost attention must be directed. The cause of a boat's broaching-to, when running before a broken sea or surf is, that her own motion being in the same direction as that of the sea, whether it be given by the force of oars or sails, or by the force of the sea itself, she opposes no resistance to it, but is carried before it. Thus, if a boat be running with her bow to the shore and her stern to the sea, the first effect of a surf or roller overtaking her is to throw up the stern, and as a consequence to depress the bow; if she then has sufficient inertia (which will be proportional to weight) to allow the sea to pass her, she will in succession pass through the descending, the horizontal, and the ascending positions, as the crest of the wave passes successively her stern, her midships, and her bow, in the reverse order in which the same positions occur to a boat propelled to seaward against a surf. This may be defined as the safe mode of running before a broken sea. But if a boat, on being overtaken by a heavy surf, has not sufficient inertia to allow it to pass her, the first of the three positions above enumerated alone occurs—her stern is raised high in the air, and the wave carries the boat before it, on its front, or unsafe side, sometimes with frightful velocity, the bow all the time deeply immersed in the hollow of the sea, where the water, being stationary, or comparatively so, offers a resistance, whilst the crest of the sea, having the actual motion which causes it to break, forces onward the stern or rear end of the boat. A boat will in this position sometimes, aided by careful oar-steerage, run a considerable distance until the wave has broken and expended itself. But it will often happen that, *if the bow be low, it will be driven under water*, when the buoyancy being lost forward, whilst the sea presses on the stern, the boat will be thrown (as it is termed) end over end; or if the bow be high, or if it be protected, as in some life-boats, by a bow airchamber, so that it does not become submerged, that the resistance forward acting on one bow will slightly turn the boat's head, and the force of the surf being transferred to the opposite quarter, *she will in a moment be turned round broadside by the sea*, and be thrown by it on her beam-ends, or altogether capsized. It is in this manner that most boats are upset in a surf, especially on flat coasts, and in this way many lives are annually lost amongst merchant seamen when attempting to land, after being compelled to desert their vessels. Hence it follows that the management of a boat, when landing through a heavy surf, must, as far as possible, be assimilated to that when proceeding to seaward against one, at least so far as to stop her progress shoreward at the moment of being overtaken by a heavy sea, and thus enabling it to pass her. There are different ways of effecting this object :—

1. *By turning the boat's head to the sea before entering the broken water, and then backing in stern foremost, pulling a few strokes ahead to meet each heavy sea, and then again backing astern.* If a sea be really heavy and a boat small, this plan will be generally the safest, as a boat can be kept more under command when the full force of the oars can be used against a heavy surf than by backing them only.

2. If rowing to shore with the stern to seaward, *by backing all the oars on the approach of a heavy sea*, and rowing ahead again as soon as it is passed to the bow of the boat, thus rowing in on the back of the wave ; or, as is practised in some lifeboats, placing the after-oarsmen with their faces forward, and making them row back at each sea on its approach.

3. If rowed in bow foremost, by towing astern a pig of ballast or large stone, or a large basket, or a canvas bag, termed a 'drogue' or drag, made for the purpose, the object of each being to hold the boat's stern back, and prevent her being turned broadside to the sea or broaching-to.

Drogues are in common use by the boatmen on the Norfolk coast; they are conical-shaped bags of about the same form and proportionate length and breadth as a candle extinguisher, about two feet wide at the mouth, and four and a half feet long. They are towed with the mouth foremost by a stout rope, a small line, termed a tripping line, being fast to the apex or pointed end. When towed with the mouth foremost they fill with water, and offer considerable resistance, thereby holding back the stern; by letting go the stouter rope and retaining the smaller line, their position is reversed, when they collapse, and can be readily hauled into the boat. Drogues are chiefly used in sailing-boats, when they both serve to check a boat's way and to keep her end on to the sea. They are, however, a great source of safety in rowing boats, and many rowing life-boats are now provided with them.

A boat's sail bent to a yard and towed astern loosed, the yard being attached to a line capable of being veered, hauled, or let go, will act in some measure as a drogue, and will tend much to break the force of the sea immediately astern of the boat.

Heavy weights should be kept out of the extreme ends of a boat; but when rowing before a heavy sea, the best trim is deepest by the stern, *which prevents the stern being readily beaten off by the sea.* A boat should be steered by an oar over the stern or on one quarter when running before a sea, as the rudder will then at times be of no use. The following general rules may therefore be depended on, when running before, or attempting to land, through a heavy surf or broken water:—

1. As far as possible avoid each sea, by placing the boat where the sea will break ahead of her.

2. If the sea be very heavy, or if the boat be small, and especially if she have a square stern, *bring her bow round to*

seaward and back her in, rowing ahead against each heavy surf sufficiently to allow it to pass the boat.

3. If it be considered safe to proceed to the shore bow foremost, *back the oars against each sea* on its approach, so as to stop the boat's way through the water as far as possible, and if there is a drogue or any other instrument in the boat which may be used as one, tow it astern to aid *in keeping the boat end on to the sea,* which is the chief object in view.

4. Bring the principal weights in the boat towards the end that is to seaward, but not to the extreme end.

5. If a boat worked by both sails and oars be running under sail for the land through a heavy sea, her crew should, under all circumstances, unless the beach be quite steep, *take down her masts and sails* before entering the broken water, and take her to land *under oars alone,* as above described. If she have sails only, her sails should be much reduced, a half-lowered foresail or other small head-sail being sufficient.

III. *Beaching, or landing through a Surf.*—The running before a surf or broken sea, and the beaching or landing of a boat, are two distinct operations : the management of boats, as above recommended, has exclusive reference to running before a surf, where the shore is so flat that the broken water extends to some distance from the beach. Thus, on a very steep beach, the first heavy fall of broken water will be on the beach itself, whilst on some very flat shores there will be broken water as far as the eye can reach, sometimes extending to even four or five miles from the land. The outermost line of broken water, on a flat shore, where the waves break in three and four fathoms water, is the heaviest, and therefore the most dangerous, and when it has been passed through in safety the danger lessens as the water shoals, until on nearing the land its force is spent and its power harmless. As the character of the sea is quite different on steep and flat shores, so is the customary management of boats on landing different in the two situations. On the flat shore, whether a boat be run or backed in, she is kept straight before or end on to the sea until she is fairly aground, when each surf takes her further in as it overtakes her, aided by the crew, who will then generally jump out to lighten her, and drag her in by her sides. As above stated, sail will in this case have been previously taken in, if set, and the boat will have been rowed or backed in by oars alone.

On the other hand, on a *steep* beach it is the general practice, in a boat of any size, to sail right on to the beach, and, in the act of landing, whether under oars or sail, to turn the boat's bow half round towards the direction in which the surf is running, so that she may be thrown on her broadside up the beach, where abundance of help is usually at hand to haul her as quickly as possible out of the reach of the sea. In such situations we believe it is nowhere the practice to back a boat in stern foremost under oars, but to row in full speed as above described.

[This agrees with my own practice during many years of open beach boating.—J. C. W.]

The following extract is also appended :—

PRACTICAL HINTS FOR THE CONSIDERATION AND GUIDANCE OF SEAMEN AND OTHERS HAVING CHARGE OR COMMAND OF BOATS.

1. Acquire the habit of sitting down in a boat, and *never stand up* to perform any work which may be done sitting.

2. *Never climb the mast of a boat*, even in smooth water, to reeve halliards, or for any other purpose, but unstep and lower the mast in preference. Many boats have been upset and very many lives lost from this cause. The smaller a boat, the more necessary this and the foregoing precaution.

3. All spare gear, such as masts, sails, oars, &c., which are stowed above the thwarts, should be lashed close to the sides of a boat; and any heavy articles on the boat's floor be secured, as well as possible, amidships, to prevent them all falling to leeward together on a heavy lurch of the sea.

[To prevent ballast shifting, see p. 253.—J. C. W.]

5. *Boats may ride out a heavy gale in the open sea in safety*, if not in comfort, by lashing their spars, oars, &c., together, and riding to leeward of them, secured to them by a span. The raft thus formed will break the sea; it may either be anchored

or drifting, according to circumstances. If the boat has a sail, the yard should be attached to the spars with the sail loosed. It will break much sea ahead. Also, a weight suspended to the clue of the sail will impede drift when requisite. In all cases of riding by spars not less than two oars should be retained in the boat, to be ready for use in case of parting from the spars.

5. When a surf breaks at only a short distance from the beach, a boat may be veered and backed through it from another boat anchored outside the surf, when two or more boats are in company; or she may be anchored and veered, or backed in from her own anchor.

APPENDIX.

INSTRUCTIONS FOR SAVING DROWNING PERSONS BY SWIMMING TO THEIR RELIEF.

1st. When you approach a person drowning in the water, assure him, with a loud and firm voice, that he is safe.

2nd. Before jumping in to save him, divest yourself as far and as quickly as possible of all clothes; tear them off, if necessary; but if there is not time, loose, at all events, the foot of your drawers, if they are tied, as if you do not do so they will fill with water and drag you.

3rd. On swimming to a person in the sea, if he be struggling, do not seize him then, but keep off for a few seconds, till he gets quiet; for it is sheer madness to take hold of a man when he is struggling in the water, and if you do you run a great risk.

4th. Then get close to him, and take fast hold of the hair of his head, turn him as quickly as possible on to his back, give him a sudden pull, and this will cause him to float; then throw yourself on your back also and swim for the shore, both hands having hold of his hair, you on your back, and he also on his, and, of course, his back to your stomach. In this way you will get sooner and safer ashore than by any other means, and you can easily thus swim with two or three persons. The writer has often, as an experiment, done it with four, and gone with them forty or fifty yards in the sea. One great advantage of this method is, that it enables you to keep your head up, and also to hold the person's head up you are trying to save. It is of primary importance that you take fast hold of the hair, and throw both the person and yourself on your backs. After many experiments, I find this vastly preferable to all the other methods. You can, in this manner, float nearly as long as you please, or until a boat or other help can be obtained.

5th. I believe there is no such thing as a death-*grasp*, at least it must be unusual, for I have seen many persons drowned, and have never witnessed it. As soon as a drowning man begins to get feeble and to lose his recollection, he gradually slackens his hold, until he quits it altogether. No apprehension need therefore be felt on that head, when attempting to rescue a drowning person.

6th. After a person has sunk to the bottom, if the water be smooth, the exact position where the body lies may be known by the air-bubbles, which will occasionally rise to the surface, allowance being made, of course, for the motion of the water if in a tideway or stream, which will have carried the bubbles out of a perpendicular course in rising to the surface. A body may be often regained from the bottom, before too late for recovery, by diving for it in the direction indicated by these bubbles.

7th. On rescuing a person by diving to the bottom, the hair of the head should be seized by one hand only, and the other used in conjunction with the feet in raising yourself and the drowning person to the surface.

8th. If in the sea, it may sometimes be a great error to try to get to land. If there be a strong 'out-setting' tide, and you are swimming, either by yourself, or having hold of a person who cannot swim, then get on to your back and float till help comes. Many a man exhausts himself by stemming the billows for the shore on a back-going tide, and sinks in the effort, when, if he had floated, a boat, or other aid, might have been obtained.

9th. These instructions apply alike to all circumstances, whether the roughest sea or smooth water.

JOSEPH R. HODGSON.

SUNDERLAND, *Dec.* 1858.

DIRECTIONS FOR RESTORING THE APPARENTLY DROWNED.

It may happen that some one or other of my readers may be present at the occurrence of an accident of the kind to which the following directions relate, and, from their perusal, be enabled to be of essential service. However this may be, it is certain that such valuable instructions as these cannot be too widely disseminated, and I therefore insert them in the work, as peculiarly *àpropos* to a volume certain to be placed in the hands of many sufficiently interested in aquatic pursuits to pass much of their time either on the bosom or the margin 'of old ocean's depths.'

These directions are issued by the Royal National Life-boat Institution.

I.

Send immediately for medical assistance, blankets, and dry clothing, but proceed to treat the patient *instantly* on the spot, in the open air, with the face downward, whether on shore or afloat ; exposing the face, neck, and chest to the wind, except in severe weather, and removing all tight clothing from the neck, and chest, especially the braces.

The points to be aimed at are—first and *immediately*, the RESTORATION OF BREATHING ; and, secondly, after breathing is restored, the PROMOTION OF WARMTH AND CIRCULATION.

The efforts to *restore Breathing* must be commenced immediately and energetically, and persevered in for one or two hours, or until a medical man has pronounced that life is extinct. Efforts to promote *Warmth* and *Circulation*, beyond removing the wet clothes and drying the skin, must not be made until the first appearance of natural breathing ; for if circulation of the blood be induced before breathing has recommenced, the restoration to life will be endangered.

II.—TO RESTORE BREATHING.

To clear the Throat.—Place the patient on the floor or ground with the face downwards and one of the arms under the forehead, in which position all fluids will more readily escape by the mouth, and the tongue itself will fall forward, leaving the entrance into the windpipe free. Assist this operation by wiping and cleansing the mouth.

If satisfactory breathing commences, use the treatment described below to promote Warmth. If there be only slight breathing—or no breathing—or if the breathing fail, then—

To Excite Breathing.—Turn the patient well and instantly on the side, supporting the head (see fig. 94), and excite the nostrils with snuff, hartshorn, or smelling salts, or tickle the throat with a feather &c., if they are at hand. Rub the chest and face warm, and dash cold water, or cold and hot water alternately, on them. If there be no success, lose not a moment, but instantly—

To Imitate Breathing.—Replace the patient on the face, raising and supporting the chest well on a folded coat or other article of dress. (See fig. 95.)

Turn the body very gently on the side, and a little beyond, and then briskly on the face, back again, repeating these measures cautiously, efficiently, and perseveringly, about fifteen times in the minute, or once every four or five seconds, occasionally varying the side.

[*By placing the patient on the chest, the weight of the body forces the air out; when turned on the side, this pressure is removed, and the air enters the chest.*]

FIG. 94.—Inspiration.

On each occasion that the body is replaced on the face make uniform but efficient pressure, with brisk movement, on the back, between and below the shoulder-blades or bones on each side, removing the pressure immediately before turning the body on the side.

During the whole of the operations let one person attend solely to the movements of the head and of the arm placed under it.

[*The first measure increases the expiration—the second commences inspiration.*]

. The result is *Respiration* or *Natural Breathing*; and, if not too late, *Life.*

Whilst the above operations are being proceeded with, dry the hands and feet, and as soon as dry clothing or blankets can be procured, strip the body, and cover or gradually reclothe it, but taking care not to interfere with the efforts to restore breathing.

III.

Should these efforts not prove successful in the course of from two to five minutes, proceed to imitate breathing by Dr. Silvester's method, as follows :—

Place the patient on the back on a flat surface, inclined a little upwards from the feet ; raise and support the head and shoulders on a small firm cushion or folded article of dress placed under the shoulder-blades.

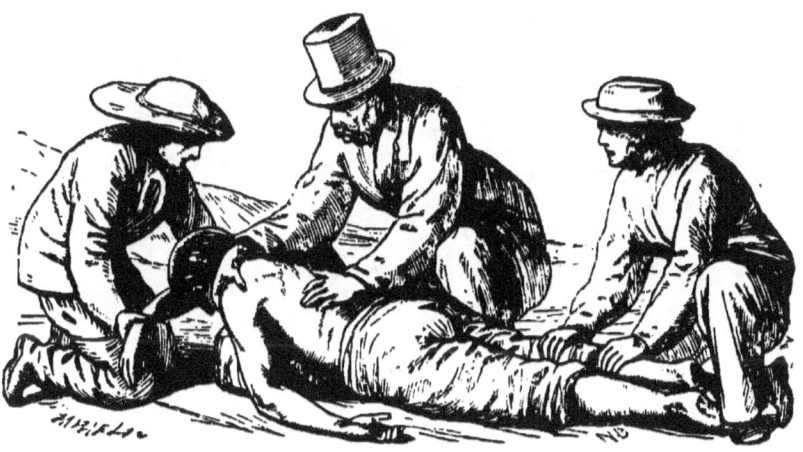

FIG. 95.—Expiration.

The foregoing two illustrations show the position of the body during the employment of Dr. Marshall Hall's method of inducing Respiration.

Draw forward the patient's tongue, and keep it projected beyond the lips : an elastic band over the tongue and under the chin will answer this purpose, or a piece of string or tape may be tied round them, or by raising the lower jaw the teeth may be made to retain the tongue in that position. Remove all tight clothing from about the neck and chest, especially the braces.

To Imitate the Movements of Breathing.—Standing at the patient's head, grasp the arms just above the elbows, and draw the arms gently and steadily upwards above the head, and *keep them stretched* upwards for two seconds as in fig. 96. (*By this means air is drawn into the lungs.*) Then turn down the patient's arms, and press them gently and firmly for two seconds against the sides of the chest as in fig. 97. (*By this means air is pressed out of the lungs.*)

Repeat these measures alternately, deliberately, and perseveringly, about fifteen times in a minute, until a spontaneous effort to respire is perceived, immediately upon which, cease to imitate the movements of breathing, and proceed to INDUCE CIRCULATION AND WARMTH.

FIG. 96.—Inspiration.

IV.—TREATMENT AFTER NATURAL BREATHING HAS BEEN RESTORED.

To Promote Warmth and Circulation.— Commence rubbing the limbs upwards, with firm grasping pressure and energy, using handkerchiefs, flannels, &c. (*By this measure the blood is propelled along the veins towards the heart.*)

The friction must be continued under the blanket or over the dry clothing.

Promote the warmth of the body by the application of hot flannels, bottles, or bladders of hot-water, heated bricks, &c., to the pit of the stomach, the arm-pits, between the thighs, and to the soles of the feet.

If the patient has been carried to a house after respiration has been restored, be careful to let the air play freely about the room.

APPENDIX.

On the restoration of life, a teaspoonful of warm water should be given; and then, if power of swallowing have returned, small quantities of wine, warm brandy and water, or coffee should be administered. The patient should be kept in bed, and a disposition to sleep encouraged.

GENERAL OBSERVATIONS.

The above treatment should be persevered in for some hours, as it is an erroneous opinion that persons are irrecoverable because life does not soon make its appearance, persons having been restored after persevering for many hours.

FIG. 97.—Expiration.

The foregoing two illustrations show the position of the body during the employment of Dr. Silvester's method of inducing Respiration.

APPEARANCES WHICH GENERALLY ACCOMPANY DEATH.

Breathing and the heart's action cease entirely; the eyelids are generally half-closed; the pupils dilated; the jaws clenched; the fingers semi-contracted; the tongue approaches to the under edges of the lips, and these, as well as the nostrils, are covered with a frothy mucus. Coldness and pallor of surface increases.

CAUTIONS.

Prevent unnecessary crowding of persons round the body, especially if in an apartment.

Avoid rough usage, and do not allow the body to remain on the back, unless the tongue is secured.

Under no circumstances hold the body up by the feet.

On no account place the body in a warm bath, unless under medical direction, and even then it should only be employed as a momentary excitant.

INDEX.

ABE	PAGE
ABERDOVEY	30
Aberrffraw, Anglesey	30
Albicore	221
Alderney	28
Anchor Bend	203
— Crowning or Scowing	202
— Drift	159
— Yoke	201
Angling for Bass	151
Appledore	30
Artificial Baits	82, 83, 199
— — for Mackerel	127
BABBACOMBE	19
Bait-tray	214
Baited Prawn-net	246
Baits—	
— Artificial	82, 83, 199
— for Dabs	117
— for Pollack	63, 96
— General, for Sea-fish	187
— for Whiffing	82
— for Whiting-Pout	100
Ballast	252
Band, India-rubber	82
Banker's Lead	41
Barking or Tanning Lines	48
Barmouth	30
Basket, the Fish	215
Bass, the	137
— Angling for	151
— Drift-line fishing for, in Bar-harbours	145
— Fly-fishing for	137
— Ground-fishing for	139
— Whiffing for	150
Beach Boat	251
— Launching	262, 263
— or Surf Boating	262
Beaching	264

BUC	PAGE
Beating to Windward	268
Beaumaris	30
Beer	16
Belaying the Main-Sheet	267
Belaying thwart	203
Bend, a	200
— the Anchor	203
Bending on Hooks	204
Boat, Beach	250
— for Fly-fishing at Sea	91
— Guernsey Spritsail	255
— Harbour	251
— Lugsail	257
— for Mackerel-fishing	123
— Management of, Hints on 267, 268, 277, 280	
— Safety Fishing	273
— Spritsail, with Mizen	254
— Shaped rig	41
— Leads, moulds for	43
Boating, Beach or Surf	262
— General Remarks on	266
— Harbour	265
Boats and Boating	250
— Centre Board	262
— Prices of	261
— Safety Fishing	269
Bobbing for Eels	183
Booms or Bobbers for Mackerel	127
Bottom for Fly-fishing	88
Bowline Knot	200
Braize or Becker, the	173
Bridlington	29
Bridport	15
Brill, the	170
Brim or Sea-Bream	170
Brixham	20
Budleigh Salterton	16
Bulter, Outhaul, for Bass	143
— Trot, or Spiller for Bass, &c.	142
Bucket, Canvas	148

INDEX.

CAB	PAGE
CABLE, slipping the	203
Capelin-net	163
Carnarvon	30
Cawsand Bay	23
Centre Board Boats	262
Chad-Brain or Sea-Bream	170
Chervin or Shrimp Ground-bait	154
— Horsehair net for	155
Christchurch	14
Clothing	219
Clotting for Eels	184
Clovelly	30
Coal or Cole, Whiting	60
Coal-fish	98
Coble, the	269
Cod, the	158
Cod-fishery, Newfoundland	159
Collar for Fly-fishing	88
Common Green Crab	196
Common or Overhand Knot	199
Conger, the	176
Cork Seats	220
Cornish Whiffing Line	81
Courge, Sand-Eel	66
— To make	67
Crab, the	219
Crab, the Common Green	196
— Hermit or Soldier	197
Cray-fish	219
Creeper Sinker	40
Crowning or Scowing Anchor	202
Cuttle-fish	193
DAB, the	115
— Bait for	117
— To salt	118
Dandy or Yawl	258
Dartmouth	21
— Rig	38
Dawlish	17
Day with the Mackerel	129
Day's Ground-fishing	101
— Whiting-fishing	55
Dew-Worm	190
Dip-Leads	34
— Mould for	41
Dipping Sand-Eels on Surface	191
Directions for Rescuing Drowning Persons	285
Disgorgers	53
Dog-fish and Sharks	182
Dome-shaped Lead	145
Dory	152
Dover	29
Drag	159
Dredge	241

GEN	PAGE
Drift-anchor	159
— fishing	55
— Line-fishing for Bass	145
— — — a Day's	74
— — on Whiting-ground	51
Drift-net	247
— Trot for Turbot	169
Drifting on Whiting-ground	55
Drowned Persons, Recovery of	286
Drowning Persons, to Rescue	285
Drum-net	249
EARTH, Dew, or Lob-Worm	190
— Worm for Bait	85
Eels, Bobbing for	183
— Clotting for	184
— Freshwater	183, 191
— — for Bait	82, 84, 191
— Hook and Line for	185
— Night-lines for	185, 186
— Spearing	186
Exmouth	17
Eye Splice	202
FALMOUTH	26
Filey	29
Fish Baits	187, 199
— Basket	215
Fisherman's Spinning-machine	216
Fishing-gear Ground	33
Fishing, Methods of	8
— Boat, Safety	269, 270, 273
— Boats, Improvement of	270
— Sheaf	92
Flamborough Head	29
Flies	88
Floating Trot	96
Floats	94
Flounder or Fluke	119
— Spearing	121
Fluking-Pick or Pike	122
Fly-fishing for Bass	137
— — at Sea	86
— — Boat for	91
Fork	121
Fowey	25
Freshwater-Eel	183
— — for Bait	82, 84, 191
GAFFS	213
Gar-fish or Long-Nose	174
Gear or Tackle required	12
General Baits for Sea-fish	187
— Remarks on Boating	266

INDEX.

GRA	PAGE
Grains, the	221
Grapnel Sinker	40
Green Crab, Common	196
Grey Mullet	153
— Chervin Ground-bait for	154
— Ground-bait for	153
Ground-bait for Sand-Smelt	157
— — for Sea-Bream	172
Ground-fishing for Bass	139
— — A Day's	101
— — Harbour	144
— — Gear	33
Guernsey	28
— Spritsail Boat	255
— Rig	36
Gurnard or Gurnet	175
Gut-links, to Knot	74
HADDOCK, the	165
Hair-links, to Knot	73
Hake, the	165
Halibut, the	122
Harbour Boat	251
— Boating	265
— Ground-fishing	144
Hartland	31
Haunts of Mackerel	136
Helford	27
Hermit or Soldier Crab	197
Herring, the	182
Herring-nets, moored	249
Hints on Boat Management	267, 280, 283
Hitch, the Timber	200
Hook and Line for Eels	185
— Holder	143
Hooks	204
— Whipping and Bending on	204
Horse-hair Lines, to make	69
— Nets for Chervin	154
Horse-Mackerel or Scad	136
Hydrography and Sea-fishing	11
ILFRACOMBE	30
Improvement of Fishing-Boats	270
India-rubber Band	82
Isle of Man	28
— — Wight	14
Isles of Scilly	27
Itchen River Rig	261
JACK for Twisting	69, 70
Jersey	28

MAI	PAGE
KENTISH Rig	37
Killick or Sling-stone	201
Knot, the Bowline	200
— Common or Overhand	199
LAMPERN or River Lamprey	192
Lamperns for Bait	82
Landing	264
— Net	214
Launce	190
Launching from flat Shore	263
— — Steep Beach	104, 262
Leads, Dip	34, 41
— Dome-shape	145
— for Mackerel-fishing	124
— Shifting	10
Leger-fishing	9
— Line for Bass	139
Lesser Lamprey	82, 192
Lester-Cock, the Sunken	218
— — Trot	218
Limpet	192
Lines	47
— Barking or Tanning	48
— Cornish Whiffing	81
— for Fly-fishing	87
— for Mackerel-fishing	124
— for Whiffing	79
Ling	164
Links, Gut, to Knot	74
— Hair, to Knot	73
Lizard, the	31
Lobsters	219
Lob-Worm	190
Long-Nose or Gar-fish	174
Looe	25
Lowestoft	29
Lugsail-Boat	257
Lug-Worm	188
Lyme	16
MACKEREL, the	123
— Artificial Baits for	127
— Booms or Bobbers for	127
— Day with the	129
— Fishing at Anchor	134
— Ground	128
— Haunts	136
— Last or Bait, to Cut	126
— Leads for	124
— Lines for	124
— Sailing-Boat for	123
— Seine	232
— Taken at Night	136
Main-Sheet, Belaying	267

MAN	PAGE	SCU	PAGE
Management of Rowing-Boats	277	Pollack, Pater-Noster.	95
Margate	30	— Rod-fishing for	93
Marks, and How to Take Them	4	— Streaming for	78
Methods of Fishing	8	Polperro	25
Mevagissey	26	Poole	14
Milford	30	Portland	15
Moored Herring-nets	249	Portsmouth	13
Mould for Boat-leads	43	— Rig	36
— for Dip-leads	41	Possible Bag	10
Mud-Worm	188	Poulpe or Sucker	195
Mullet, Grey	153	Pout or Whiting-Pout	99
— Red	174	Power or Poor-Cod	101
Mussel	50, 51, 106, 187	Practical Hints, &c.	283
		Prawn	196, 242, 245
		— Net	242
NETS, Capelin	163	— — Baited	246
— Drift	247	Prices of Boats	261
— Drum	249		
— Hand	91		
— Landing	91, 214	RACE or Rauning Pollack	98
— Moored Herring	249	Rag-Worm	188
— Peter	249	Ramsgate	30
— Prawn, Baited	246	Razor-fish	197
— Remarks on	222	Recovery of Persons apparently	
— Sand-Smelt	157	Drowned	286
— Shrimp and Prawn	242	Red Mullet	174
— Strand	243	Reels	212
— Tanning	249	Remarks on Nets	222
— Trammel	222	— on Beach or Surf Boating	262
— Triangle	222	Rig, the Itchen River	260
Newfoundland Cod-fishery	159	Rock-fish or Wrasse	170
— Lead	41	— Ling	165
Night-lines for Eels	185, 186	— Worm	188
Night or Small Seine	231	Rod for Fly-fishing	86
Nossil-cock	216	Rod-fishing for Pollack	93
		Rowing-Boats, Management of	277
OCEAN-FISHING	220	Running	268
Otter	218		
— Trawl	240	SAFETY Fishing-Boats	269–273
Outhaul Bulter	143	Sailing-Boat for Mackerel-fishing	123
Overhand or Common Knot	199	St. Ives	31
		Salcombe Harbour	22
		Sand-Eel	190
PADSTOW	31	— Cut for Mackerel Bait	135
Pater-Noster for Pollack	95	— Dipping on Surface	191
— for Smelt	156	— Living to Bait with	63, 64, 65
Penzance	31	— Courge	66
Peter Nets	249	— Seine	228
Pilchard, the	182	Sand-Smelt	155
— Bait	48, 49	— — Ground-Bait	157
— Seine	235	— — Nets for	157
Pipe-leads and Moulds	72	Sand-Worm, the White	189
Plaice, the	168	Scad or Horse-Mackerel	136
Plymouth	22	Scarborough	29
Pollack-Whiting	60	Scowing the Anchor	202
— Baits for	63, 82	Scudding	268

SEA	PAGE	WHI	PAGE
Sea-Bream	170	Staying	267
— Ground-Bait for	172	Strand-net	243
Sea-fish, General Baits for	187	Streaming for Pollack	78
Sea-fishing Gear	12	Sucker or Poulpe	195
— and Hydrography	11	Sunken Lester-Cock	218
Sea-Loach	165	Swanage	14
Sea Tape-Worm or Varm	190		
Seaton	16		
Seats, Cork	220	TACKING or Staying	267
Seine	227	Tackle or Gear required	12
— Mackerel	232	Tanning Lines	48
— Night or Small	231	— Nets	249
— the Pilchard	235	Teignmouth	18
— the Sand-Eel	228	Tenby	30
Sharks and Dog-fish	182	Thwart, Belaying	203
Sheaf-fishing	92	Tide-way Fishing	61
Shifting leads	10	Timber Hitch	200
Shore-fishing	24, 93, 139	Tobacco-Pipe Bait	127
Short-splice, Two-strand	202	Torquay	19
Shrewsbury Thread	223, 227	Trace for Fly-fishing	88
Shrimps	196, 242, 245	Train-Oil	163
— to Bait with	69	Trammel	222
Ground Bait for Mullet	154	Trawl	237
— and Prawn-nets	242, 243	— the Otter	240
Sidmouth	16	Triangle-net	222
Sillock	98	Trot-Basket	143
Silver-Whiting	32	Trot, Bulter, or Spiller	142
Sinkers, Weight of	46	— Floating	96
Skate	181	— The Lester-Cock	218
Slapton Lake	21	Turbot, the	169
Sling-stone or Killick	201	— Drift-Trot	169
Slipping the Cable	203	Twisting-Lines, Jack for	69
Smelt	155	Tynemouth	29
— Pater-Noster for	155		
Snatch-blocks	46		
Snooding	34, 205	VARM or Sea Tape-Worm	190
Snoods	205		
Soft Iron Hooks	204		
Soldier Crab	197	WEARING	267
Soldier-line	77	Weight of Sinkers	46
Sole	167	Weymouth	15
Solen	197	Whelk, the	193
Solent	14	Whiffing	9, 79
Southampton	14	— Baits for	82
— Dip	34	— for Bass	150
South-West Coast Fishing, &c.	13–31	— Line, Cornish	81
Spearing Eels	186	— Lines for	79
— Flounders	121, 122	Whipping on Hooks	204
Spinner, Fisherman's	216	Whitby	29
Splice, Eye	202	White Bait	182
— Short	202	— Sand-Worm	189
Sprat	182	Whiting	31
Spritsail-Boat with Mizen	254	— Coal or Cole	60
— — Guernsey	255	— Fishing, a Day's	55
Sprool Rig	37	— Ground, Drifting on	55
Squid	194	— Pollack	60
Start Point	22	— Pout	95

WHI	PAGE	YOK	PAGE
Whiting-Pout, Baits for	100	Worms Varm, or Sea Tape	190
— To Cure	53	— White Sand	189
Winch for Fly-fishing	87	Wrasse or Rock-fish	170
Windward, Beating to	268		
Worms, Earth	190	YARMOUTH	29
— Lug	188	Yawl or Dandy	258
— Mud, Rag, or Rock	188	Yoke Anchor	201

LONDON: PRINTED BY
SPOTTISWOODE AND CO., NEW-STREET SQUARE
AND PARLIAMENT STREET

Advertisements. 1

SEA FISHING TACKLE FOR ALL PARTS OF THE WORLD.

HEARDER & SON,

General Fishing Tackle Makers

TO

**H.M. GOVERNMENT AND EXPLORING EXPEDITIONS,
H R.H. THE PRINCE OF WALES, H.R.H. THE DUKE OF EDINBURGH,
AND THE LEADING NOBILITY AND GENTRY OF THE WORLD,**

MAKERS OF

IMPROVED NATURALIST and other DREDGES, SIEVES, and SWABS,

AS USED BY THE PRINCIPAL SCIENTIFIC SOCIETIES,
Were awarded the GOLD MEDAL, the DIPLOMA OF HONOUR, and the

ONLY SPECIAL PRIZE (MONEY)

That was given for COLLECTION of MOUNTED HAND LINES and GEAR, and which was competed for by the whole world at the International Fisheries Exhibition, 1883;
And also were awarded the ONLY MEDAL and DIPLOMA OF HONOUR at the Cork Exhibition, 1883.

This makes 15 Medals and 3 Special Awards for excellence and superiority.

Their collection of Gear and Baits, which have so long been the admiration of all practical fishermen, is suited for Surface, Drift or Railing. or Deep Sea Fishing, and their Trots, Spillers, and Boulters have received special awards for their completeness.

INVENTORS OF THE

CELEBRATED OTTER TRAWLS,

Which have received that greatest amount of praise and flattery, viz., imitation.

BEAM TRAWLS, with jointed Pipe or Wood Beams, TRAMMELS, SEINES. and all kinds of Nets, of best material only, and made by practical fishermen who know how to use them.
MOUNTED HAND LINES for Pollack, Bass, Lythe. Billett, Cod Fish, and all surface fish. Also for Cod, Conger, Turbot, Hake, Halibut, Snook, Cape Salmon, Bonito, Albacore, Dolphin, or Shark.
Inventors and Manufacturers of the celebrated SILVER SPINNERS. SAND EELS, and other India Rubber Baits; also the world-renowned

PLANO CONVEX MINNOW,

For Fresh or Salt Water Fishing.

HEARDER & SON could fill pages with newspaper notices and encomiums, but space will not permit.
Illustrated Guide to Sea and River Fishing gratis; including a copious list of all Tackle suited for fishing or scientific research in any part of the world.

Only Address—195 UNION STREET, PLYMOUTH.

NO AGENTS.

Yacht Smiths, Plumbers, and Galvanizers, Electrical Bells, &c.

HERBERT E. HOUNSELL

(LIMITED),

GOLD MEDAL, LONDON, 1883.   GOLD MEDAL, PARIS, 1875.   SILVER MEDAL, LONDON, 1883.

MANUFACTURERS OF

NETS,
LINES,
AND
TWINES

OF EVERY DESCRIPTION, IN

HEMP, FLAX, AND COTTON.

ADDRESS FULLY,

'PELICAN' WORKS,

BRIDPORT, DORSET.

Advertisements.

HERBERT E. HOUNSELL, LIMITED,
PELICAN WORKS, BRIDPORT, ENGLAND.
PRICE LIST, 1884.

TERMS—Nett Cash. Delivered in London, Liverpool, Plymouth, Cowes, or Bristol.

TRAMMEL OR FLUE NETS.
(FITTED COMPLETE FOR USE).

				£	s.	d.
10 fms. long by	6 ft. deep	..	1	10	0
15 fms. ,, ,,	6 ft. ,,	..	2	5	0
20 fms. ,, ,,	6 ft. ,,	..	3	0	0
25 fms. ,, ,,	6 ft. ,,	..	3	15	0
30 fms. ,, ,,	6 ft. ,,	..	4	10	0
35 fms. ,, ,,	6 ft. ,,	..	5	5	0
40 fms. ,, ,,	6 ft. ,,	..	6	0	0
50 fms. ,, ,,	6 ft. ,,	..	7	10	0
20 fms. ,, ,,	9 ft. ,,	..	3	16	0
25 fms ,, ,,	9 ft. ,,	..	4	16	0
30 fms. ,, ,,	9 ft. ,,	..	5	15	0
40 fms. ,, ,,	9 ft. ,,	..	7	13	6
50 fms. ,, ,,	9 ft. ,,	..	9	12	0
20 fms. ,, ,,	12 ft. ,,	..	4	10	0
25 fms. ,, ,,	12 ft. ,,	..	5	12	6
30 fms. ,, ,,	12 ft. ,,	..	6	15	0
40 fms. ,, ,,	12 ft. ,,	..	9	0	0
50 fms. ,, ,,	12 ft. ,,	..	11	5	0

If Barked, 1s. 6d. per 10 fms. by 6 ft., extra.

OTTER TRAWLS.
Barked, Corked, and Leaded, with Pockets.

	Including Otter Boards.		Complete with Warp, &c.	
	£ s.	d.	£ s.	d.
Width 30 ft. mouth ..	4 5	0	7 5	0
,, 42 ft. ,, ..	7 0	0	10 10	0
,, 60 ft. ,, ..	9 10	0	15 0	0
,, 90 ft. ,, ..	17 10	0	27 10	0

DEEP-SEA BEAM TRAWLS.

Made from 3-strand best quality twine, barked and fitted complete, with pockets to prevent the escape of fish (but no beam, irons, or hauling ropes).

For a

10, 12, 16, 20, 24, 30, 35, 40 ft. Beam, 48/- 50/- 57/6 70/- 90/- 115/- 145/- 170/- each.

FISHING LINES AND NETS.

All sorts, Mounted or not Mounted in Stock. Seines, Drift and other Fishing Nets of all descriptions, made and mounted to order.

Full Size Net Hammocks complete, especially recommended for Yachts and Camping Out, 7s. 6d. each ; if barked, 8s. 6d. each.

PLEASE ADDRESS ALL COMMUNICATIONS TO THE WORKS.

'To set a trammel from a boat the following arrangements are required. Two buoy-lines with corks at intervals, and stones at the ends about 25 lbs. weight, must be provided ; to one of which, close to the stone, make fast the lead-line of the trammel, and at the breadth of the net, above the stone, make fast the cork-line, being careful not to stretch the trammel up too high, lest the strain be taken by the network instead of the buoy-line, which, being stronger than the netting, ought to take the whole of the strain. Place the buoy-line carefully in a coil with the stone by itself in the middle of the boat, and proceed to drop the cork-line in the stern-sheets as near the stern as possible, but the lead-line should be in advance of it about 3 feet when the slack net will naturally take its place between the two. When you have thus arranged the whole net, place the second buoy-line conveniently, together with the stone, on one side of the net, having first secured the head and foot-lines, as mentioned above, to the buoy-line, which you are now to throw overboard, and then proceed to lower the stone with care and deliberation to the bottom. It is always better that two should be in the boat on these occasions, as one can pull slowly whilst the other pays out the net. A trammel should always be shot with, and not across, the tide, for if the latter mode were adopted the force of the current would tend to depress the net towards the ground, and thereby injure its efficiency. A trammel of 40 fathoms length will be found quite large enough for general use, and if two of these nets be required they should not exceed 30 fathoms each, as they are then very convenient for river fishing, and for sea work a long net is at once made by joining the two together. Trammels and others should be spread on a clean shingle beach or grass field, or hoisted up to dry after using, and all weeds picked carefully out; they should likewise be barked, in common with other nets, at least once a season. All broken meshes should be at once repaired, as "a stitch in time saves nine." Many yachts on coming to anchor of an evening in a roadstead set this net : it should, if shot at six or seven, be hauled at about half-past nine p.m., it may then be shot again and hauled at daylight. If left the whole time without examination the fish will probably be devoured by Crabs, Squid, &c., to which the Red Mullet generally are the first to fall victims.'—Wilcock's "Sea Fisherman," page 225, 4th Edition.

THE
EGERIA FISHING BASKET

Prize Medal Awarded at

NATIONAL FISHERIES EXHIBITION,

NORWICH 1881.

REGISTERED.

A COMPLETE OUTFIT

FOR

SEA-FISHING,

DESIGNED AND ARRANGED BY

BRADDELL & SON,

CASTLE PLACE (Next the Ulster Club),

BELFAST.

SOLE MANUFACTURERS.

B. & SON respectfully invite the attention of the Nobility and Gentry of the United Kingdom, and Yachtsmen &c. in all parts of the globe, to the above Outfit for Sea-Fishing, which has already elicited the warmest encomiums from a large circle of the aristocracy of the country, and has been most favourably noticed by THE FIELD and other first-class newspapers.

From the Editor of 'The Field.'

'EGERIA FISHING BASKET.—This is a flat basket containing everything in the way of tackle which a sea-fisherman can be expected to require, and filled up of the best possible materials, and in the most perfect manner. When going a-fishing with this basket, the sea-fisherman has not to think whether he has these leads, or those hooks, or that disgorger. They are all here under his hand, stowed away as conveniently as possible—nothing missing, nothing wanting. Lines for streaming, whiffing, bottom fishing &c., beautifully rigged and fitted with spare chop-sticks, hooks, leads, horse line, bass flies, bait knife, gaff &c., all are there, many of them fitted into neat japanned cases, and complete in their way, not forgetting a big sandwich box, and, last but not least, what a Scotch miller once described to us as a "muckle whusky flask." This useful collection of conveniently stowed sea-fishing tackle is the production of Messrs. BRADDELL & SON, of Belfast. It is intended for the use of yachtsmen, and will no doubt be largely patronised by them, as containing in a moderate space all that they can require.'—THE FIELD, June 1871.

UNDER THE PATRONAGE OF HER MAJESTY THE QUEEN
H.R.H. THE PRINCE OF WALES,
H.R.H. THE DUKE OF EDINBURGH, &c. &c. &c.

PASCALL ATKEY & SON,
YACHT FITTERS,
Compass Adjusters and Galvanizers,
WEST COWES, ISLE OF WIGHT,
Inventors and Sole Manufacturers of the

R.Y.S. YACHT'S COOKING APPARATUS,
Which they have supplied to over 3,500 Yachts; also their Registered

R.Y.S., Wilton, and other Porcelain Saloon Stoves,
As awarded a Prize at the Vienna Exhibition.

PATENT MAINSHEET BUFFERS,
SIR W. THOMSON'S AND OTHER PATENT COMPASSES.
PATENT MAST WINCHES.
PATENT VECTIS DECK AND SIDE LIGHTS.
SALOON LAMPS IN GREAT VARIETY.

WATCH CLOCKS to strike Watch and Dog Watch,
ALSO BAROMETERS TO MATCH,
AND
YACHT FITTINGS OF ALL DESCRIPTIONS
TO SUIT ALL TONNAGES.

Illustrated Catalogues on application, also Estimates.

ISLE OF WIGHT AND SOUTH OF ENGLAND GALVANIZING WORKS AND YACHT FITTING DEPÔT.
(OPPOSITE THE STEAMBOAT LANDING STAGE).

Advertisements.

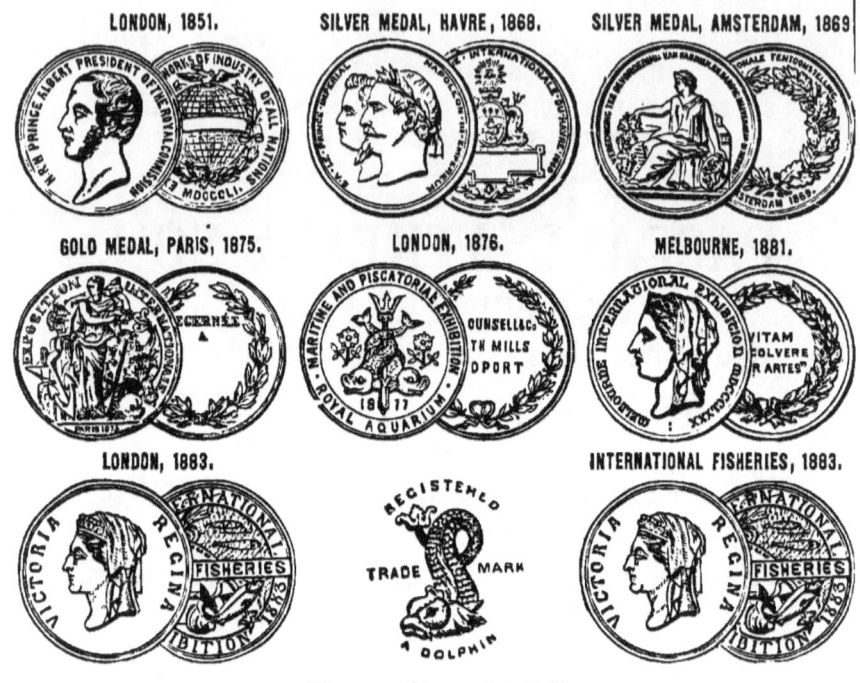

Please address in full.

WILLIAM HOUNSELL & CO.,
NORTH MILLS, BRIDPORT.

LINES, TWINES, SEINES, TRAMMELS, TRAWLS, DRIFT NETS, &c.

Machinery secured by Royal Letters Patent.

PRICE SHEETS AND SAMPLES FORWARDED ON APPLICATION.

C. & R. BROOKS,
PRACTICAL FISHERMEN,

INVENTORS AND MANUFACTURERS OF

The Most Advanced Marine Tackle in the World.

GOLD MEDAL
AND
GOVERNMENT DIPLOMA
OF HONOUR.

HAND LINES,
LONG LINES,

BAITS, AND GENERAL TACKLE OUTFITS.

HORSEHAIR DRIFT LINES,
AND
SAND EEL COURGES FOR BAR HARBOUR & OTHER FISHING

With all Tackle and Appliances mentioned in this Work.

Red and Grey Eels for Pollack, 4s. 2d. per doz. post free.

ESTIMATES FREE.
ILLUSTRATED PRICE LIST, 6d. FREE TO CUSTOMERS.

ONLY ADDRESS:

C. & R. BROOKS,
SOUTH OF ENGLAND TACKLE MANUFACTORY,
STONEHOUSE, PLYMOUTH.

x

STEPHEN WHETHAM & SONS,
𝔐erchants,
SPINNERS AND MANUFACTURERS,
BRIDPORT.

SAIL CLOTH, TWINES, NETS, LINES, &c.

Warehouse: 40 GRACECHURCH STREET, LONDON.

LONDON YACHT AGENCY,
V. WING, Proprietor.
10 DUKE STREET, ST. JAMES'S, S.W.

Southampton Branch {OFFICE—100 HIGH STREET. / LYING-UP STATION—NORTHAM.}

About 300 Steam and Sailing Yachts are usually on this Agency's Books for Sale or Hire.

Insurances, Surveys, Inspections, and Valuations; Altering, Repairing, or Building Superintended. Estimates, Specifications, and Plans furnished or advice given thereon. Accounts audited, and all matters connected with Yachting attended to.

YACHTS TAKEN CHARGE OF FOR SALE OR OTHERWISE.

Agents or Correspondents at the various Yachting Stations and Ports of Call at Home or Abroad.

SALMON, COD, CONGER, PIKE, TROUT.

THESE noted FISHLINE CORDS in 16-plait or cable laid dressed or undressed; also Waterproof TAPERED, at one or both ends, Salmon and Trout Lines, as used and recommended by the Editor of 'THE FIELD,' and all noted anglers. Also Improved LANDING NETS that do not catch the hooks, made from hard-cabled waterproof cord, as used and recommended by Col. Whyte and others. Black Prince Salmon Line, so called from its many excellent qualities in spinning, casting, trolling, and mahseer fishing, as used and recommended by Sir Wm. Elliott and officers of the Indian Army. See catalogue. For samples, address, inclosing two stamps,

Cotton Twine Spinning Company,
51 CORPORATION STREET, MANCHESTER.

ALFRED H. BROWN
(Associate of the Institute of Naval Architects),
WATERLOO PLACE, PALL MALL, LONDON, S.W.

Steam and Sailing Yachts of every description Designed, and their Building superintended.

𝔈stimates for 𝔅uilding and 𝔑epairs free of all cost.

DETAILED DRAWINGS AND MODELS OF NEW YACHTS SUBMITTED FOR APPROVAL FREE OF ALL CHARGE.

Steam and Sailing Yachts of all sizes for Sale or Charter.

THE
PISCATORIAL ATLAS
OF THE
NORTH SEA, ENGLISH AND ST. GEORGE'S CHANNELS.
BY
O. T. OLSEN, F.L.S., F.R.G.S.

Showing at a glance the Fishing-ports, Harbours, Species of Fish (how, where, and when caught), Boats and Fishing-gear, and other special information concerning Fish and Fisheries.

Dedicated, by special permission, to H.R.H. the Duke of Edinburgh, K.G., &c. &c.

Complete in One Volume of 50 Charts, 22 × 18 in., Printed in Colours, and bound in cloth.

	£	s.	d.
Boats and Fish Coloured by hand	2	12	6
Do. do. Uncoloured	2	2	0

BY THE SAME AUTHOR.

	s.	d.
The Fisherman's Nautical Almanac	1	0
The Fisherman's Seamanship	2	0
The Fisherman's Seamanship, with Fishing-grounds and Sailing directions	3	6
Signal System for the Fisheries	0	6
The Mariner's Compass	0	2
Chart of the Fishing Banks	8	0
Chart of the River Humber	0	6

TAYLOR & FRANCIS, Red Lion Court, London.
O. T. OLSEN, Fish Dock Road, Grimsby.

FISHING NETS, FISHING LINES, FISHING GEAR
FOR ALL PURPOSES OF THE
SEA-FISHERMAN,
Both for the Amateur and the Professional, made from the finest Qualities of Hemp, Cotton, and Flax; made in the best form, and that most suitable to their various uses.

Also,
CABLE-LAID and PLAITED WATERPROOFED CORDAGES, ROPES, LINES in Flax, Silk, Cotton, Hemp.
SALMON and FISHING NETTINGS, for Sea and Inland Fisheries.
RABBIT NETTINGS.—Our improved first-class FLAX NETS, with Patent Kink-proof Runners, from 10s. per 100 yards complete.
GAME and FOWLING NETTINGS. PHEASANT and POULTRY NETS.
LAWN TENNIS NETS. GARDEN NETTINGS.

DAVID REID & CO., 7 Hilton Street, Manchester.

ESTABLISHED TWO CENTURIES.

EATON & DELLER,
6 & 7 CROOKED LANE, LONDON BRIDGE,
MANUFACTURERS OF ALL KINDS OF
TACKLE FOR SEA AND RIVER FISHING.

Nets and all kinds of Fishing Gear made to Order.

FRANCIS FRANCIS'S FISHING BOOK.
With a Portrait of the Author and his Gillie, and 17 other Plates (11 Plain and 6 Coloured), in post 8vo. price 15s.

A BOOK ON ANGLING;
Being a complete Treatise on the Art of Angling in every branch.
By FRANCIS FRANCIS, of *The Field*.
The Fifth Edition, revised and improved by the Author.

London, LONGMANS & CO.

The Ninth Edition, in 8vo., with 20 Coloured Plates, price 14s.

THE FLY-FISHER'S ENTOMOLOGY:
Including Coloured Representations of the Natural and Artificial Insect, and a few Observations and Instructions on Trout and Grayling Fishing.

By ALFRED RONALDS.

London, LONGMANS & CO.

A CATALOGUE OF WORKS

IN

GENERAL LITERATURE

PUBLISHED BY

MESSRS. LONGMANS, GREEN, & CO.,

39 PATERNOSTER ROW, LONDON, E.C.,

15 EAST 16TH STREET, NEW YORK.

MESSRS. LONGMANS, GREEN, & CO.

Issue the undermentioned Lists of their Publications, which may be had post free on application:—

1. MONTHLY LIST OF NEW WORKS AND NEW EDITIONS.
2. QUARTERLY LIST OF ANNOUNCEMENTS AND NEW WORKS.
3. NOTES ON BOOKS; BEING AN ANALYSIS OF THE WORKS PUBLISHED DURING EACH QUARTER.
4. CATALOGUE OF SCIENTIFIC WORKS.
5. CATALOGUE OF MEDICAL AND SURGICAL WORKS.
6. CATALOGUE OF SCHOOL BOOKS AND EDUCATIONAL WORKS.
7. CATALOGUE OF BOOKS FOR ELEMENTARY SCHOOLS AND PUPIL TEACHERS.
8. CATALOGUE OF THEOLOGICAL WORKS BY DIVINES AND MEMBERS OF THE CHURCH OF ENGLAND.
9. CATALOGUE OF WORKS IN GENERAL LITERATURE.

ABBEY (Rev. C. J.) and OVERTON (Rev. J. H.).—THE ENGLISH CHURCH IN THE EIGHTEENTH CENTURY. Cr. 8vo. 7s. 6d.

ABBOTT (Evelyn).—A HISTORY OF GREECE. In Two Parts.
Part I.—From the Earliest Times to the Ionian Revolt. Cr. 8vo. 10s. 6d.
Part II.—500-445 B.C. 10s. 6d.

—— A SKELETON OUTLINE OF GREEK HISTORY. Chronologically Arranged. Crown 8vo. 2s. 6d.

—— HELLENICA. A Collection of Essays on Greek Poetry, Philosophy, History, and Religion. Edited by EVELYN ABBOTT. 8vo. 16s.

ACLAND (A. H. Dyke) and RANSOME (Cyril).—A HANDBOOK IN OUTLINE OF THE POLITICAL HISTORY OF ENGLAND TO 1890. Chronologically Arranged. Crown 8vo. 6s.

ACTON (Eliza).—MODERN COOKERY. With 150 Woodcuts. Fcp. 8vo. 4s. 6d.

A. K. H. B.—THE ESSAYS AND CONTRIBUTIONS OF. Crown 8vo. 3s. 6d. each.

Autumn Holidays of a Country Parson.
Changed Aspects of Unchanged Truths.
Commonplace Philosopher.
Counsel and Comfort from a City Pulpit.
Critical Essays of a Country Parson.
East Coast Days and Memories.
Graver Thoughts of a Country Parson. Three Series.
Landscapes, Churches, and Moralities.
Leisure Hours in Town.
Lessons of Middle Age.
Our Little Life. Two Series.
Our Homely Comedy and Tragedy.
Present Day Thoughts.
Recreations of a Country Parson. Three Series. Also 1st Series. 6d.
Seaside Musings.
Sunday Afternoons in the Parish Church of a Scottish University City.

——— 'To Meet the Day' through the Christian Year; being a Text of Scripture, with an Original Meditation and a Short Selection in Verse for Every Day. Crown 8vo. 4s. 6d.

——— TWENTY-FIVE YEARS OF ST. ANDREWS. 1865-1890. 2 vols. 8vo. Vol. I. 12s. Vol. II. 15s.

AMOS (Sheldon).—A PRIMER OF THE ENGLISH CONSTITUTION AND GOVERNMENT. Crown 8vo. 6s.

ANNUAL REGISTER (The). A Review of Public Events at Home and Abroad, for the year 1891. 8vo. 18s.

*** Volumes of the 'Annual Register' for the years 1863-1890 can still be had.

ANSTEY (F.).—THE BLACK POODLE, and other Stories. Crown 8vo. 2s. boards.; 2s. 6d. cloth.

——— VOCES POPULI. Reprinted from *Punch*. With Illustrations by J. BERNARD PARTRIDGE. First Series, Fcp. 4to. 5s. Second Series. Fcp. 4to. 6s.

——— THE TRAVELLING COMPANIONS. Reprinted from *Punch*. With Illustrations by J. BARNARD PARTRIDGE. Post 4to. 5s.

ARISTOTLE—The Works of.

——— THE POLITICS, G. Bekker's Greek Text of Books I. III. IV. (VII.), with an English Translation by W. E. BOLLAND, and short Introductory Essays by ANDREW LANG. Crown 8vo. 7s. 6d.

——— THE POLITICS, Introductory Essays. By ANDREW LANG. (From Bolland and Lang's ' Politics'.) Crown 8vo. 2s. 6d.

——— THE ETHICS, Greek Text, illustrated with Essays and Notes. By Sir ALEXANDER GRANT, Bart. 2 vols. 8vo. 32s.

——— THE NICOMACHEAN ETHICS, newly translated into English. By ROBERT WILLIAMS. Crown 8vo. 7s. 6d.

ARMSTRONG (Ed.).—ELISABETH FARNESE: the Termagant of Spain. 8vo. 16s.

ARMSTRONG (G. F. Savage-).—POEMS: Lyrical and Dramatic. Fcp. 8vo. 6s.

BY THE SAME AUTHOR. Fcp. 8vo.

King Saul. 5s.
King David. 6s.
King Solomon. 6s.
Ugone; a Tragedy. 6s.
A Garland from Greece. Poems. 7s. 6d.
Stories of Wicklow. Poems. 7s. 6d.
Mephistopheles in Broadcloth; a Satire. 4s.
One in the Infinite; a Poem. Crown 8vo. 7s. 6d.
The Life and Letters of Edmond J. Armstrong. 7s. 6d.

ARMSTRONG (E. J.).—POETICAL WORKS. Fcp. 8vo. 5s.

——— ESSAYS AND SKETCHES. Fcp. 8vo. 5s.

ARNOLD (Sir Edwin).—THE LIGHT OF THE WORLD, or the Great Consummation. A Poem. Crown 8vo. 7s. 6d. net.

Presentation Edition. With Illustrations by W. HOLMAN HUNT. 4to. 20s. net.
[*In the Press.*

————— POTIPHAR'S WIFE, and other Poems. Crown 8vo. 5s. net.

————— SEAS AND LANDS. With 71 Illustrations. Crown 8vo. 7s. 6d.

————— ADZUMA; OR, THE JAPANESE WIFE. A Play. Cr. 8vo. 6s. 6d. net.

ARNOLD (Dr. T.).—INTRODUCTORY LECTURES ON MODERN HISTORY. 8vo. 7s. 6d.

————— MISCELLANEOUS WORKS. 8vo. 7s. 6d.

ASHLEY (J. W.).—ENGLISH ECONOMIC HISTORY AND THEORY. Part I.—The Middle Ages. Crown 8vo. 5s.

ATELIER (The) du Lys; or, An Art Student in the Reign of Terror. By the Author of 'Mademoiselle Mori'. Crown 8vo. 2s. 6d.

BY THE SAME AUTHOR. Crown 2s. 6d. each.

MADEMOISELLE MORI.   A CHILD OF THE REVOLU-
THAT CHILD.   TION.
UNDER A CLOUD.   HESTER'S VENTURE.
THE FIDDLER OF LUGAU.   IN THE OLDEN TIME.

————— THE YOUNGER SISTER : a Tale. Crown 8vo. 6s.

BACON.—COMPLETE WORKS. Edited by R. L. ELLIS, J. SPEDDING, and D. D. HEATH. 7 vols. 8vo. £3 13s. 6d.

————— LETTERS AND LIFE, INCLUDING ALL HIS OCCASIONAL WORKS. Edited by J. SPEDDING. 7 vols. 8vo. £4 4s.

————— THE ESSAYS; with Annotations. By Archbishop WHATELY. 8vo. 10s. 6d.

————— THE ESSAYS; with Introduction, Notes, and Index. By E. A. ABBOTT. 2 vols. Fcp. 8vo. 6s. Text and Index only. Fcp. 8vo. 2s. 6d.

BADMINTON LIBRARY (The), edited by the DUKE OF BEAUFORT, assisted by ALFRED E. T. WATSON.

ATHLETICS AND FOOTBALL. By MONTAGUE SHEARMAN. With 41 Illustrations. Crown 8vo. 10s. 6d.

BOATING. By W. B. WOODGATE. With 49 Illustrations. Crown 8vo. 10s. 6d.

COURSING AND FALCONRY. By HARDING COX and the Hon. GERALD LASCELLES. With 76 Illustrations. Crown 8vo. 10s. 6d.

CRICKET. By A. G. STEEL and the Hon. R. H. LYTTELTON. With 63 Illustrations. Crown 8vo. 10s. 6d.

CYCLING. By VISCOUNT BURY (Earl of Albemarle) and G. LACY HILLIER. With 89 Illustrations. Crown 8vo. 10s. 6d.

DRIVING. By the DUKE OF BEAUFORT. With 65 Illustrations. Crown 8vo. 10s. 6d.

FENCING, BOXING, AND WRESTLING. By WALTER H. POLLOCK, F. C. GROVE, C. PREVOST, E. B. MICHELL, and WALTER ARMSTRONG. With 42 Illustrations. Crown 8vo. 10s. 6d.

FISHING. By H. CHOLMONDELEY-PENNELL.
Vol. I. Salmon, Trout, and Grayling. 158 Illustrations. Crown 8vo. 10s. 6d.
Vol. II. Pike and other Coarse Fish. 132 Illustrations. Crown 8vo. 10s. 6d.

[*Continued.*

4   *A CATALOGUE OF BOOKS IN GENERAL LITERATURE*

BADMINTON LIBRARY (The)—*(continued).*
GOLF. By HORACE HUTCHINSON, the Rt. Hon. A. J. BALFOUR, M.P., ANDREW LANG, Sir W. G. SIMPSON, Bart., &c. With 88 Illustrations. Crown 8vo. 10s. 6d.

HUNTING. By the DUKE OF BEAUFORT, and MOWBRAY MORRIS. With 53 Illustrations. Crown 8vo. 10s. 6d.

MOUNTAINEERING. By C. T. DENT, Sir F. POLLOCK, Bart., W. M. CONWAY, DOUGLAS FRESHFIELD, C. E. MATHEWS, C. PILKINGTON, and other Writers. With Illustrations by H. G. WILLINK.

RACING AND STEEPLECHASING. By the EARL OF SUFFOLK AND BERKSHIRE, W. G. CRAVEN, &c. 56 Illustrations. Crown 8vo. 10s. 6d.

RIDING AND POLO. By Captain ROBERT WEIR, Riding-Master, R.H.G., J. MORAY BROWN, &c. With 59 Illustrations. Crown 8vo. 10s. 6d.

SHOOTING. By LORD WALSINGHAM, and Sir RALPH PAYNE-GALLWEY, Bart.
Vol. I. Field and Covert. With 105 Illustrations. Crown 8vo. 10s. 6d.
Vol. II. Moor and Marsh. With 65 Illustrations. Crown 8vo. 10s. 6d.

SKATING, CURLING, TOBOGGANING, &c. By J. M. HEATHCOTE, C. G. TEBBUTT, T. MAXWELL WITHAM, the Rev. JOHN KERR, ORMOND HAKE, and Colonel BUCK. With 284 Illustrations. Crown 8vo. 10s. 6d.

TENNIS, LAWN TENNIS, RACKETS, AND FIVES. By J. M. and C. G. HEATHCOTE, E. O. PLEYDELL-BOUVERIE, and A. C. AINGER. With 79 Illustrations. Crown 8vo. 10s. 6d.

BAGEHOT (Walter).—BIOGRAPHICAL STUDIES. 8vo. 12s.
——— ECONOMIC STUDIES. 8vo. 10s. 6d.
——— LITERARY STUDIES. 2 vols. 8vo. 28s.
——— THE POSTULATES OF ENGLISH POLITICAL ECONOMY. Crown 8vo. 2s. 6d.

BAGWELL (Richard).—IRELAND UNDER THE TUDORS. (3 vols.) Vols. I. and II. From the first invasion of the Northmen to the year 1578. 8vo. 32s. Vol. III. 1578-1603. 8vo. 18s.

BAIN (Alex.).—MENTAL AND MORAL SCIENCE. Crown 8vo. 10s. 6d.
——— SENSES AND THE INTELLECT. 8vo. 15s.
——— EMOTIONS AND THE WILL. 8vo. 15s.
——— LOGIC, DEDUCTIVE AND INDUCTIVE. Part I., *Deduction*, 4s. Part II., *Induction*, 6s. 6d.
——— PRACTICAL ESSAYS. Crown 8vo. 2s.

BAKER (Sir S. W.).—EIGHT YEARS IN CEYLON. With 6 Illustrations. Crown 8vo. 3s. 6d.
——— THE RIFLE AND THE HOUND IN CEYLON. With 6 Illustrations. Crown 8vo. 3s. 6d.

BALL (The Rt. Hon. T. J.).—THE REFORMED CHURCH OF IRELAND (1537-1889). 8vo. 7s. 6d.
——— HISTORICAL REVIEW OF THE LEGISLATIVE SYSTEMS OPERATIVE IN IRELAND (1172-1800). 8vo. 6s.

BARING-GOULD (Rev. S.).—CURIOUS MYTHS OF THE MIDDLE AGES. Crown 8vo. 3s. 6d.
——— ORIGIN AND DEVELOPMENT OF RELIGIOUS BELIEF. 2 vols. Crown 8vo. 3s. 6d. each.

PUBLISHED BY MESSRS. LONGMANS, GREEN, & CO. 5

BEACONSFIELD (The Earl of).—NOVELS AND TALES. The Hughenden Edition. With 2 Portraits and 11 Vignettes. 11 vols. Crown 8vo. 42s.

Endymion.
Lothair.
Coningsby.
Tancred. Sybil.
Venetia.
Henrietta Temple.
Contarini Fleming, &c.
Alroy, Ixion, &c.
The Young Duke, &c.
Vivian Grey.

NOVELS AND TALES. Cheap Edition. 11 vols. Crown 8vo. 1s. each, boards; 1s. 6d. each, cloth.

BECKER (Professor).—GALLUS; or, Roman Scenes in the Time of Augustus. Illustrated. Post 8vo. 7s. 6d.

——— CHARICLES; or, Illustrations of the Private Life of the Ancient Greeks. Illustrated. Post 8vo. 7s. 6d.

BELL (Mrs. Hugh).—CHAMBER COMEDIES. Crown 8vo. 6s.

——— NURSERY COMEDIES. Fcp. 8vo. 1s. 6d.

BENT (J. Theodore).—THE RUINED CITIES OF MASHONALAND: being a Record of Excavations and Explorations, 1891-2. With numerous Illustrations and Maps. 8vo. 18s.

BRASSEY (Lady).—A VOYAGE IN THE 'SUNBEAM,' OUR HOME ON THE OCEAN FOR ELEVEN MONTHS.
Library Edition. With 8 Maps and Charts, and 118 Illustrations, 8vo. 21s.
Cabinet Edition. With Map and 66 Illustrations, Crown 8vo. 7s. 6d.
'Silver Library' Edition. With 66 Illustrations, Crown 8vo. 3s. 6d.
School Edition. With 37 Illustrations, Fcp. 2s. cloth, or 3s. white parchment.
Popular Edition. With 60 Illustrations, 4to. 6d. sewed, 1s. cloth.

——— SUNSHINE AND STORM IN THE EAST.
Library Edition. With 2 Maps and 114 Illustrations, 8vo. 21s.
Cabinet Edition. With 2 Maps and 114 Illustrations, Crown 8vo. 7s. 6d.
Popular Edition. With 103 Illustrations, 4to. 6d. sewed, 1s. cloth.

——— IN THE TRADES, THE TROPICS, AND THE 'ROARING FORTIES'.
Cabinet Edition. With Map and 220 Illustrations, Crown 8vo. 7s. 6d.
Popular Edition. With 183 Illustrations, 4to. 6d. sewed, 1s. cloth.

——— THE LAST VOYAGE TO INDIA AND AUSTRALIA IN THE 'SUNBEAM'. With Charts and Maps, and 40 Illustrations in Monotone (20 full-page), and nearly 200 Illustrations in the Text. 8vo. 21s.

——— THREE VOYAGES IN THE 'SUNBEAM'. Popular Edition. With 346 Illustrations, 4to. 2s. 6d.

"BRENDA."—WITHOUT A REFERENCE. A Story for Children. Crown 8vo. 3s. 6d.

——— OLD ENGLAND'S STORY. In little Words for little Children. With 29 Illustrations by SIDNEY P. HALL, &c. Imperial 16mo. 3s. 6d.

BRIGHT (Rev. J. Franck).—A HISTORY OF ENGLAND. 4 vols. Cr. 8vo.
Period I.—Mediæval Monarchy: The Departure of the Romans to Richard III. From A.D. 449 to 1485. 4s. 6d.
Period II.—Personal Monarchy: Henry VII. to James II. From 1485 to 1688. 5s.
Period III.—Constitutional Monarchy: William and Mary to William IV. From 1689 to 1837. 7s. 6d.
Period IV.—The Growth of Democracy: Victoria. From 1837 to 1880. 6s.

BUCKLE (Henry Thomas).—HISTORY OF CIVILISATION IN ENGLAND AND FRANCE, SPAIN AND SCOTLAND. 3 vols. Cr. 8vo. 24s.

BULL (Thomas).—HINTS TO MOTHERS ON THE MANAGEMENT OF THEIR HEALTH during the Period of Pregnancy. Fcp. 8vo. 1s. 6d.

—— THE MATERNAL MANAGEMENT OF CHILDREN IN HEALTH AND DISEASE. Fcp. 8vo. 1s. 6d.

BUTLER (Samuel).—EREWHON. Crown 8vo. 5s.

—— THE FAIR HAVEN. A Work in Defence of the Miraculous Element in our Lord's Ministry. Crown 8vo. 7s. 6d.

—— LIFE AND HABIT. An Essay after a Completer View of Evolution. Cr. 8vo. 7s. 6d.

—— EVOLUTION, OLD AND NEW. Crown 8vo. 10s. 6d.

—— UNCONSCIOUS MEMORY. Crown 8vo. 7s. 6d.

—— ALPS AND SANCTUARIES OF PIEDMONT AND THE CANTON TICINO. Illustrated. Pott 4to. 10s. 6d.

—— SELECTIONS FROM WORKS. Crown 8vo. 7s. 6d.

—— LUCK, OR CUNNING, AS THE MAIN MEANS OF ORGANIC MODIFICATION? Crown 8vo. 7s. 6d.

—— EX VOTO. An Account of the Sacro Monte or New Jerusalem at Varallo-Sesia. Crown 8vo. 10s. 6d.

—— HOLBEIN'S 'LA DANSE'. 3s.

CARLYLE (Thomas).—THOMAS CARLYLE: a History of his Life. By J. A. FROUDE. 1795-1835, 2 vols. Cr. 8vo. 7s. 1834-1881, 2 vols. Cr. 8vo. 7s.

LAST WORDS OF THOMAS CARLYLE—Wotton Reinfred—Excursion (Futile enough) to Paris—Letters to Varnhagen von Ense, &c. Cr. 8vo. 6s. 6d. net.

CHETWYND (Sir George).—RACING REMINISCENCES AND EXPERIENCES OF THE TURF. 2 vols. 8vo. 21s.

CHILD (Gilbert W.).—CHURCH AND STATE UNDER THE TUDORS. 8vo. 15s.

CHILTON (E.).—THE HISTORY OF A FAILURE, and other Tales. Fcp. 8vo. 3s. 6d.

CHISHOLM (G. G.).—HANDBOOK OF COMMERCIAL GEOGRAPHY. New Edition. With 29 Maps. 8vo. 10s. net.

CLERKE (Agnes M.).—FAMILIAR STUDIES IN HOMER. Crown 8vo. 7s. 6d.

CLODD (Edward).—THE STORY OF CREATION: a Plain Account of Evolution. With 77 Illustrations. Crown 8vo. 3s. 6d.

CLUTTERBUCK (W. J.).—ABOUT CEYLON AND BORNEO. With 47 Illustrations. Crown 8vo. 10s. 6d.

COLENSO (J. W.).—THE PENTATEUCH AND BOOK OF JOSHUA CRITICALLY EXAMINED. Crown 8vo. 6s.

COMYN (L. N.).—ATHERSTONE PRIORY: a Tale. Crown 8vo. 2s. 6d.

CONINGTON (John).—THE ÆNEID OF VIRGIL. Translated into English Verse. Crown 8vo. 6s.

—— THE POEMS OF VIRGIL. Translated into English Prose. Cr. 8vo. 6s.

COPLESTON (Reginald Stephen, D.D., Bishop of Colombo).—BUDDHISM, PRIMITIVE AND PRESENT, IN MAGADHA AND IN CEYLON. 8vo. 16s.

COX (Rev. Sir G. W.).—A HISTORY OF GREECE, from the Earliest Period to the Death of Alexander the Great. With 11 Maps. Cr. 8vo 7s. 6d.

CRAKE (Rev. A. D.).—HISTORICAL TALES. Cr. 8vo. 5 vols. 2s. 6d. each.

Edwy the Fair; or, The First Chronicle of Æscendune.
Alfgar the Dane; or, The Second Chronicle of Æscendune.
The Rival Heirs: being the Third and Last Chronicle of Æscendune.
The House of Walderne. A Tale of the Cloister and the Forest in the Days of the Barons' Wars.
Brian Fitz-Count. A Story of Wallingford Castle and Dorchester Abbey.

——— HISTORY OF THE CHURCH UNDER THE ROMAN EMPIRE, A.D. 30-476. Crown 8vo. 7s. 6d.

CREIGHTON (Mandell, D.D.)—HISTORY OF THE PAPACY DURING THE REFORMATION. 8vo. Vols. I. and II., 1378-1464, 32s.; Vols. III. and IV., 1464-1518, 24s.

CROZIER (John Beattie, M.D.).—CIVILISATION AND PROGRESS. Revised and Enlarged, and with New Preface. More fully explaining the nature of the New Organon used in the solution of its problems. 8vo. 14s.

CRUMP (A.).—A SHORT ENQUIRY INTO THE FORMATION OF POLITICAL OPINION, from the Reign of the Great Families to the Advent of Democracy. 8vo. 7s. 6d.

——— AN INVESTIGATION INTO THE CAUSES OF THE GREAT FALL IN PRICES which took place coincidently with the Demonetisation of Silver by Germany. 8vo. 6s.

CURZON (George N., M.P.).—PERSIA AND THE PERSIAN QUESTION. With 9 Maps, 96 Illustrations, Appendices, and an Index. 2 vols. 8vo. 42s.

DANTE.—LA COMMEDIA DI DANTE. A New Text, carefully Revised with the aid of the most recent Editions and Collations. Small 8vo. 6s.

DE LA SAUSSAYE (Prof. Chantepie).—A MANUAL OF THE SCIENCE OF RELIGION. Translated by Mrs. COLYER FERGUSSON (née MAX MÜLLER). Crown 8vo. 12s. 6d.

DEAD SHOT (THE); or, Sportman's Complete Guide. Being a Treatise on the Use of the Gun, with Rudimentary and Finishing Lessons on the Art of Shooting Game of all kinds, also Game Driving, Wild-Fowl and Pigeon Shooting, Dog Breaking, &c. By MARKSMAN. Crown 8vo. 10s. 6d.

DELAND (Margaret, Author of 'John Ward ').—THE STORY OF A CHILD. Crown 8vo. 5s.

DE SALIS (Mrs.).—Works by :—

Cakes and Confections à la Mode. Fcp. 8vo. 1s. 6d.
Dressed Game and Poultry à la Mode. Fcp. 8vo. 1s. 6d.
Dressed Vegetables à la Mode. Fcp. 8vo. 1s. 6d.
Drinks à la Mode. Fcp. 8vo. 1s. 6d.
Entrées à la Mode. Fcp. 1s. 8vo. 6d.
Floral Decorations. Fcp. 8vo. 1s. 6d.
Oysters à la Mode. Fcp. 8vo. 1s. 6d.
Puddings and Pastry à la Mode. Fcp. 8vo. 1s. 6d.

Savouries à la Mode. Fcp. 8vo. 1s. 6d.
Soups and Dressed Fish à la Mode. Fcp. 8vo. 1s. 6d.
Sweets and Supper Dishes à la Mode. Fcp. 8vo. 1s. 6d.
Tempting Dishes for Small Incomes. Fcp. 8vo. 1s. 6d.
Wrinkles and Notions for every Household. Crown 8vo. 1s. 6d.
New-Laid Eggs: Hints for Amateur Poultry Rearers. Fcp. 8vo. 1s. 6d.

DE TOCQUEVILLE (Alexis).—DEMOCRACY IN AMERICA Translated by HENRY REEVE, C.B. 2 vols. Crown 8vo. 16s.

DOROTHY WALLIS: an Autobiography. With Preface by WALTER BESANT. Crown 8vo. 6s.

DOUGALL (L.).—BEGGARS ALL; a Novel. Crown 8vo. 3s. 6d.

DOWELL (Stephen).—A HISTORY OF TAXATION AND TAXES IN ENGLAND. 4 vols. 8vo. Vols. I. and II., The History of Taxation, 21s. Vols. III. and IV., The History of Taxes, 21s.

DOYLE (A. Conan).—MICAH CLARKE: a Tale of Monmouth's Rebellion. With Frontispiece and Vignette. Crown 8vo. 3s. 6d.

——— THE CAPTAIN OF THE POLESTAR; and other Tales. Cr. 8vo. 3s. 6d.

EWALD (Heinrich).—THE ANTIQUITIES OF ISRAEL. 8vo. 12s. 6d.

——— THE HISTORY OF ISRAEL. 8vo. Vols. I. and II. 24s. Vols. III. and IV. 21s. Vol. V. 18s. Vol. VI. 16s. Vol. VII. 21s. Vol. VIII. 18s.

FALKENER (Edward).—GAMES, ANCIENT AND ORIENTAL, AND HOW TO PLAY THEM. Being the Games of the Ancient Egyptians, the Hiera Gramme of the Greeks, the Ludus Latrunculorum of the Romans, and the Oriental Games of Chess, Draughts, Backgammon, and Magic Squares. With numerous Photographs, Diagrams, &c. 8vo. 21s.

FARNELL (G. S.).—GREEK LYRIC POETRY. 8vo. 16s.

FARRAR (F. W.).—LANGUAGE AND LANGUAGES. Crown 8vo. 6s.

——— DARKNESS AND DAWN; or, Scenes in the Days of Nero. An Historic Tale. Crown 8vo. 7s. 6d.

FITZPATRICK (W. J.).—SECRET SERVICE UNDER PITT. 8vo. 14s.

FITZWYGRAM (Major-General Sir F.).—HORSES AND STABLES. With 19 pages of Illustrations. 8vo. 5s.

FORD (Horace).—THE THEORY AND PRACTICE OF ARCHERY. New Edition, thoroughly Revised and Re-written by W. BUTT. 8vo. 14s.

FOUARD (Abbé Constant).—THE CHRIST THE SON OF GOD. With Introduction by Cardinal Manning. 2 vols. Crown 8vo. 14s.

——— ST. PETER AND THE FIRST YEARS OF CHRISTIANITY. Translated from the Second Edition, with the Author's sanction, by GEORGE F. X. GRIFFITH. With an Introduction by Cardinal GIBBONS. Cr. 8vo. 9s.

FOX (C. J.).—THE EARLY HISTORY OF CHARLES JAMES FOX. By the Right Hon. Sir. G. O. TREVELYAN, Bart.
Library Edition. 8vo. 18s. | Cabinet Edition. Crown 8vo. 6s.

FRANCIS (Francis).—A BOOK ON ANGLING: including full Illustrated Lists of Salmon Flies. Post 8vo. 15s.

FREEMAN (E. A.).—THE HISTORICAL GEOGRAPHY OF EUROPE. With 65 Maps. 2 vols. 8vo. 31s. 6d.

FROUDE (James A.).—THE HISTORY OF ENGLAND, from the Fall of Wolsey to the Defeat of the Spanish Armada. 12 vols. Crown 8vo. £2 2s.

——— THE DIVORCE OF CATHERINE OF ARAGON: The Story as told by the Imperial Ambassadors resident at the Court of Henry VIII. *In Usum Laicorum.* Crown 8vo. 6s.

——— THE ENGLISH IN IRELAND IN THE EIGHTEENTH CENTURY. 3 vols. Crown 8vo. 18s.

——— SHORT STUDIES ON GREAT SUBJECTS.
Cabinet Edition. 4 vols. Cr. 8vo. 24s. | Cheap Edit. 4 vols. Cr. 8vo. 3s. 6d. ea.

——— THE SPANISH STORY OF THE ARMADA, and other Essays, Historical and Descriptive. Crown 8vo. 6s.   [*Continued.*

FROUDE (James A.)—(*Continued*).
——— CÆSAR: a Sketch. Crown 8vo. 3s. 6d.
——— OCEANA; OR, ENGLAND AND HER COLONIES. With 9 Illustrations. Crown 8vo. 2s. boards, 2s. 6d. cloth.
——— THE ENGLISH IN THE WEST INDIES; or, the Bow of Ulysses. With 9 Illustrations. Crown 8vo. 2s. boards, 2s. 6d. cloth.
——— THE TWO CHIEFS OF DUNBOY; an Irish Romance of the Last Century. Crown 8vo. 3s. 6d.
——— THOMAS CARLYLE, a History of his Life. 1795 to 1835. 2 vols. Crown 8vo. 7s. 1834 to 1881. 2 vols. Crown 8vo. 7s.

GALLWEY (Sir Ralph Payne-).—LETTERS TO YOUNG SHOOTERS. First Series. Crown 8vo. 7s. 6d. Second Series. Crown 8vo. 12s. 6d.

GARDINER (Samuel Rawson).—HISTORY OF ENGLAND, 1603-1642. 10 vols. Crown 8vo. price 6s. each.
——— A HISTORY OF THE GREAT CIVIL WAR, 1642-1649. (3 vols.) Vol. I. 1642-1644. With 24 Maps. 8vo. (*out of print*). Vol. II. 1644-1647. With 21 Maps. 8vo. 24s. Vol. III. 1647-1649. With 8 Maps. 28s.
——— THE STUDENT'S HISTORY OF ENGLAND. Vol. I. B.C. 55-A.D. 1509, with 173 Illustrations, Crown 8vo. 4s. Vol. II. 1509-1689, with 96 Illustrations. Crown 8vo. 4s. Vol. III. 1689-1885, with 109 Illustrations. Crown 8vo. 4s. Complete in 1 vol. With 378 Illustrations. Crown 8vo. 12s.
——— A SCHOOL ATLAS OF ENGLISH HISTORY. A Companion Atlas to 'Student's History of England'. 66 Maps and 22 Plans. Fcap. 4to. 5s.

GOETHE.—FAUST. A New Translation chiefly in Blank Verse; with Introduction and Notes. By JAMES ADEY BIRDS. Crown 8vo. 6s.
——— FAUST. The Second Part. A New Translation in Verse. By JAMES ADEY BIRDS. Crown 8vo. 6s.

GREEN (T. H.)—THE WORKS OF THOMAS HILL GREEN. (3 Vols.) Vols. I. and II. 8vo. 16s. each. Vol. III. 8vo. 21s.
——— THE WITNESS OF GOD AND FAITH: Two Lay Sermons. Fcp. 8vo. 2s.

GREVILLE (C. C. F.).—A JOURNAL OF THE REIGNS OF KING GEORGE IV., KING WILLIAM IV., AND QUEEN VICTORIA. Edited by H. REEVE. 8 vols. Crown 8vo. 6s. each.

GWILT (Joseph).—AN ENCYCLOPÆDIA OF ARCHITECTURE. With more than 1700 Engravings on Wood. 8vo. 52s. 6d.

HAGGARD (H. Rider).—SHE. With 32 Illustrations. Crown 8vo. 3s. 6d.
——— ALLAN QUATERMAIN. With 31 Illustrations. Crown 8vo. 3s. 6d.
——— MAIWA'S REVENGE. Crown 8vo. 1s. boards, 1s. 6d. cloth.
——— COLONEL QUARITCH, V.C. Crown 8vo. 3s. 6d.
——— CLEOPATRA: With 29 Illustrations. Crown 8vo. 3s. 6d.
——— BEATRICE. Crown 8vo. 3s. 6d.
——— ERIC BRIGHTEYES. With 51 Illustrations. Crown 8vo. 6s.
——— NADA THE LILY. With 23 Illustrations by C. H. M. KERR. Cr. 8vo. 6s.

HAGGARD (H. Rider) and LANG (Andrew).—THE WORLD'S DESIRE. Crown 8vo. 6s.

HALLIWELL-PHILLIPPS (J. O.)—A CALENDAR OF THE HALLIWELL-PHILLIPPS COLLECTION OF SHAKESPEAREAN RARITIES. Second Edition. Enlarged by Ernest E. Baker. 8vo. 10s. 6d.

—— OUTLINES OF THE LIFE OF SHAKESPEARE. With numerous Illustrations and Facsimiles. 2 vols. Royal 8vo. 21s.

HARRISON (Jane E.).—MYTHS OF THE ODYSSEY IN ART AND LITERATURE. Illustrated with Outline Drawings. 8vo. 18s.

HARRISON (Mary).—COOKERY FOR BUSY LIVES AND SMALL INCOMES. Fcp. 8vo. 1s.

HARTE (Bret).—IN THE CARQUINEZ WOODS. Fcp. 8vo. 1s. bds., 1s. 6d. cloth.

—— BY SHORE AND SEDGE. 16mo. 1s.

—— ON THE FRONTIER. 16mo. 1s.

*** Complete in one Volume. Crown 8vo. 3s. 6d.

HARTWIG (Dr.).—THE SEA AND ITS LIVING WONDERS. With 12 Plates and 303 Woodcuts. 8vo. 7s. net.

THE TROPICAL WORLD. With 8 Plates and 172 Woodcuts. 8vo. 7s. net.

THE POLAR WORLD. With 3 Maps, 8 Plates and 85 Woodcuts. 8vo. 7s. net.

THE SUBTERRANEAN WORLD. With 3 Maps and 80 Woodcuts. 8vo. 7s. net.

THE AERIAL WORLD. With Map, 8 Plates and 60 Woodcuts. 8vo. 7s. net.

HAVELOCK.—MEMOIRS OF SIR HENRY HAVELOCK, K.C.B. By JOHN CLARK MARSHMAN. Crown 8vo. 3s. 6d.

HEARN (W. Edward).—THE GOVERNMENT OF ENGLAND: its Structure and its Development. 8vo. 16s.

—— THE ARYAN HOUSEHOLD: its Structure and ts Development. An Introduction to Comparative Jurisprudence. 8vo. 16s.

HISTORIC TOWNS. Edited by E. A. FREEMAN and Rev. WILLIAM HUNT. With Maps and Plans. Crown 8vo 3s. 6d. each.

Bristol. By Rev. W. Hunt.
Carlisle. By Dr. Mandell Creighton.
Cinque Ports. By Montagu Burrows.
Colchester. By Rev. E. L. Cutts.
Exeter. By E. A. Freeman.
London. By Rev. W. J. Loftie.

Oxford. By Rev. C. W. Boase.
Winchester. By Rev. G. W. Kitchin.
New York. By Theodore Roosevelt.
Boston (U.S.). By Henry Cabot Lodge.
York. By Rev. James Raine.

HODGSON (Shadworth H.).—TIME AND SPACE: a Metaphysical Essay. 8vo. 16s.

—— THE THEORY OF PRACTICE: an Ethical Enquiry. 2 vols. 8vo. 24s.

—— THE PHILOSOPHY OF REFLECTION. 2 vols. 8vo. 21s.

—— OUTCAST ESSAYS AND VERSE TRANSLATIONS. Crown 8vo. 8s. 6d.

HOOPER (George).—ABRAHAM FABERT: Governor of Sedan, Marshall of France. His Life and Times, 1599-1662. With a Portrait. 8vo. 10s. 6d.

HOWITT (William).—VISITS TO REMARKABLE PLACES. 80 Illustrations. Crown 8vo. 3s. 6d.

HULLAH (John).—COURSE OF LECTURES ON THE HISTORY OF MODERN MUSIC. 8vo. 8s. 6d.

—— COURSE OF LECTURES ON THE TRANSITION PERIOD OF MUSICAL HISTORY. 8vo. 10s. 6d.

HUME.—THE PHILOSOPHICAL WORKS OF DAVID HUME. Edited by T. H. GREEN and T. H. GROSE. 4 vols. 8vo. 56s.

HUTH (Alfred H.).—THE MARRIAGE OF NEAR KIN, considered with respect to the Law of Nations, the Result of Experience, and the Teachings of Biology. Royal 8vo. 21s.

INGELOW (Jean).—POETICAL WORKS. 2 vols. Fcp. 8vo. 12s.

—— LYRICAL AND OTHER POEMS. Selected from the Writings of JEAN INGELOW. Fcp. 8vo. 2s. 6d. cloth plain, 3s. cloth gilt.

—— VERY YOUNG and QUITE ANOTHER STORY: Two Stories. Crown 8vo. 6s.

INGRAM (T. Dunbar).—ENGLAND AND ROME: a History of the Relations between the Papacy and the English State and Church from the Norman Conquest to the Revolution of 1688. 8vo. 14s.

JAMESON (Mrs.).—SACRED AND LEGENDARY ART. With 19 Etchings and 187 Woodcuts. 2 vols. 8vo. 20s. net.

—— LEGENDS OF THE MADONNA, the Virgin Mary as represented in Sacred and Legendary Art. With 27 Etchings and 165 Woodcuts. 8vo. 10s. net.

—— LEGENDS OF THE MONASTIC ORDERS. With 11 Etchings and 88 Woodcuts. 8vo. 10s. net.

—— HISTORY OF OUR LORD. His Types and Precursors. Completed by LADY EASTLAKE. With 31 Etchings and 281 Woodcuts. 2 vols. 8vo. 20s. net.

JEFFERIES (Richard).—FIELD AND HEDGEROW. Last Essays. Crown 8vo. 3s. 6d.

—— THE STORY OF MY HEART: My Autobiography. Crown 8vo. 3s. 6d.

—— RED DEER. With 17 Illustrations by J. CHARLTON and H. TUNALY. Crown 8vo. 3s. 6d.

—— THE TOILERS OF THE FIELD. With autotype reproduction of bust of Richard Jefferies in Salisbury Cathedral. Crown 8vo. 6s.

JENNINGS (Rev. A. C.).—ECCLESIA ANGLICANA. A History of the Church of Christ in England. Crown 8vo. 7s. 6d.

JEWSBURY.—A SELECTION FROM THE LETTERS OF GERALDINE JEWSBURY TO JANE WELSH CARLYLE. Edited by Mrs. ALEXANDER IRELAND, and Prefaced by a Monograph on Miss Jewsbury by the Editor. 8vo. 6s.

JOHNSON (J. & J. H.).—THE PATENTEE'S MANUAL; a Treatise on the Law and Practice of Letters Patent. 8vo. 10s. 6d.

JORDAN (William Leighton).—THE STANDARD OF VALUE. 8vo. 6s.

JUSTINIAN.—THE INSTITUTES OF JUSTINIAN; Latin Text, with English Introduction, &c. By THOMAS C. SANDARS. 8vo. 18s.

12 *A CATALOGUE OF BOOKS IN GENERAL LITERATURE*

KANT (Immanuel).—CRITIQUE OF PRACTICAL REASON, AND OTHER WORKS ON THE THEORY OF ETHICS. 8vo. 12*s*. 6*d*.
—— INTRODUCTION TO LOGIC. Translated by T. K. Abbott. Notes by S. T. Coleridge. 8vo. 6*s*.

KEITH DERAMORE. A Novel. By the Author of 'Miss Molly'. Cr. 8vo. 6*s*.

KILLICK (Rev. A. H.).—HANDBOOK TO MILL'S SYSTEM OF LOGIC. Crown 8vo. 3*s*. 6*d*.

KNIGHT (E. F.).—THE CRUISE OF THE 'ALERTE'; the Narrative of a Search for Treasure on the Desert Island of Trinidad. With 2 Maps and 23 Illustrations. Crown 8vo. 3*s*. 6*d*.
—— WHERE THREE EMPIRES MEET. A Narrative of Recent Travel in Kashmir, Western Tibet, Gilgit, and the adjacent countries. 8vo.

LADD (George T.).—ELEMENTS OF PHYSIOLOGICAL PSYCHOLOGY. 8vo. 21*s*.
—— OUTLINES OF PHYSIOLOGICAL PSYCHOLOGY. A Text-Book of Mental Science for Academies and Colleges. 8vo. 12*s*.

LANG (Andrew).—CUSTOM AND MYTH: Studies of Early Usage and Belief. With 15 Illustrations. Crown 8vo. 7*s*. 6*d*.
—— HOMER AND THE EPIC. Crown 8vo. 9*s*. *net*.
—— BOOKS AND BOOKMEN. With 2 Coloured Plates and 17 Illustrations. Fcp. 8vo. 2*s*. 6*d*. *net*.
—— LETTERS TO DEAD AUTHORS. Fcp. 8vo. 2*s*. 6*d*. *net*.
—— OLD FRIENDS. Fcp. 8vo. 2*s*. 6*d*. *net*.
—— LETTERS ON LITERATURE. Fcp. 8vo. 2*s*. 6*d*. *net*.
—— GRASS OF PARNASSUS. Fcp. 8vo. 2*s*. 6*d*. *net*.
—— BALLADS OF BOOKS. Edited by ANDREW LANG. Fcp. 8vo. 6*s*.
—— THE BLUE FAIRY BOOK. Edited by ANDREW LANG. With 8 Plates and 130 Illustrations in the Text. Crown 8vo. 6*s*.
—— THE RED FAIRY BOOK. Edited by ANDREW LANG. With 4 Plates and 96 Illustrations in the Text. Crown 8vo. 6*s*.
—— THE BLUE POETRY BOOK. With 12 Plates and 88 Illustrations in the Text. Crown 8vo. 6*s*.
—— THE BLUE POETRY BOOK. School Edition, without Illustrations. Fcp. 8vo. 2*s*. 6*d*.
—— THE GREEN FAIRY BOOK. Edited by ANDREW LANG. With 13 Plates and 88 Illustrations in the Text by H. J. Ford. Crown 8vo. 6*s*.
—— ANGLING SKETCHES. With Illustrations by W. G. BURN-MURDOCH. Crown 8vo. 7*s*. 6*d*.

LAVISSE (Ernest).—GENERAL VIEW OF THE POLITICAL HISTORY OF EUROPE. Crown 8vo. 5*s*.

LECKY (W. E. H.).—HISTORY OF ENGLAND IN THE EIGHTEENTH CENTURY. Library Edition, 8vo. Vols. I. and II. 1700-1760. 36*s*. Vols. III. and IV. 1760-1784. 36*s*. Vols. V. and VI. 1784-1793. 36*s*. Vols. VII. and VIII. 1793-1800. 36*s*. Cabinet Edition, 12 vols. Crown 8vo. 6*s*. each.
—— THE HISTORY OF EUROPEAN MORALS FROM AUGUSTUS TO CHARLEMAGNE. 2 vols. Crown 8vo. 16*s*.
—— HISTORY OF THE RISE AND INFLUENCE OF THE SPIRIT OF RATIONALISM IN EUROPE. 2 vols. Crown 8vo. 16*s*.
—— POEMS. Fcap. 8vo. 5*s*.

LEES (J. A.) and CLUTTERBUCK (W. J.).—B.C. 1887, A RAMBLE IN BRITISH COLUMBIA. With Map and 75 Illusts. Cr. 8vo. 3s. 6d.

LEWES (George Henry).—THE HISTORY OF PHILOSOPHY, from Thales to Comte. 2 vo s. 8vo. 32s.

LEYTON (Frank).—THE SHADOWS OF THE LAKE, and other Poems. Crown 8vo. 7s. 6d. Cheap Edition. Crown 8vo. 3s. 6d.

LLOYD (F. J.).—THE SCIENCE OF AGRICULTURE. 8vo. 12s.

LONGMAN (Frederick W.).—CHESS OPENINGS. Fcp. 8vo. 2s. 6d.

―――― FREDERICK THE GREAT AND THE SEVEN YEARS' WAR. Fcp. 8vo. 2s. 6d.

LONGMORE (Sir T.).—RICHARD WISEMAN, Surgeon and Sergeant-Surgeon to Charles II. A Biographical Study. With Portrait. 8vo. 10s. 6d.

LOUDON (J. C.).—ENCYCLOPÆDIA OF GARDENING. With 1000 Woodcuts. 8vo. 21s.

―――― ENCYCLOPÆDIA OF AGRICULTURE; the Laying-out, Improvement, and Management of Landed Property. With 1100 Woodcuts. 8vo. 21s.

―――― ENCYCLOPÆDIA OF PLANTS; the Specific Character, &c., of all Plants found in Great Britain. With 12,000 Woodcuts. 8vo. 42s.

LUBBOCK (Sir J.).—THE ORIGIN OF CIVILISATION and the Primitive Condition of Man. With 5 Plates and 20 Illustrations in the Text. 8vo. 18s.

LYALL (Edna).—THE AUTOBIOGRAPHY OF A SLANDER. Fcp. 8vo. 1s. sewed.
Presentation Edition, with 20 Illustrations by L. SPEED. Crown 8vo. 5s.

LYDEKKER (R., B.A.).—PHASES OF ANIMAL LIFE, PAST AND PRESENT. With 82 Illustrations. Crown 8vo. 6s.

LYDE (Lionel W.).—AN INTRODUCTION TO ANCIENT HISTORY. With 3 Coloured Maps. Crown 8vo. 3s.

LYONS (Rev. Daniel).—CHRISTIANITY AND INFALLIBILITY— Both or Neither. Crown 8vo. 5s.

LYTTON (Earl of).—MARAH.—By OWEN MEREDITH (the late Earl of Lytton). Fcp. 8vo. 6s. 6d.

―――― KING POPPY; a Fantasia. Crown 8vo. 10s. 6d.

MACAULAY (Lord).—COMPLETE WORKS OF LORD MACAULAY.
Library Edition, 8 vols. 8vo. £5 5s. | Cabinet Edition, 16 vols. post 8vo. £4 16s.

―――― HISTORY OF ENGLAND FROM THE ACCESSION OF JAMES THE SECOND.
Popular Edition, 2 vols. Crown 8vo. 5s. | People's Edition, 4 vols. Crown 8vo. 16s.
Student's Edition, 2 vols. Crown 8vo. 12s. | Cabinet Edition, 8 vols. Post 8vo. 48s.
| Library Edition, 5 vols. 8vo. £4.

―――― CRITICAL AND HISTORICAL ESSAYS, WITH LAYS OF ANCIENT ROME, in 1 volume.
Popular Edition, Crown 8vo. 2s. 6d. | 'Silver Library' Edition. With Portrait and Illustrations to the 'Lays'.
Authorised Edition, Crown 8vo. 2s. 6d., or 3s. 6d. gilt edges. | Crown 8vo. 3s. 6d.

―――― CRITICAL AND HISTORICAL ESSAYS.
Student's Edition. Crown 8vo. 6s. | Trevelyan Edition, 2 vols. Crown 8vo. 9s.
People's Edition, 2 vols. Crown 8vo. 8s. | Cabinet Edition, 4 vols. Post 8vo. 24s.
| Library Edition, 3 vols. 8vo. 36s.

[*Continued.*

14   A CATALOGUE OF BOOKS IN GENERAL LITERATURE

MACAULAY (Lord)—(*Continued*).

——— ESSAYS which may be had separately, price 6*d.* each sewed.   1*s.* each cloth.

Addison and Walpole.
Frederic the Great.
Croker's Boswell's Johnson.
Hallam's Constitutional History.
Warren Hastings (3*d.* sewed, 6*d.* cloth).
The Earl of Chatham (Two Essays).

Ranke and Gladstone.
Milton and Machiavelli.
Lord Bacon.
Lord Clive.
Lord Byron, and the Comic Dramatists of the Restoration.

The Essay on Warren Hastings, annotated by S. Hales. Fcp. 8vo. 1*s.* 6*d.*

The Essay on Lord Clive, annotated by H. Courthope Bowen. Fcp. 8vo. 2*s.* 6*d.*

——— SPEECHES.   People's Edition, Crown 8vo. 3*s.* 6*d.*

——— LAYS OF ANCIENT ROME, &c. Illustrated by G. Scharf. Library Edition. Fcp. 4to. 10*s.* 6*d.*

Bijou Edition, 18mo. 2*s.* 6*d.* gilt top.

Popular Edition, Fcp. 4to. 6*d* sewed, 1*s.* cloth.

——————————————— Illustrated by J. R. Weguelin. Crown 8vo. 3*s.* 6*d.* gilt edges.

Cabinet Edition, Post 8vo. 3*s.* 6*d.*

Annotated Edition, Fcp. 8vo. 1*s.* sewed, 1*s.* 6*d.* cloth.

——— MISCELLANEOUS WRITINGS.
People's Edition.   Crown 8vo. 4*s.* 6*d.* | Library Edition, 2 vols. 8vo. 21*s.*

——— MISCELLANEOUS WRITINGS AND SPEECHES.
Popular Edition.   Crown 8vo. 2*s.* 6*d.* | Cabinet Edition, Post 8vo. 24*s.*
Student's Edition.   Crown 8vo. 6*s.*

——— SELECTIONS FROM THE WRITINGS OF LORD MACAULAY.
Edited, with Notes, by the Right Hon. Sir G. O. TREVELYAN. Crown 8vo. 6*s.*

——— THE LIFE AND LETTERS OF LORD MACAULAY.   By the Right Hon. Sir G. O. TREVELYAN.

Popular Edition.   Crown. 8vo. 2*s.* 6*d.* | Cabinet Edition, 2 vols. Post 8vo. 12*s.*
Student's Edition.   Crown 8vo. 6*s.* | Library Edition, 2 vols. 8vo. 36*s.*

MACDONALD (George).—UNSPOKEN SERMONS.   Three Series.
Crown 8vo. 3*s.* 6*d.* each.

——— THE MIRACLES OF OUR LORD.   Crown 8vo. 3*s.* 6*d.*

——— A BOOK OF STRIFE, IN THE FORM OF THE DIARY OF AN OLD SOUL : Poems. 12mo. 6*s.*

MACFARREN (Sir G. A.).—LECTURES ON HARMONY. 8vo. 12*s.*

MACKAIL (J. W.).—SELECT EPIGRAMS FROM THE GREEK ANTHOLOGY. With a Revised Text, Introduction, Translation, &c. 8vo. 16*s.*

MACLEOD (Henry D.).—THE ELEMENTS OF BANKING.   Crown 8vo. 3*s.* 6*d.*

——— THE THEORY AND PRACTICE OF BANKING.   Vol. I. 8vo. 12*s.*, Vol. II. 14*s.*

——— THE THEORY OF CREDIT.   8vo. Vol. I. [*New Edition in the Press*] ; Vol. II. Part I. 4*s.* 6*d.* ; Vol. II. Part II. 10*s.* 6*d.*

MANNERING (G. E.).—WITH AXE AND ROPE IN THE NEW ZEALAND ALPS. Illustrated. 8vo. 12*s.* 6*d.*

MANUALS OF CATHOLIC PHILOSOPHY (*Stonyhurst Series*).

Logic. By Richard F. Clarke. Crown 8vo. 5s.
First Principles of Knowledge. By John Rickaby. Crown 8vo. 5s.
Moral Philosophy (Ethics and Natural Law). By Joseph Rickaby. Crown 8vo. 5s.

General Metaphysics. By John Rickaby. Crown 8vo. 5s.
Psychology. By Michael Maher. Crown 8vo. 6s. 6d.
Natural Theology. By Bernard Boedder. Crown 8vo. 6s. 6d.
A Manual of Political Economy. By C. S. Devas. 6s. 6d.

MARBOT (Baron de).—THE MEMOIRS OF. Translated from the French. Crown 8vo. 7s. 6d.

MARTINEAU (James).—HOURS OF THOUGHT ON SACRED THINGS. Two Volumes of Sermons. 2 vols. Crown 8vo. 7s. 6d. each.

———— ENDEAVOURS AFTER THE CHRISTIAN LIFE. Discourses. Crown 8vo. 7s. 6d.

———— HOME PRAYERS. Crown 8vo. 3s. 6d.

———— THE SEAT OF AUTHORITY IN RELIGION. 8vo. 14s.

———— ESSAYS, REVIEWS, AND ADDRESSES. 4 vols. Crown 8vo. 7s. 6d. each.

I. Personal: Political.
II. Ecclesiastical: Historical.
III. Theological: Philosophical.
IV. Academical: Religious.

MATTHEWS (Brander).—A FAMILY TREE, and other Stories. Crown 8vo. 6s.

———— PEN AND INK—Selected Papers. Crown 8vo. 5s.

———— WITH MY FRIENDS: Tales told in Partnership. Crown 8vo. 6s.

MAUNDER'S TREASURIES. Fcp. 8vo. 6s. each volume

Biographical Treasury.
Treasury of Natural History. With 900 Woodcuts.
Treasury of Geography. With 7 Maps and 16 Plates.
Scientific and Literary Treasury.
Historical Treasury.
Treasury of Knowledge.

The Treasury of Bible Knowledge. By the Rev. J. AYRE. With 5 Maps, 15 Plates, and 300 Woodcuts. Fcp. 8vo. 6s.
The Treasury of Botany. Edited by J. LINDLEY and T. MOORE. With 274 Woodcuts and 20 Steel Plates. 2 vols.

MAX MÜLLER (F.).—SELECTED ESSAYS ON LANGUAGE, MYTHOLOGY, AND RELIGION. 2 vols. Crown 8vo. 16s.

———— THREE LECTURES ON THE SCIENCE OF LANGUAGE. Cr. 8vo. 3s.

———— THE SCIENCE OF LANGUAGE, founded on Lectures delivered at the Royal Institution in 1861 and 1863. 2 vols. Crown 8vo. 21s.

———— HIBBERT LECTURES ON THE ORIGIN AND GROWTH OF RELIGION, as illustrated by the Religions of India. Crown 8vo. 7s. 6d.

———— INTRODUCTION TO THE SCIENCE OF RELIGION; Four Lectures delivered at the Royal Institution. Crown 8vo. 7s. 6d.

———— NATURAL RELIGION. The Gifford Lectures, delivered before the University of Glasgow in 1888. Crown 8vo. 10s. 6d.

———— PHYSICAL RELIGION. The Gifford Lectures, delivered before the University of Glasgow in 1890. Crown 8vo. 10s. 6d.

———— ANTHROPOLOGICAL RELIGION: The Gifford Lectures delivered before the University of Glasgow in 1891. Crown 8vo. 10s. 6d.

[*Continued.*

MAX MÜLLER (F.)—(*Continued*).
—— THEOSOPHY OR PSYCHOLOGICAL RELIGION: the Gifford Lectures delivered before the University of Glasgow in 1892. Crown 8vo.
—— THE SCIENCE OF THOUGHT. 8vo. 21s.
—— THREE INTRODUCTORY LECTURES ON THE SCIENCE OF THOUGHT. 8vo. 2s. 6d.
—— BIOGRAPHIES OF WORDS, AND THE HOME OF THE ARYAS. Crown 8vo. 7s. 6d.
—— INDIA, WHAT CAN IT TEACH US? Crown 8vo. 3s. 6d.
—— A SANSKRIT GRAMMAR FOR BEGINNERS. New and Abridged Edition. By A. A. MACDONELL. Crown 8vo. 6s.

MAY (Sir Thomas Erskine).—THE CONSTITUTIONAL HISTORY OF ENGLAND since the Accession of George III. 3 vols. Crown 8vo. 18s.

MEADE (L. T.).—DADDY'S BOY. With Illustrations. Crown 8vo. 3s. 6d.
—— DEB AND THE DUCHESS. Illust. by M. E. Edwards. Cr. 8vo. 3s. 6d.
—— THE BERESFORD PRIZE. Illustrated by M. E. Edwards. Cr. 8vo. 5s.

MEATH (The Earl of).—SOCIAL ARROWS: Reprinted Articles on various Social Subjects. Crown 8vo. 5s.
—— PROSPERITY OR PAUPERISM? Physical, Industrial, and Technical Training. Edited by the EARL OF MEATH. 8vo. 5s.

MELVILLE (G. J. Whyte).—Novels by. Crown 8vo. 1s. each, boards; 1s. 6d. each, cloth.

The Gladiators.	The Queen's Maries.	Digby Grand.
The Interpreter.	Holmby House.	General Bounce.
Good for Nothing.	Kate Coventry.	

MENDELSSOHN.—THE LETTERS OF FELIX MENDELSSOHN. Translated by Lady Wallace. 2 vols. Crown 8vo. 10s.

MERIVALE (Rev. Chas.).—HISTORY OF THE ROMANS UNDER THE EMPIRE. Cabinet Edition, 8 vols. Crown 8vo. 48s. Popular Edition, 8 vols. Crown 8vo. 3s. 6d. each.
—— THE FALL OF THE ROMAN REPUBLIC: a Short History of the Last Century of the Commonwealth. 12mo. 7s. 6d.
—— GENERAL HISTORY OF ROME FROM B.C. 753 TO A.D. 476. Cr. 8vo. 7s. 6d.
—— THE ROMAN TRIUMVIRATES. With Maps. Fcp. 8vo. 2s. 6d.

MILL (James).—ANALYSIS OF THE PHENOMENA OF THE HUMAN MIND. 2 vols. 8vo. 28s.

MILL (John Stuart).—PRINCIPLES OF POLITICAL ECONOMY.
Library Edition, 2 vols. 8vo. 30s. | People's Edition, 1 vol. Crown 8vo. 3s. 6d.
—— A SYSTEM OF LOGIC. Crown 8vo. 3s. 6d.
—— ON LIBERTY. Crown 8vo. 1s. 4d.
—— ON REPRESENTATIVE GOVERNMENT. Crown 8vo. 2s.
—— UTILITARIANISM. 8vo. 5s.
—— EXAMINATION OF SIR WILLIAM HAMILTON'S PHILOSOPHY. 8vo. 16s.
—— NATURE, THE UTILITY OF RELIGION AND THEISM. Three Essays, 8vo. 5s.

PUBLISHED BY MESSRS. LONGMANS, GREEN, & CO.   17

MILNER (George).—COUNTRY PLEASURES; the Chronicle of a Year chiefly in a Garden. Crown 8vo. 3s. 6d.

MOLESWORTH (Mrs.).—SILVERTHORNS. With Illustrations by F. Noel Paton. Cr. 8vo. 5s.
—— THE PALACE IN THE GARDEN. With Illustrations. Cr. 8vo. 5s.
—— THE THIRD MISS ST. QUENTIN. Crown 8vo. 6s.
—— NEIGHBOURS. With Illustrations by M. Ellen Edwards. Cr. 8vo. 6s.
—— THE STORY OF A SPRING MORNING. With Illustrations. Cr.8vo. 5s.
—— STORIES OF THE SAINTS FOR CHILDREN: the Black Letter Saints. With Illustrations. Royal 16mo. 5s.

MOORE (Edward).—DANTE AND HIS EARLY BIOGRAPHERS. Crown 8vo. 4s. 6d.

MULHALL (Michael G.).—HISTORY OF PRICES SINCE THE YEAR 1850. Crown 8vo. 6s.

NANSEN (Dr. Fridtjof).—THE FIRST CROSSING OF GREENLAND. With numerous Illustrations and a Map. Crown 8vo. 7s. 6d.

NAPIER.—THE LIFE OF SIR JOSEPH NAPIER, BART., EX-LORD CHANCELLOR OF IRELAND. By ALEX. CHARLES EWALD. 8vo. 15s.
—— THE LECTURES, ESSAYS, AND LETTERS OF THE RIGHT HON. SIR JOSEPH NAPIER, BART. 8vo. 12s. 6d.

NESBIT (E.).—LEAVES OF LIFE: Verses. Crown 8vo. 5s.
—— LAYS AND LEGENDS. FIRST Series. Crown 8vo. 3s. 6d. SECOND Series. With Portrait. Crown 8vo. 5s.

NEWMAN (Cardinal).—Works by:—

Discourses to Mixed Congregations. Cabinet Edition, Crown 8vo. 6s. Cheap Edition, 3s. 6d.
Sermons on Various Occasions. Cabinet Edition, Cr. 8vo. 6s. Cheap Edition, 3s. 6d.
The Idea of a University defined and Illustrated. Cabinet Edition, Cr. 8vo. 7s. Cheap Edition, Cr. 8vo. 3s. 6d.
Historical Sketches. Cabinet Edition, 3 vols. Crown 8vo. 6s. each. Cheap Edition, 3 vols. Cr. 8vo. 3s. 6d. each.
The Arians of the Fourth Century. Cabinet Edition, Crown 8vo. 6s. Cheap Edition, Crown 8vo. 3s. 6d.
Select Treatises of St. Athanasius in Controversy with the Arians. Freely Translated. 2 vols. Crown 8vo. 15s.
Discussions and Arguments on Various Subjects. Cabinet Edition, Crown 8vo. 6s. Cheap Edition, Crown 8vo. 3s. 6d.
Apologia Pro Vita Sua. Cabinet Ed., Crown 8vo. 6s. Cheap Ed. 3s. 6d.
Development of Christian Doctrine. Cabinet Edition, Crown 8vo. 6s. Cheap Edition, Cr. 8vo. 3s. 6d.

Certain Difficulties felt by Anglicans in Catholic Teaching Considered. Cabinet Edition. Vol. I. Crown 8vo. 7s. 6d.; Vol. II. Crown 8vo. 5s. 6d. Cheap Edition, 2 vols. Crown 8vo. 3s. 6d. each.
The Via Media of the Anglican Church, Illustrated in Lectures, &c. Cabinet Edition, 2 vols. Cr. 8vo. 6s. each. Cheap Edition, 2 vols. Crown 8vo. 3s. 6d. each.
Essays, Critical and Historical. Cabinet Edition, 2 vols. Crown 8vo. 12s. Cheap Edition, 2 vols. Cr. 8vo. 7s.
Biblical and Ecclesiastical Miracles. Cabinet Edition, Crown 8vo. 6s. Cheap Edition, Crown 8vo. 3s. 6d.
Present Position of Catholics in England. Cabinet Edition, Crown 8vo. 7s. 6d. Cheap Edition, Crown 8vo. 3s. 6d.
Tracts. 1. Dissertatiunculæ. 2. On the Text of the Seven Epistles of St. Ignatius. 3. Doctrinal Causes of Arianism. 4. Apollinarianism. 5. St. Cyril's Formula. 6. Ordo de Tempore. 7. Douay Version of Scripture. Crown 8vo. 8s.

[Continued.

NEWMAN (Cardinal).—Works by :—(*continued*).

An Essay in Aid of a Grammar of Assent. Cabinet Edition, Crown 8vo. 7s. 6d. Cheap Edition, Crown 8vo. 3s. 6d.

Callista: a Tale of the Third Century. Cabinet Edition, Crown 8vo. 6s. Cheap Edition, Crown 8vo. 3s. 6d.

Loss and Gain: a Tale. Cabinet Edition, Crown 8vo. 6s. Cheap Edition, Crown 8vo. 3s. 6d.

The Dream of Gerontius. 16mo. 6d. sewed, 1s. cloth.

Verses on Various Occasions. Cabinet Edition, Crown 8vo. 6s. Cheap Edition, Crown 8vo. 3s. 6d.

⁎ *For Cardinal Newman's other Works see Messrs. Longmans & Co.'s Catalogue of Theological Works.*

NORTON (Charles L.).—A HANDBOOK OF FLORIDA. 49 Maps and Plans. Fcp. 8vo. 5s.

O'BRIEN (William).—WHEN WE WERE BOYS: A Novel. Cr. 8vo. 2s. 6d.

OLIPHANT (Mrs.).—MADAM. Crown 8vo. 1s. boards; 1s. 6d. cloth.

———— IN TRUST. Crown 8vo. 1s. boards; 1s. 6d. cloth.

OMAN (C. W. C.).—A HISTORY OF GREECE FROM THE EARLIEST TIMES TO THE MACEDONIAN CONQUEST. With Maps. Crown 8vo. 4s. 6d.

PARKES (Sir Henry).—FIFTY YEARS IN THE MAKING OF AUSTRALIAN HISTORY. With Portraits. 2 vols. 8vo. 32s.

PAUL (Hermann).—PRINCIPLES OF THE HISTORY OF LANGUAGE. Translated by H. A. Strong. 8vo. 10s. 6d.

PAYN (James).—THE LUCK OF THE DARRELLS. Cr. 8vo. 1s. bds.; 1s. 6d. cl.

———— THICKER THAN WATER. Crown 8vo. 1s. boards; 1s. 6d. cloth.

PERRING (Sir Philip).—HARD KNOTS IN SHAKESPEARE. 8vo. 7s. 6d.

———— THE 'WORKS AND DAYS' OF MOSES. Crown 8vo. 3s. 6d.

PHILLIPPS-WOLLEY (C.).—SNAP: a Legend of the Lone Mountain. With 13 Illustrations by H. G. Willink. Crown 8vo. 3s. 6d.

POLE (W.).—THE THEORY OF THE MODERN SCIENTIFIC GAME OF WHIST. Fcp. 8vo. 2s. 6d.

POOLE (W. H. and Mrs.).—COOKERY FOR THE DIABETIC. Fcp. 8vo. 2s. 6d.

PRAEGER (F.).—WAGNER AS I KNEW HIM. Crown 8vo. 7s. 6d.

PRATT (A. E., F.R.G.S.).—TO THE SNOWS OF TIBET THROUGH CHINA. With 33 Illustrations and a Map. 8vo. 18s.

PRENDERGAST (John P.).—IRELAND, FROM THE RESTORATION TO THE REVOLUTION, 1660-1690. 8vo. 5s.

PROCTOR (R.A.).—Works by:—

Old and New Astronomy. 4to. 36s.
The Orbs Around Us. Crown 8vo. 5s.
Other Worlds than Ours. With 14 Illustrations. Crown 8vo. 5s. Cheap Edition, 3s. 6d.
The Moon. Crown 8vo. 5s.
Universe of Stars. 8vo. 10s. 6d.
Larger Star Atlas for the Library, in 12 Circular Maps, with Introduction and 2 Index Pages. Folio, 15s. or Maps only, 12s. 6d.
The Student's Atlas. In 12 Circular Maps. 8vo. 5s.
New Star Atlas. In 12 Circular Maps. Crown 8vo. 5s.
Light Science for Leisure Hours. 3 vols. Crown 8vo. 5s. each.
Chance and Luck. Crown 8vo. 2s. boards; 2s. 6d. cloth.
Pleasant Ways in Science. Cr. 8vo. 5s. Cheap Edition, 3s. 6d.
How to Play Whist: with the Laws and Etiquette of Whist. Crown 8vo. 3s. 6d.
Home Whist: an Easy Guide to Correct Play. 16mo. 1s.
The Stars in their Season. 12 Maps. Royal 8vo. 5s.
Star Primer. Showing the Starry Sky Week by Week, in 24 Hourly Maps. Crown 4to. 2s. 6d.
The Seasons Pictured in 48 Sun-Views of the Earth, and 24 Zodiacal Maps, &c. Demy 4to. 5s.
Strength and Happiness. With 9 Illustrations. Crown 8vo. 5s.
Strength: How to get Strong and keep Strong. Crown 8vo. 2s.
Rough Ways Made Smooth. Essays on Scientific Subjects. Crown 8vo. 5s. Cheap Edition, 3s. 6d.
Our Place among Infinities. Cr. 8vo. 5s.
The Expanse of Heaven. Cr. 8vo. 5s.
The Great Pyramid. Crown 8vo. 5s.
Myths and Marvels of Astronomy Crown 8vo. 5s.
Nature Studies. By Grant Allen, A. Wilson, T. Foster, E. Clodd, and R. A. Proctor. Crown 8vo. 5s.
Leisure Readings. By E. Clodd, A. Wilson, T. Foster, A. C. Ranyard, and R. A. Proctor. Crown 8vo. 5s.

RANSOME (Cyril).—THE RISE OF CONSTITUTIONAL GOVERNMENT IN ENGLAND: being a Series of Twenty Lectures. Crown 8vo. 6s.

READER (Emily E.).—VOICES FROM FLOWER-LAND: a Birthday Book and Language of Flowers. Illustrated by ADA BROOKE. Royal 16mo. Cloth, 2s. 6d.; vegetable vellum, 3s. 6d.

REPLY (A) TO DR. LIGHTFOOT'S ESSAYS. By the Author of 'Supernatural Religion'. 8vo. 6s.

RIBOT (Th.).—THE PSYCHOLOGY OF ATTENTION. Crown 8vo. 3s.

RICH (A.).—A DICTIONARY OF ROMAN AND GREEK ANTIQUITIES. With 2000 Woodcuts. Crown 8vo. 7s. 6d.

RICHARDSON (Dr. B. W.).—NATIONAL HEALTH. A Review of the Works of Sir Edwin Chadwick, K.C.B. Crown 4s. 6d.

RIVERS (T. and T. F.).—THE MINIATURE FRUIT GARDEN; or, The Culture of Pyramidal and Bush Fruit Trees. With 32 Illustrations. Crown 8vo. 4s.

RIVERS (T.).—THE ROSE AMATEUR'S GUIDE. Fcp. 8vo. 4s. 6d.

ROBERTSON (A.).—THE KIDNAPPED SQUATTER, and other Australian Tales. Crown 8vo. 6s.

ROGET (John Lewis).—A HISTORY OF THE 'OLD WATER COLOUR' SOCIETY. 2 vols. Royal 8vo. 42s.

ROGET (Peter M.).—THESAURUS OF ENGLISH WORDS AND PHRASES. Crown 8vo. 10s. 6d.

ROMANES (George John, M.A., LL.D., F.R.S.).—DARWIN, AND AFTER DARWIN: an Exposition of the Darwinian Theory and a Discussion of Post-Darwinian Questions. Part I.—The Darwinian Theory. With a Portrait of Darwin and 125 Illustrations. Crown 8vo. 10s. 6d.

RONALDS (Alfred).—THE FLY-FISHER'S ETYMOLOGY. With 20 Coloured Plates. 8vo. 14s.

ROSSETTI (Maria Francesca).—A SHADOW OF DANTE: being an Essay towards studying Himself, his World, and his Pilgrimage. Cr.8vo. 10s.6d.

ROUND (J. H., M.A.).—GEOFFREY DE MANDEVILLE: a Study of the Anarchy. 8vo. 16s.

RUSSELL.—A LIFE OF LORD JOHN RUSSELL. By SPENCER WALPOLE. 2 vols. 8vo. 36s. Cabinet Edition, 2 vols. Crown 8vo. 12s.

SEEBOHM (Frederick).—THE OXFORD REFORMERS — JOHN COLET, ERASMUS, AND THOMAS MORE. 8vo. 14s.
——— THE ENGLISH VILLAGE COMMUNITY Examined in its Relations to the Manorial and Tribal Systems, &c. 13 Maps and Plates. 8vo. 16s.
——— THE ERA OF THE PROTESTANT REVOLUTION. With Map. Fcp. 8vo. 2s. 6d.

SEWELL (Elizabeth M.).—STORIES AND TALES. Crown 8vo. 1s. 6d. each, cloth plain ; 2s. 6d. each, cloth extra, gilt edges :—

Amy Herbert.	Katharine Ashton.	Gertrude.
The Earl's Daughter.	Margaret Percival.	Ivors.
The Experience of Life.	Laneton Parsonage.	Home Life.
A Glimpse of the World.	Ursula.	After Life.
Cleve Hall.		

SHAKESPEARE.—BOWDLER'S FAMILY SHAKESPEARE. 1 vol. 8vo. With 36 Woodcuts, 14s., or in 6 vols. Fcp. 8vo. 21s.
——— OUTLINES OF THE LIFE OF SHAKESPEARE. By J. O. HALLIWELL-PHILLIPPS. With Illustrations. 2 vols. Royal 8vo. £1 1s.
——— SHAKESPEARE'S TRUE LIFE. By JAMES WALTER. With 500 Illustrations. Imp. 8vo. 21s.
——— THE SHAKESPEARE BIRTHDAY BOOK. By MARY F. DUNBAR. 32mo. 1s. 6d. cloth. With Photographs, 32mo. 5s. Drawing-Room Edition, with Photographs, Fcp. 8vo. 10s. 6d.

SHERBROOKE (Viscount).—LIFE AND LETTERS OF THE RIGHT HON. ROBERT LOWE, VISCOUNT SHERBROOKE, G.C.B. By A. PATCHETT MARTIN. With 5 Copper-plate Portraits, &c. 2 vols. 8vo.

SHIRRES (L. P.).—AN ANALYSIS OF THE IDEAS OF ECONOMICS. Crown 8vo. 6s.

SIDGWICK (Alfred).—DISTINCTION : and the Criticism of Beliefs. Cr. 8vo. 6s.

SILVER LIBRARY, The.—Crown 8vo. price 3s. 6d. each volume.

BAKER'S (Sir S. W.) Eight Years in Ceylon. With 6 Illustrations.
——— Rifle and Hound in Ceylon. With 6 Illustrations.
BARING-GOULD'S (S.) Curious Myths of the Middle Ages.
——— Origin and Development of Religious Belief. 2 vols.
BRASSEY'S (Lady) A Voyage in the 'Sunbeam'. With 66 Illustrations.

CLODD'S (E.) Story of Creation : a Plain Account of Evolution. With 77 Illustrations.
CONYBEARE (Rev. W. J.) and HOWSON'S (Very Rev. J. S.) Life and Epistles of St. Paul. 46 Illustrations.
DOUGALL'S (L.) Beggars All ; a Novel.
DOYLE'S (A. Conan) Micah Clarke : a Tale of Monmouth's Rebellion.

[Continued.

SILVER LIBRARY, The.—(*Continued*).

DOYLE'S (A. Conan) The Captain of the Polestar, and other Tales.

FROUDE'S (J. A.) Short Studies on Great Subjects. 4 vols.

———— The History of England, from the Fall of Wolsey to the Defeat of the Spanish Armada. 12 vols.

———— Cæsar : a Sketch.

———— Thomas Carlyle : a History of his Life. 1795-1835. 2 vols. 1834-1881. 2 vols.

———— The Two Chiefs of Dunboy : an Irish Romance of the Last Century.

GLEIG'S (Rev. G. R.) Life of the Duke of Wellington. With Portrait.

HAGGARD'S (H. R.) She : A History of Adventure. 32 Illustrations.

———— Allan Quatermain. With 20 Illustrations.

———— Colonel Quaritch, V.C. : a Tale of Country Life.

———— Cleopatra. With 29 Full-page Illustrations.

———— Beatrice.

HARTE'S (Bret) In the Carquinez Woods, and other Stories.

HELMHOLTZ'S (Professor) Popular Lectures on Scientific Subjects. With 68 Woodcuts. 2 vols.

HOWITT'S (W.) Visits to Remarkable Places. 80 Illustrations.

JEFFERIES' (R.) The Story of My Heart. With Portrait.

———— Field and Hedgerow. Last Essays of. With Portrait.

———— Red Deer. With 17 Illust.

KNIGHT'S (E. F.) Cruise of the 'Alerte,' a Search for Treasure. With 2 Maps and 23 Illustrations.

LEES (J. A.) and CLUTTERBUCK'S (W. J.) B.C. 1887. British Columbia. 75 Illustrations.

MACAULAY'S (Lord) Essays—Lays of Ancient Rome. In 1 vol. With Portrait and Illustrations to the 'Lays'.

MACLEOD'S (H. D.) The Elements of Banking.

MARSHMAN'S (J. C.) Memoirs of Sir Henry Havelock.

MAX MÜLLER'S (F.) India, What can it teach us ?

———— Introduction to the Science of Religion.

MERIVALE'S (Dean) History of the Romans under the Empire. 8 vols.

MILL'S (J. S.) Principles of Political Economy.

———— System of Logic.

MILNER'S (G.) Country Pleasures.

NEWMAN'S (Cardinal) Historical Sketches. 3 vols.

———— Fifteen Sermons Preached before the University of Oxford.

———— Apologia Pro Vita Sua.

———— Callista : a Tale of the Third Century.

———— Loss and Gain : a Tale.

———— Essays, Critical and Historical. 2 vols.

———— Sermons on Various Occasions.

———— Lectures on the Doctrine of Justification.

———— Fifteen Sermons Preached before the University of Oxford.

———— An Essay on the Development of Christian Doctrine.

———— The Arians of the Fourth Century.

———— Verses on Various Occasions.

———— Difficulties felt by Anglicans in Catholic Teaching Considered. 2 vols.

———— The Idea of a University defined and Illustrated.

———— Biblical and Ecclesiastical Miracles.

———— Discussions and Arguments on Various Subjects.

———— Grammar of Assent.

———— The Via Media of the Anglican Church. 2 vols.

———— Parochial and Plain Sermons. 8 vols.

———— Selection from 'Parochial and Plain Sermons'.

———— Discourses Addressed to Mixed Congregations.

———— Present Position of Catholics in England.

———— Sermons bearing upon Subjects of the Day.

PHILLIPPS-WOLLEY'S (C.) Snap : a Legend of the Lone Mountains. 13 Illustrations.

PROCTOR'S (R. A.) Other Worlds than Ours.

[*Continued.*

SILVER LIBRARY, The.—*(Continued.)*

PROCTOR'S (R. A.) Rough Ways made Smooth.
———— Pleasant Ways in Science.
STANLEY'S (Bishop) Familiar History of Birds. With 160 Illustrations.
STEVENSON (Robert Louis) and OSBOURNE'S (Lloyd) The Wrong Box.
WEYMAN'S (Stanley J.) The House of the Wolf: a Romance.
WOOD'S (Rev. J. G.) Petland Revisited. With 33 Illustrations.
———— Strange Dwellings. With 60 Illustrations.
———— Out of Doors. With 11 Illustrations.

SMITH (R. Bosworth).—CARTHAGE AND THE CARTHAGINIANS. Maps, Plans, &c. Crown 8vo. 6s.

STANLEY (E.).—A FAMILIAR HISTORY OF BIRDS. With 160 Woodcuts. Crown 8vo. 3s. 6d.

STEPHEN (Sir James).— ESSAYS IN ECCLESIASTICAL BIOGRAPHY. Crown 8vo. 7s. 6d.

STEPHENS (H. Morse).—A HISTORY OF THE FRENCH REVOLUTION. 3 vols. 8vo. Vol. I. 18s. Vol. II. 18s. [*Vol. III. in the press.*

STEVENSON (Robt. Louis).—A CHILD'S GARDEN OF VERSES. Small Fcp. 8vo. 5s.
———— A CHILD'S GARLAND OF SONGS, Gathered from 'A Child's Garden of Verses'. Set to Music by C. VILLIERS STANFORD, Mus. Doc. 4to. 2s. sewed, 3s. 6d. cloth gilt.
———— THE DYNAMITER. Fcp. 8vo. 1s. sewed, 1s. 6d. cloth.
———— STRANGE CASE OF DR. JEKYLL AND MR. HYDE. Fcp. 8vo. 1s. sewed, 1s. 6d. cloth.

STEVENSON (Robert Louis) and OSBOURNE (Lloyd).—THE WRONG BOX. Crown 8vo. 3s. 6d.

STOCK (St. George).—DEDUCTIVE LOGIC. Fcp. 8vo. 3s. 6d.

STRONG (Herbert A.), LOGEMAN (Willem S.) and WHEELER (B. I.).—INTRODUCTION TO THE STUDY OF THE HISTORY OF LANGUAGE. 8vo. 10s. 6d.

SULLY (James).—THE HUMAN MIND. 2 vols. 8vo. 21s.
———— OUTLINES OF PSYCHOLOGY. 8vo. 9s.
———— THE TEACHER'S HANDBOOK OF PSYCHOLOGY. Cr. 8vo. 5s.

SUPERNATURAL RELIGION; an Inquiry into the Reality of Divine Revelation. 3 vols. 8vo. 36s.

REPLY (A) TO DR. LIGHTFOOT'S ESSAYS. By the Author of 'Supernatural Religion'. 8vo. 7s. 6d.

SUTTNER (Bertha Von).—LAY DOWN YOUR ARMS (*Die Waffen Nieder*): The Autobiography of Martha Tilling. Translated by T. HOLMES. Crown 8vo. 7s. 6d.

SYMES (J. E.).—PRELUDE TO MODERN HISTORY: a Brief Sketch of the World's History from the Third to the Ninth Century. Cr. 8vo. 2s. 6d.

TAYLOR (Colonel Meadows).—A STUDENT'S MANUAL OF THE HISTORY OF INDIA. Crown 8vo. 7s. 6d.

THOMPSON (Annie).—A MORAL DILEMMA: a Novel. Cr. 8vo. 6s.

THOMPSON (D. Greenleaf).—THE PROBLEM OF EVIL: an Introduction to the Practical Sciences. 8vo. 10s. 6d.
———— A SYSTEM OF PSYCHOLOGY. 2 vols. 8vo. 36s.
———— THE RELIGIOUS SENTIMENTS OF THE HUMAN MIND. 8vo. 7s. 6d.
———— SOCIAL PROGRESS: an Essay. 8vo. 7s. 6d.
———— THE PHILOSOPHY OF FICTION IN LITERATURE: an Essay. Crown 8vo. 6s.

THOMSON (Most Rev. William, D.D., late Archbishop of York).—
OUTLINES OF THE NECESSARY LAWS OF THOUGHT : a Treatise
on Pure and Applied Logic. Post 8vo. 6s.
THREE IN NORWAY. By Two of THEM. With a Map and 59 Illustrations.
Crown 8vo. 2s. boards ; 2s. 6d. cloth.

TOYNBEE (Arnold).—LECTURES ON THE INDUSTRIAL REVO-
LUTION OF THE 18th CENTURY IN ENGLAND. 8vo. 10s. 6d.

TREVELYAN (Sir G. O., Bart.).—THE LIFE AND LETTERS OF
LORD MACAULAY.
Popular Edition. Crown 8vo. 2s. 6d. | Cabinet Edition, 2 vols. Cr. 8vo. 12s.
Student's Edition. Crown 8vo. 6s. | Library Edition, 2 vols. 8vo. 36s.
——— THE EARLY HISTORY OF CHARLES JAMES FOX. Library
Edition, 8vo. 18s. Cabinet Edition, Crown 8vo. 6s.

TROLLOPE (Anthony).—THE WARDEN. Cr. 8vo. 1s. bds., 1s. 6d. cl.
——— BARCHESTER TOWERS. Crown 8vo. 1s. boards, 1s. 6d. cloth.

VERNEY (Frances Parthenope).—MEMOIRS OF THE VERNEY
FAMILY DURING THE CIVIL WAR. Compiled from the Letters and
Illustrated by the Portraits at Claydon House, Bucks. With 38 Portraits,
Woodcuts, and Facsimile. 2 vols. Royal 8vo. 42s.

VILLE (G.).—THE PERPLEXED FARMER : How is he to meet Alien
Competition ? Crown 8vo. 5s.

VIRGIL. — PUBLI VERGILI MARONIS BUCOLICA, GEORGICA,
ÆNEIS ; the Works of VIRGIL, Latin Text, with English Commentary and
Index. By B. H. KENNEDY. Crown 8vo. 10s. 6d.
——— THE ÆNEID OF VIRGIL. Translated into English Verse. By
John Conington. Crown 8vo. 6s.
——— THE POEMS OF VIRGIL. Translated into English Prose. By
John Conington. Crown 8vo. 6s.
——— THE ECLOGUES AND GEORGICS OF VIRGIL. Translated from
the Latin by J. W. Mackail. Printed on Dutch Hand-made Paper. 16mo. 5s.
——— THE ÆNEID OF VERGIL. Books I. to VI. Translated into English
Verse by JAMES RHOADES. Crown 8vo. 5s.

WAKEMAN (H. O.) and HASSALL (A.).—ESSAYS INTRODUC-
TORY TO THE STUDY OF ENGLISH CONSTITUTIONAL HISTORY.
Edited by H. O. WAKEMAN and A. HASSALL. Crown 8vo. 6s.

WALFORD (Mrs. L. B.).—THE MISCHIEF OF MONICA. Cr. 8vo. 2s. 6d.
——— THE ONE GOOD GUEST. Crown 8vo. 6s.
——— TWELVE ENGLISH AUTHORESSES. With Portrait of HANNAH
MORE. Crown 8vo. 4s. 6d.

WALKER (A. Campbell-).—THE CORRECT CARD; or, How to Play
at Whist ; a Whist Catechism. Fcp. 8vo. 2s. 6d.

WALPOLE (Spencer).—HISTORY OF ENGLAND FROM THE CON-
CLUSION OF THE GREAT WAR IN 1815 to 1858. 6 vols. Crown 8vo.
6s. each.
——— THE LAND OF HOME RULE : being an Account of the History and
Institutions of the Isle of Man. Crown 8vo. 6s.

WELLINGTON.—LIFE OF THE DUKE OF WELLINGTON. By the
Rev. G. R. GLEIG. Crown 8vo. 3s. 6d.

WEYMAN (Stanley J.).—THE HOUSE OF THE WOLF : a Romance.
Crown 8vo. 3s. 6d.

WHATELY (Archbishop).—ELEMENTS OF LOGIC. Cr. 8vo. 4s. 6d.
——— ELEMENTS OF RHETORIC. Crown 8vo. 4s. 6d.
——— LESSONS ON REASONING. Fcp. 8vo. 1s. 6d.
——— BACON'S ESSAYS, with Annotations. 8vo. 10s. 6d.

WHISHAW (Fred. J.).—OUT OF DOORS IN TSAR LAND: a Record of the Seeings and Doings of a Wanderer in Russia. With Frontispiece and Vignette by CHARLES WHYMPER.

WILCOCKS (J. C.).—THE SEA FISHERMAN. Comprising the Chief Methods of Hook and Line Fishing in the British and other Seas, and Remarks on Nets, Boats, and Boating. Profusely Illustrated. Crown 8vo. 6s.

WILLICH (Charles M.).—POPULAR TABLES for giving Information for ascertaining the value of Lifehold, Leasehold, and Church Property, the Public Funds, &c. Edited by H. BENCE JONES. Crown 8vo. 10s. 6d.

WITT (Prof.)—Works by. Translated by Frances Younghusband.
—————— THE TROJAN WAR. Crown 8vo. 2s.
—————— MYTHS OF HELLAS; or, Greek Tales. Crown 8vo. 3s. 6d.
—————— THE WANDERINGS OF ULYSSES. Crown 8vo. 3s. 6d.
—————— THE RETREAT OF THE TEN THOUSAND; being the Story of Xenophon's 'Anabasis'. With Illustrations. Crown 8vo. 3s. 6d.

WOLFF (Henry W.).—RAMBLES IN THE BLACK FOREST. Crown 8vo. 7s. 6d.
—————— THE WATERING PLACES OF THE VOSGES. With Map. Crown 8vo. 4s. 6d.
—————— THE COUNTRY OF THE VOSGES. With a Map. 8vo. 12s.
—————— PEOPLE'S BANKS: a Record of Social and Economic Success. 8vo. 7s. 6d.

WOOD (Rev. J. G.).—HOMES WITHOUT HANDS; a Description of the Habitations of Animals. With 140 Illustrations. 8vo. 7s. net.
—————— INSECTS AT HOME; a Popular Account of British Insects, their Structure, Habits, and Transformations. With 700 Illustrations. 8vo. 7s. net.
—————— INSECTS ABROAD; a Popular Account of Foreign Insects, their Structure, Habits, and Transformations. With 600 Illustrations. 8vo. 7s. net.
—————— BIBLE ANIMALS; a Description of every Living Creature mentioned in the Scriptures. With 112 Illustrations. 8vo. 7s. net.
—————— STRANGE DWELLINGS; abridged from 'Homes without Hands'. With 60 Illustrations. Crown 8vo. 3s. 6d.
—————— OUT OF DOORS; a Selection of Original Articles on Practical Natural History. With 11 Illustrations. Crown 8vo. 3s. 6d.
—————— PETLAND REVISITED. With 33 Illustrations. Crown 8vo. 3s. 6d.

WORDSWORTH (Bishop Charles).—ANNALS OF MY LIFE. First Series, 1806-1846. 8vo. 15s. Second Series, 1847-1856. 8vo.

WYLIE (J. H.).—HISTORY OF ENGLAND UNDER HENRY THE FOURTH. Crown 8vo. Vol. I. 10s. 6d.; Vol. II.

ZELLER (Dr. E.).—HISTORY OF ECLECTICISM IN GREEK PHILOSOPHY. Translated by Sarah F. Alleyne. Crown 8vo. 10s. 6d.
—————— THE STOICS, EPICUREANS, AND SCEPTICS. Translated by the Rev. O. J. Reichel. Crown 8vo. 15s.
—————— SOCRATES AND THE SOCRATIC SCHOOLS. Translated by the Rev. O. J. Reichel. Crown 8vo. 10s. 6d.
—————— PLATO AND THE OLDER ACADEMY. Translated by Sarah F. Alleyne and Alfred Goodwin. Crown 8vo. 18s.
—————— THE PRE-SOCRATIC SCHOOLS. Translated by Sarah F. Alleyne. 2 vols. Crown 8vo. 30s.
—————— OUTLINES OF THE HISTORY OF GREEK PHILOSOPHY. Translated by Sarah F. Alleyne and Evelyn Abbott. Crown 8vo. 10s. 6d.

www.ingramcontent.com/pod-product-compliance
Lightning Source LLC
Chambersburg PA
CBHW032048220426
43664CB00008B/912